TEXTBOOK OF LITHOLOGY

TEXTBOOK OF LITHOLOGY

KERN C. JACKSON

Professor of Geology
University of Arkansas

McGRAW-HILL BOOK COMPANY

New York
St. Louis
San Francisco
Düsseldorf
London
Mexico
Panama
Sydney
Toronto

TEXTBOOK OF LITHOLOGY

Library of Congress Catalog Card Number 72-95810

ISBN 07-032143-4

11 12 13 14 15 16   KPKP   78321098

PREFACE

The science of geology is based on the developed art of personal observation of detail. Modern culture has of necessity trained us not to observe detail, except when it is essential for a specific set of responses. For instance, if every passing car on a highway registered on a driver, his mind would be so distracted with detail that he would soon be in trouble. Instead, a driver becomes conscious of an individual car only if it behaves in an erratic or dangerous manner. Then that one car stands out from the background of dozens of others which are behaving in a normal manner. Similarly, passengers in a car will be conscious of different elements of the scenery. A geologist will be conscious of the great majority of outcrops or geomorphic features he passes, whereas a nongeologist will be conscious of only the overpowering features or those which are landmarks for his navigation. After a trip the motor-minded teenage boy will be able to recite the make and model of each unusual car passed, and a woman will be able to describe the regional architectural style of houses or the number of antique shops passed up. Thus each person is "programmed" to observe the detail in which he is interested, and the vast volume of other detail does not register.

It is with this background that the average undergraduate major in geology arrives at college, having had no training in observation of the details of rocks and minerals. A rock has been, heretofore, merely something to kick while walking, and on a beach, it could be classified as a "skipping rock," a "chucking rock," or merely a "rock." Occasionally, a student has had a rock collection, but even these students have generally made little detailed observation. It is during the undergraduate course work that a foundation for observation of geological detail must be laid. This text is designed for this purpose. The course level at which it is aimed should have as prerequisites general geology, mineralogy, and college chemistry. An ability to identify the more typical examples of common rock-forming minerals by their physical properties is assumed, as is knowledge of the chemical composition and normal compositional variations in the major mineral groups.

The text is divided into two areas: Chapters 1 to 4 are designed for classroom lecture material on the genesis of the major rock types; Chapters 5 to 7 are designed for use in the laboratory in conjunction with the detailed study of specimens. In Chapters 1 to 4 the classical concepts of rock genesis are developed without bringing in the details of modern experimental techniques. The data of structural mineralogy, hydrothermal and high-pressure experimental data, isotope and trace-element chemistry, and so forth, are not considered, but this is not intended to minimize their importance. The purpose is to give a framework of basic concepts to which the results of those advanced techniques can be added by graduate-level courses.

Chapter 2, on igneous-rock petrology, essentially develops the concept of crystal differentiation following Bowen. Only the simplest two-component phase diagrams are employed in the discussion. However, a more complete discussion of the principles of interpretation of two- and three-component phase diagrams is given in the Appendix for those who wish to go further in the application of this data. Chapter 3, on sedimentary-rock petrology, examines the origin, transportation, and deposition of sediments and the influence of each successive step on the ultimate product — a sedimentary rock. The influence of climate on the character of clastic sediments is particularly emphasized. Chapter 4, on metamorphic-rock petrology, emphasizes processes and products of metamorphism. In this chapter only the classical Barrovian concept of regional metamorphism and typical low-pressure contact metamorphism are considered. The transitional phases between these extremes are not included. Thus for each major rock type a base is laid on which refinements and modifications can be built in subsequent course work. Many terms are introduced in these chapters without definitions. These terms are defined in Chapters 5 to 7, and it is assumed that the laboratory study will proceed concurrently with the classroom discussion.

Chapters 5 to 7, on the petrography of rocks, attempt to begin the programming of the student in the observation of details. A separate chapter is devoted to the major features of each rock group: igneous, sedimentary, and metamorphic. Textural and structural features are defined and their significance is discussed. Basic mineralogy and typical characteristics of these minerals are described for each major rock type in an attempt to make the student conscious of exactly what he is seeing. With very few exceptions all the features described are observable in selected specimens with a hand lens or, at most, a binocular microscope. Thin sections and the petrographic microscope are not necessary for the observation of most of the detail described, but of course, many of the features are more readily recognized in thin section. More advanced instrumental techniques of mineral identification have been omitted, thus eliminating mineral identification based on composition in solid-solution series or based on the nature of clay minerals present; but this is in accord with the necessities of field identification of rocks. Chapters 5 to 7 are written with an emphasis on a laboratory involving the detailed study of a restricted number of ideal specimens rather than on the cursory examination of a large number of specimens.

The title Textbook of Lithology was expressly selected to emphasize the nature of the laboratory study. Lithology is defined in part by the

1960 edition of the AGI Glossary of Geology* as "the study of rocks based on megascopic examination of samples." Petrology, a term most commonly applied to courses for which this text is designed, is defined in the same volume as "a general term for the study *by all available methods* of the natural history of rocks, including their origins, present conditions, alterations and decay." (Italics are the author's.) Inasmuch as chemical, petrographic, and advanced instrumental techniques, such as X-ray and DTA, are not included in this text, "all available methods" are not considered, and lithology is the correct term.

All the photographs in this book have been prepared by Mr. James E. Edson, currently a graduate student at the University of Arkansas. The photographs in the field were taken largely by the author. The individual rock specimens illustrated are from the author's collection, from "Ward's Universal Rock Collection" of Ward's Natural Science Establishment (300 specimens), or from the natural accretion of material around a university Geology Department. Photographs of thin sections were prepared by placing the thin section directly in an enlarger and producing a film negative. No polarizing devices were used, and the magnification was kept low so that, except for color and opacity, the photographs have the appearance of a rock under a hand lens.

This book has evolved to its present form over a period of years, and the author wishes to express his appreciation to the many students who have helped form it through classroom discussions and testing of approaches. The author is also indebted to his colleagues, Drs. J. H. Quinn, R. H. Konig, H. F. Garner, and R. J. Willard, for many ideas hashed out in discussions. The basic approach of Chapter 3 is that of Dr. J. H. Quinn, and many ideas expressed there are both his and those of Dr. H. F. Garner. Finally, the author is deeply indebted to a long line of excellent instructors, including Professors Chester B. Slawson, Willys A. Seaman, Vincent L. Ayres, Reynolds M. Denning, Stanley A. Tyler, and R. Conrad Emmons, each of whom contributed much to the development of ideas and, particularly, to the development of an ability to observe details in rocks.

Kern C. Jackson

*Glossary of Geology and Related Sciences, 2d ed., The American Geological Institute, Washington, D.C., 1960.

CONTENTS

1 THE EARTH

Before we can consider the various rocks of the surficial layers of the earth, we must have some knowledge of the earth as a unit, particularly of the earth's interior. A clear distinction must be made between *facts* about the earth's interior and the *interpretation* of those facts. The facts are limited to the average density of the earth, the velocity of seismic waves, and the temperature measurements in a thin surficial zone. Any interpretation of the conditions and materials within the earth must be based on this information and must satisfy all the facts.

FACTS ABOUT THE EARTH'S INTERIOR

Density

In the visible portions of the earth's crust the density of major rock types varies from 2.0 or less for unconsolidated sediments to 3.4 for the densest igneous rocks (Table 1-1). The average density of all surface rocks has been estimated at 2.7 to 2.8. The material of the earth's interior must be denser than the surface materials, because the density of the earth is about twice that of the surficial materials.

The earth's mean density can be measured by comparing the attraction of the earth with the attraction of a body of known mass for the same object. In 1878, Von Jolly first measured the density of the earth by application of this principle. He mounted a beam balance with pans at the top of a high tower and suspended a second pair of scale pans from the first, 69 ft below. Four identical glass globes of equal volume and weight were then prepared. Two were each filled with 5 kg of mercury and sealed, and the other two were sealed empty. The mercury-filled globes were placed on the upper pans, the empty globes were placed on the lower pans, and the system was brought to balance. The two globes on one side of the balance arm were then interchanged, bringing one mercury-filled globe closer to the center of the earth. Gravitational force, being inversely proportional to the square of the distance, increased the weight of the mercury-filled sphere. Since there was no change other than distance from the center of the earth, it was possible to measure the increased attraction of the earth on the mass of mercury. The gain in weight was 0.031 g.

Table 1-1.   Average Densities of Rock Materials*

Igneous rocks	No. of samples	Mean density	Range of density
Granite	155	2.667	2.516–2.809
Granodiorite	11	2.716	2.668–2.785
Syenite	24	2.757	2.630–2.899
Quartz diorite	21	2.806	2.680–2.960
Diorite	13	2.839	2.721–2.960
Norite	11	2.984	2.720–3.020
Gabbro	27	2.976	2.850–3.120
Diabase	40	2.965	2.804–3.110
Anorthosite	12	2.734	2.640–2.920
Peridotite	3	3.234	3.152–3.276
Pyroxenite	8	3.231	3.10–3.318

Sediments and sedimentary rocks	Porosity, %	Dry density	H_2O-saturated density
Soils	40–50	1.00–2.00	1.50–2.40
Loess	20–60	0.80–1.60	1.40–1.80
Sand	30–48	1.37–1.81	1.85–2.14
Gravels	20–37.7	1.36–2.19	1.65–2.39
Clays (less than 100 ft deep)	40–50	1.30–1.60	1.80–2.00
Atlantic Ocean bottom clay	48.6–51.9	1.21–1.34	1.73–1.83
Globigerina ooze, Pacific Ocean	1.8	2.79	2.81
Dakota sandstone, Iowa (upper Cretaceous)	33.8–38.2	1.65–1.76	2.03–2.10
Eagle sandstone, Montana (upper Cretaceous)	23.6–27.1	1.93–2.06	2.20–2.30
Hale sandstone, Oklahoma (Pennsylvanian)	13.8–18.1	2.24–2.28	2.42
Savanna sandstone, Oklahoma (Pennsylvanian)	5.4–8.4	2.42–2.50	2.50–2.55
Graneros shale, Wyoming (upper Cretaceous)	32.5–33.3	1.78–1.87	2.11–2.20
Graneros shale, Kansas (upper Cretaceous)	24.6–25.2	1.98–1.99	2.23–2.24
Weston shale, Kansas (Pennsylvanian)	15.5–16.0	2.28–2.29	2.44–2.45
Hamilton shale, Missouri (Devonian)	11.3–11.6	2.32	2.43–2.44
Stanley shale, Oklahoma (Pennsylvanian)	2.5–6.8	2.45–2.60	2.52–2.62
Chalk, England	17.6–42.8	1.53–2.22	1.96–2.40
Greenhorn limestone, Wyoming (upper Cretaceous)	37.6	1.74	2.12
Portland limestone, England	18.1	2.21	2.39
Limestone, Buston, England	14.1	2.31	2.45
Caddo lime, Texas (Pennsylvanian)	4.4	2.57	2.61
Ophir formation limestone, Utah (Cambrian)	0.0–1.0	2.76	2.76–2.77
Anhydrite (torsion balance)		2.9	
Gypsum (torsion balance)		2.2	
Rock salt (torsion balance)		2.1	
Chert, England	11.1	2.29	2.40
Dolomite		2.23–2.68	

Table 1-1.   Average Densities of Rock Materials (Continued)

Metamorphic rocks	Porosity, %	Dry density	H₂O-saturated density
Quartzite, England	6.4	2.48	2.54
Quartzite (auriferous), Africa	1.9–2.8	2.59–2.61	2.62–2.63
Quartzite, England	0.8	2.63	2.64
Slate, England	0.6–6.0	2.59–2.83	2.64–2.65
Schist, England (Archean)	2.3	2.68	2.70
Schist, Tasmania	6.0	2.64	2.70
Marble, 34 samples from 12 states	0.4–2.1	2.66–2.86	2.68–2.86
Eclogite		3.415	

*Data from F. Birch, *Handbook of physical constants, Geol. Soc. America Spec. Paper 36*, 1942.

The experiment was then repeated after a lead sphere had been placed under the lower pan on the same side of the balance where the globes were to be interchanged. This time the mercury-filled sphere gained 0.03159 g, 0.59 mg more than previously. The gravitative attraction of the mass of lead, whose center was 57 cm below the center of the mercury sphere, caused the additional gain of weight. If the lead sphere at an effective distance of 57 cm exerted a pull of 0.59 mg on the mercury, and the earth at an effective distance of its radius (approximately 637,000,000 cm) exerted a pull of 5 kg, or 5,000,000 mg, then the mass of the earth could be calculated from that of the lead sphere. Von Jolly's result was about 6,100 billion billion (6.1×10^{21}) metric tons. Dividing this by the known volume of the earth gave a density of 5.69. Subsequent refinements of the technique give 5.516 as the best figure to date.

The mean density of the earth can also be calculated from the known radius and the measured acceleration of gravity, making a correction for the ellipticity of the earth. The formula for mean density ρ_m of a sphere is

$$\rho_m = \frac{3g}{4rG}$$

where g is the acceleration of gravity, r is the radius, and G is the newtonian gravity constant (approximately 6.668×10^{-8}). To correct for ellipticity, use the equatorial radius r_e, the gravity at the equator at sea level g_e, the ellipticity e, and the angular velocity of the earth:

$g_e = 978.409$ gal*

$r_e = 6,378,388$ m

$$e = \frac{r_e - r_{polar}}{r_e} = \frac{1}{297}$$

*One gal is an acceleration of 1 cm/sec².

The corrected value gives a mean density of 5.5144, which is not markedly different from the experimentally determined value.

Seismology

The most valuable tools for the exploration of the interior of the earth are the various seismic waves. When earthquake-generating movements take place within the earth, they initiate a series of waves of differing types which move outward from their source as spherical fronts. The different waves move at different velocities in different types of material. In addition, they are reflected and refracted within the earth at contacts between zones of contrasting physical characteristics. Thus if the exact time of an earthquake movement is known, the velocity of each wave type can be determined from the time of arrival of each wave at distant seismograph stations. If a series of stations is placed at varying distances from the source, the velocity of waves at different depths within the earth can be determined. From these velocities and from the character of the waves transmitted through various zones in the earth, a concept of the earth's interior can start to be developed.

Three fundamental waves are generated by earthquakes: primary, or P, waves (Fig. 1-1a), secondary, or S, waves (Fig. 1-1b), and surface waves. P and S waves travel through the body of the earth, whereas surface waves travel through only the surficial layers of the earth. Each of these waves involves a different type of particle motion in the mass of the earth through which it passes.

The P wave has the highest velocity, since it is the first to arrive at a distant seismograph station. It is a compressional wave, in that as it passes, each particle moves back and forth along the direction of wave propagation (Fig. 1-1a). Thus P waves are analogous to sound waves in their transmission. Seismographs record P waves quite distinctly at any distance from the focus of the earthquake up to a distance of 11,300 to 11,500 km (103° of arc) on the surface. Beyond this distance P waves fade out and are not received at stations between 11,500 and 15,900 km (143° of arc) from the earthquake locus. At 15,900 km they reappear on the seismograph and are recorded at stations up to 20,000 km (180° of arc) from the focus. The zone in which P waves are not recorded or are very weak is called the *shadow zone*, and it indicates an abrupt change in the physical properties of earth materials at a depth of 2,900 km. The change of properties at this depth is a major discontinuity within the earth called the *Gutenberg discontinuity*, named after the man who first determined its depth. The shadow zone can be explained by the refraction of waves at the boundary between contrasting materials. Seismic waves go out from the locus in all possible directions. Those entering the earth on a path nearly parallel to the surface return to the surface a short arc distance from the focus and penetrate the earth to only a shallow depth (paths 1 and 2, Fig. 1-2). Waves entering the earth at a

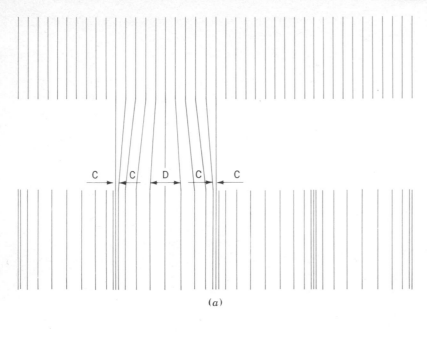

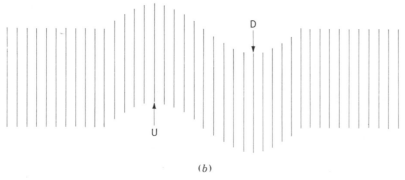

Fig. 1-1. (*a*) The nature of *P*-wave motion shown by the schematic displacement of equally spaced lines perpendicular to the direction of wave travel. The lines are connected through to their original positions for one wavelength. C–compressional zone; D–dilational zone. (*b*) The nature of *S*-wave motion shown by the schematic displacement of equally spaced lines perpendicular to the direction of wave travel. A single wavelength of *S* motion is shown.

steep angle relative to the surface penetrate deeply into the earth. At a specific angle of incidence (path 4, Fig. 1-2) the wave will just reach the 2,900 km discontinuity and then emerge at the surface 11,300 to 11,500 km from the focus (102° to 104° of arc). Waves with paths closer to the radius

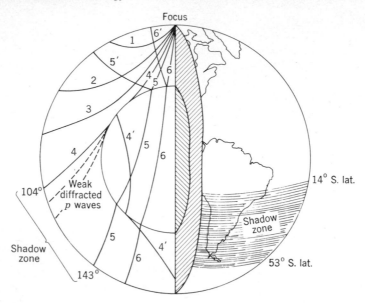

Fig. 1-2. The shadow zone of an earthquake originating near the North Pole. Rays 1, 2, and 3 are high-incident waves and emerge at the surface without reaching the Gutenberg discontinuity. Ray 4 just grazes the Gutenberg discontinuity and emerges at the surface 104° from the focus. It is also refracted into the core and emerges 187° from the focus (4′). Rays 5 and 6 are low-incident rays and are refracted into the core; they are also reflected back to the surface by the discontinuity (5′ and 6′). (*Modified from B. Gutenberg, "Internal Constitution of the Earth," Dover Publications, 1951.*)

(paths 5 and 6, Fig. 1-2) will not merely graze the 2,900 discontinuity but will strike it at such an angle that they will be refracted into the lower zone and will come to the surface at distances greater than 15,900 km (143° of arc) from the focus.

In 1909, A. Mohorovicic discovered that the record of a near earthquake at least 100 to 150 km distant showed two types of initial P waves. They arrived at the station at slightly different times and with different characteristics. This can be explained by assuming that one P impulse travels directly from the earthquake focus through upper layers of the earth to the station and that the other travels in a lower, higher-velocity, level after being refracted at the boundary between two contrasting rock types. The travel time of the two waves indicates that the P-wave velocity is approximately 5.5 km/sec in the upper level and approximately 8.2 km/sec in the lower level. The break in seismic velocities is called the *Mohorovicic discontinuity* (the "Moho").

S waves have a velocity approximately six-tenths that of P waves. Therefore the S wave arrives at a distant seismograph station later than the P wave, and the time lapse between the arrival of these two waves can be used to determine the distance of the earthquake from the station. The motion of any particle as the S wave passes is perpendicular to the direction of wave propagation (Fig. 1-1b). The S wave is thus a shear, or shaking, wave and is analogous to light waves in its transmission. S waves, like P waves, are recorded up to an arc distance of 102 to 104° from the focus of the earthquake, but unlike the P waves, they are not recorded beyond the shadow zone through the arc distance of 143 to 180°. This fact indicates that the material below the Gutenberg discontinuity cannot transmit shear waves and that it therefore has practically no rigidity. Absence of rigidity is a characteristic of fluids, and therefore most seismologists refer to the material of this deep zone as a liquid.

Surface waves are of many types, involving complex movements and varying velocities. Several types penetrate the earth to a depth of only one wavelength. Since the waves have different wavelengths, they can be used to study the characteristics of the surficial layer of the earth above the Mohorovicic discontinuity. Surface waves show clearly that the surficial layers under continents and oceanic basins are markedly different, and a study of these waves can delimit structural units. It has been found without exception that the continental type of structure is present in any large area where water depth is less than about 1,000 fathoms (1.8 km) and that the oceanic type of structure is present under water depths greater than about 2,000 fathoms (3.6 km). The Mohorovicic discontinuity under continental masses seems to be at a depth of approximately 33 km, whereas under oceanic sectors it seems to be at a depth of about 10 km below sea level. The velocity of surface waves under the ocean is somewhat higher than it is under continents, a fact which indicates some difference of material. Some variations exist in the depth of the Moho: under some young mountain ranges the discontinuity is deeper than average for the continents.

On the evidence of seismic waves, therefore, it is known that the earth is divided into three major units which are separated by discontinuities. The upper, or surficial, unit above the Mohorovicic discontinuity is known as the *crust*. This zone varies in thickness and, apparently, in the character of its rock material. The two fundamental types of crust are continental and oceanic. Between the Moho and Gutenberg discontinuities is the main bulk of the earth's mass known as the *mantle*. The velocity of earthquake waves increases with depth in the mantle, except in certain zones which probably mark the minor discontinuities. Below the Gutenberg discontinuity is the inner portion of the earth called the *core*. The outer portion of the core behaves like a liquid in the transmission of seismic waves but seismic evidence indicates the presence of an inner core which has differing characteristics below a discontinuity at a depth of about 5,000 km. Internal sub-divisions are outlined in Table 1-2.

Table 1-2.   Dimensions and Descriptions of Internal Layers of the Earth, Based on Seismic Evidence*

Layer	Depth to boundaries, km	Velocity P wave, km/sec	Fraction of volume	Features of regions
Crust A†	0	5–6.5	0.0155	Fairly heterogeneous
	— 33 —			
B		8.2	0.1667	Probably homogeneous
	413			
Mantle C			0.2131	Transition region
	984			
D		13.5	0.4428	Probably homogeneous
	— 2,898 —			
E		8	0.1516	Homogeneous fluid
	4,982			
Core F		10	0.0026	Transition layer
	5,121			
G		11.5	0.0076	Inner core (solid?)
	6,371			Center of the earth

*From J. A. Jacobs, R. D. Russell, and J. T. Wilson, "Physics and Geology," p. 24, McGraw–Hill, 1959.
†The thickness of the crust under the continents is not constant, averaging about 33 km. Under the oceans the crust is much thinner, only about 5 or 6 km thick.

INTERPRETATIONS OF THE EARTH'S INTERIOR
Earth Models

A model of the interior of the earth must first of all fit the facts. It must be made up of three major units of contrasting materials to correspond to the physical properties indicated by seismic waves. In addition, it must conform with the average specific gravity of the earth and the energy of the earth's motion. Several different models have been built based on these data, but none apparently is completely adequate. As new information is obtained from refined observation and new techniques, especially with respect to high-pressure experiments, the models will be revised and refined. Still, until man is able to penetrate to great depth with the drill, no one will "know" what the interior is actually like.

An interpretation of the material of the inner earth can be approached through comparison with the materials of meteorites. Meteorites and the earth are usually assumed to have had the same origin, since meteorites can be considered fragments of a sister planet. Most meteorites are composed of three types of materials: ultramafic silicate rock, iron-nickel alloys, and minor amounts of iron sulfides. Meteorites are classified according to the structure, the relative proportions of silicate and metallic phases, and the mineralogy of the silicate phase. This classification, with the mineralogical composition of each, is given in Table 1-3.

Table 1-3. Types and Mineral Compositions of Meteorites*

Type	Mineral	Number	Olivine	Pyroxene	Feldspar	Fe-Ni alloy	Sulfide
Achondrite	Pyroxene	17	<5	~90	~5	<1	Tr
	Pyroxene-plagioclase	39	<5	~30	~65	<1	Tr
Chondrites†	Enstatite	11	.....	50–60	5–10	20–28	7–15
	Bronzite	~300	25–40	20–35	5–10	16–21	~5
	Hypersthene	~600	35–60	25–35	5–10	1–10	~5
Stony irons	Pallasites	40	45–70	.....	.....	30–55	Tr
	Mesosiderites	22	<2	~30	~20	~50	Tr
Irons		545	Tr	Tr	.....	98	<1

*From B. Mason, "Meteorites," Wiley, 1962.
†Carbonaceous chondrites are a quantitatively minor type composed of amorphous hydrous silicates or serpentine with hydrocarbons.

The relative abundance of the various types of meteorites is significant in correlating meteorites with earth materials. The mantle makes up more than 82 percent of the volume of the earth, and the core makes up slightly more than 16 percent. Thus the most common meteorite types are presumably equivalent to the mantle. The relative abundance of meteorites is given in Table 1-4. Meteorites seen to fall and then recovered are the most significant, for they give the best sampling of the influx of larger fragments into the atmosphere. The markedly different proportion of stones and irons found accidentally is probably accounted for by the fact that stones are not always recognized because of their similarity to many terrestrial rocks.

The idea that the interior of the earth is composed of materials similar to meteorites predates seismic evidence. It apparently originated with Daubree in 1866. Subsequently, the concept of an iron-nickel-alloy core became

Table 1-4.   Relative Abundance of Meteorites*

	Accidentally found		Seen to fall and found	
	Number	%	Number	%
Irons	503	59	42	6
Stony irons	55	6	12	2
Stones	304	35	628	92

*From B. Mason, "Meteorites," Wiley, 1962.

firmly entrenched; however, new ideas have been put forward. The two most familiar earth models based on meteorites are those of Buddington (1943)* and Daly (1943 and 1946).† Both assume that the crustal layers are composed of normal silicate rocks, grading from granitic to gabbroic or peridotitic composition from the surface downward to a depth of approximately 70 km. Immediately below this zone, both authors assume a zone of peridotitic material having peculiar physical characteristics: the material is plastic enough to yield under the long-term stresses of isostasy but rigid enough to rupture under the short-term stresses of deep-focus earthquakes. According to Buddington this zone extends to a depth of 500 km, and the plasticity is due to enrichment of volatiles in the crystalline peridotite. Daly, in his model, extends this zone to a depth of 450 km and obtains the necessary plasticity by assuming that the material is composed of crystalline peridotite containing an interstitial glass phase. Both models assume that the lower portion of the mantle is composed of meteoritic types of materials, ranging with depth from achondritic to palasitic composition at the Gutenberg discontinuity. The core in both models is assumed to be an iron-nickel alloy similar to siderite meteorites.

In 1941, Kuhn and Rittmann‡ suggested that seismic discontinuities may be due not to changes in types of material within the earth but to changes in the physical state of materials of uniform composition. They also suggested that the core may be composed of "undifferentiated solar material" rich in hydrogen. This second suggestion met with much objection and has been generally discredited, but it opened anew the question of the nature of the core materials. One result was the suggestion of Ramsey (1948)§ that the

*A. F. Buddington, Some petrological concepts and the interior of the earth, *Am. Mineralogist*, vol. 28, pp. 119–140, 1943.
†R. A. Daly, Meteorites and an earth-model, *Geol. Soc. America Bull.*, vol. 54, pp. 401–455, 1943; Nature of the asthenosphere, *Geol. Soc. America Bull.*, vol. 57, pp. 707–726, 1946.
‡W. Kuhn and A. Rittmann, Summarized in: K. Rankama and T. G. Sahama, "Geochemistry," pp. 82–83, The University of Chicago Press, 1950.
§H. W. Ramsey, On the constitution of the terrestrial planets, *Royal Astron. Soc. Monthly Notices*, vol. 108, pp. 406–413 (1948).

core is not an iron-nickel alloy but, rather, a silicate material identical to the mantle that is broken down by high pressure to a "metallic" state. Breaking down involves the destruction of atomic structures and crystal lattices accompanied by a great decrease in volume; such changes have been effected in some materials subjected to pressures now attainable in the laboratory.

An interpretation of the relationship at the Mohorovicic discontinuity has recently been proposed by Kennedy (1959)[*] and involves a change of state from low-density to high-density rock as a result of pressure and temperature. Kennedy assumes the inversion of basalt or gabbro at the base of the crust to the chemically equivalent high-pressure rock eclogite at the top of the mantle. Eclogite has a mean density of 3.3 g/cm³, is composed of a jadeitic pyroxene and garnet, and contains no feldspar. Gabbro has a mean density of 2.95 g/cm³. The density contrast of about 10 percent between gabbro and eclogite is almost the same as that indicated by seismic velocities at the Mohorovicic discontinuity.

Experimental studies of the basalt-eclogite relationship by Yoder and Tilley (1962)[†] show that this inversion is possible at pressures equivalent to those at the depth of the Moho under the continents. However, their results show that this inversion is not probable at the oceanic Moho (Fig. 1-3), since the pressure is insufficient. Wyllie (1963)[‡] suggests that the upper mantle, except immediately under the continents, is a feldspar or garnet peridotite. The physical properties of peridotites and eclogite are very similar so that the peridotite-basalt contact under the oceans is equivalent to the eclogite-basalt contact under the continents and gives the seismic discontinuity of the Moho (Fig. 1-4).

Inclusions of ultramafic rocks are common in many basaltic volcanic regions. Kushiro and Kuno (1963)[§] point out that these inclusions from 70 known occurrences are consistently similar both chemically and mineralogically, in spite of worldwide distribution and variation in the host volcanic rocks. They thus consider this material to represent the upper mantle and they point out that all varieties of basalt can be produced from material of the composition of these masses by melting from 2 to 9 percent of the material. It is also possible that these masses represent the residue from some more fundamental material from which basaltic magmas have been selectively melted.

[*]G. C. Kennedy, The origin of continents, mountain ranges and ocean basins, *Am. Scientist*, vol. 47, pp. 491–504, 1959.
[†]H. S. Yoder and C. E. Tilley, Origin of basalt magmas, *Jour. Petrology*, vol. 3, pp. 342–532, 1962.
[‡]P. J. Wyllie, The nature of the Mohorovicic discontinuity, a compromise, *Jour. Geophys. Research*, vol. 68, pp. 4611–4619, 1963.
[§]I. Kushiro and H. Kuno, Origin of primary basalt magmas and classification of basaltic rocks, *Jour. Petrology*, vol. 4, pp. 75–89, 1963.

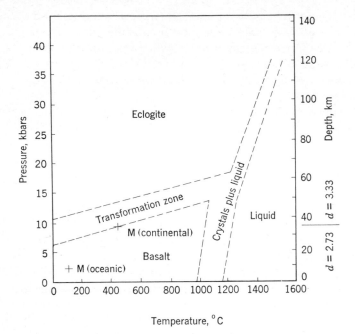

Fig. 1-3. Temperature-pressure relationships of the basalt-eclogite transition and estimated pressure-temperature conditions at the average oceanic and continental positions of the Mohorovicic discontinuity M. (*After H. S. Yoder and C. E. Tilley, Jour. Petrology, vol. 3, 1962.*)

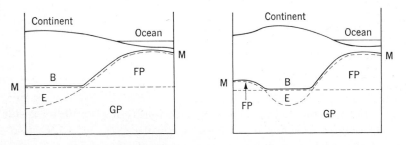

Fig. 1-4. Possible arrangements within the earth of a chemical discontinuity (dashed line) and a phase-density inversion (dotted line) and the Mohorovicic discontinuity. B–basalt; E–eclogite; GP–garnet peridotite; FP–feldspar peridotite. (*After P. J. Wyllie, Jour. Geophys. Research, vol. 68, 1963.*)

MacDonald (1959)* presented a model of the earth which has an average composition of chondritic meteorites. In his model the mantle and core are differentiated into three major units. The upper mantle to a depth of 600 to 700 km contains the major portion of radioactive heat-producing elements and consists of material which is eclogite to peridotite in composition; the lower mantle is approximately a dunite; and the core is an iron-nickel alloy containing appreciable metallic silicon. This model is based on the close similarity of occurrence of the elements in chondrites and in the solar atmosphere as well as on analyses of heat generation by radioactive decay and heat loss from the earth.

These models of the earth, then, are all based on the assumption that meteorites represent the parental material from which the earth was formed. The most generally accepted models assume a thin crust composed of normal igneous rocks, ranging from granitic to gabbroic or peridotitic composition; a mantle composed dominantly of silicate material of mafic to ultramafic composition (eclogite, peridotite, and dunite), perhaps containing minor amounts of an iron-nickel-alloy phase; and a core of metallic character dominated by iron and nickel, perhaps containing important amounts of silicon. Such models fit the requirements of seismic evidence and density.

Composition of the Earth

Based on the above models, it is now possible to estimate the average composition of the earth. Its bulk composition is essentially determined by the relative amounts of mantle and core, since the crust composes less than 0.5 percent of the earth's mass. Mason (1958)† computes the composition of the earth from the following assumptions:

1. The iron-alloy core has the composition of the average meteoritic nickel-iron.
2. The silicate mantle has peridotitic composition.
3. The earth contains 8 percent sulfide of the composition of meteoritic troilite distributed arbitrarily between the core and mantle.
4. The weight percentages are 28.8 percent iron-nickel, 8 percent sulfide, and 63.2 percent silicate, on the basis of the above assumptions.

The results of such a calculation are given in Table 1-5. Other results would be obtained if other assumptions were made. If we assume that a portion of the mantle is eclogite, the proportion of aluminum, calcium, sodium, potassium, and titanium would be increased somewhat at the

*G. J. F. MacDonald, Chondrites and the chemical composition of the earth, in: P. H. Abelson (ed.), "Researches in Geochemistry," vol. 1, pp. 476–494, Wiley, 1959.
†B. Mason, "Principles of Geochemistry," 2d ed., p. 50, Wiley, 1958.

Table 1-5.   Calculation of the Bulk Composition of the Earth*

	Metal × 0.288		Silicate × 0.632		Sulfide × 0.080		Total
Fe	90.78	26.14	6.90	4.36	61.1	4.89	35.39
Ni	8.59	2.47	. . . . .	. . . . .	2.88	0.23	2.70
Co	0.63	0.18	. . . . .	. . . . .	0.21	0.02	0.20
O			43.97	27.79			27.79
Si			20.00	12.64			12.64
Al			0.69	0.44			0.44
Mg			26.90	17.00			17.00
Ca			0.97	0.61			0.61
Na			0.22	0.14			0.14
K			0.11	0.07			0.07
Ti			0.07	0.04			0.04
Cr			. . . . .	. . . . .	0.12	0.01	0.01
P			0.02	0.01	0.31	0.02	0.03
S			. . . . .	. . . . .	34.3	2.74	2.74
Mn			0.15	0.09			0.09

*From B. Mason, "Principles of Geochemistry," 2d ed., p. 50, Wiley, 1958.

expense of magnesium, but the general order of magnitude of the abundance of the elements would not be changed appreciably.

Mason (1962) calculates the average composition of meteoritic matter on the assumption that it is composed of 74.7 percent silicate, 5.7 percent troilite, and 19.6 percent nickel-iron alloys, as shown in Table 1-6. The discrepancies between the two estimates shown in Table 1-6 arise primarily from the difference of composition in the silicate phase assumed. The meteorites used for the estimation are high-iron chondrites, all of which contain pyroxene in appreciable amounts and feldspar in minor amounts. This will account for the appreciably larger estimate for oxygen, silicon, calcium, aluminum, and sodium and the smaller estimate for iron and magnesium over the estimate assuming the mantle to be average peridotite.

With all the assumptions in such calculations, the exact numerical proportions of each element have limited significance. However, some semiquantitative interpretations can be made. Four elements—iron, oxygen, magnesium, and silicon—make up more than 90 percent of the earth's bulk; four more—nickel, calcium, aluminum, and sulfur—may occur in amounts exceeding 1 percent; seven more—sodium, potassium, chromium, cobalt, phosphorus, manganese, and titanium—do occur in amounts exceeding 0.1 percent; all other elements probably total less than 0.1 percent. Therefore 15 elements dominate the entire bulk of the earth.

Using similar approximations regarding the composition of the earth, and making the further assumption that the earth was entirely fluid

**Table 1-6.   Average Compositions of
Meteorites and the Earth***

	Meteorites	*Earth*
O	33.0	27.79
Fe	28.6	35.39
Si	16.95	12.64
Mg	13.86	17.00
S	2.07	2.74
Ni	1.68	2.70
Ca	1.39	0.61
Al	1.10	0.44
Na	0.68	0.14
Cr	0.30	0.01
Mn	0.20	0.09
P	0.13	0.03
K	0.10	0.07
Co	0.10	0.20
Ti	0.08	0.04

*From B. Mason, "Meteorites," table 18,
pp. 151–152, Wiley, 1962.

early in its history, Goldschmidt (1922)* reasoned that all elements would
have been distributed into four major phases according to their chemical
activity. In the fluid earth, oxygen exceeded sulfur, but oxygen and sulfur
combined were insufficient to combine with all the metals. Therefore three
fluid phases would have developed: a metallic phase, a sulfide phase, and an
oxide phase. The oxide phase can be considered a silicate phase, because
silicon would have been most stable as an oxide and would have formed the
major bulk. Iron, since it is the most abundant metal, would have been
dominant in all three phases, which can thus be considered as consisting of
iron, iron sulfide, and iron silicate. Each metal would then have distributed
itself among the phases according to its stability relative to the iron phases
in reactions of the following types:

$$\text{Metal} + \text{FeS} \rightleftharpoons \text{metal S} + \text{Fe}$$
$$\text{Metal} + \text{FeSiO} \rightleftharpoons \text{metal SiO} + \text{Fe}$$

Thus a metal which was more stable as the sulfide than iron would have
reacted with iron sulfide to form sulfide plus free iron, or a metal which was
more stable as the silicate than iron would have reacted with iron silicate to
form metal silicate plus free iron. For example,

$$\text{Cu} + \text{FeS} \rightarrow \text{CuS} + \text{Fe}$$
$$\text{K} + \text{Fe silicate} \rightarrow \text{K silicate} + \text{Fe}$$

*V. M. Goldschmidt, "Geochemistry," chap. II, pp. 11–26, Oxford, 1954.

Reverse reactions are also possible:

$$Fe + NiS \rightarrow FeS + Ni$$
$$Fe + Ni \text{ silicate} \rightarrow Fe \text{ silicate} + Ni$$
$$Fe \text{ silicate} + KS \rightarrow FeS + K \text{ silicate}$$
$$FeS + Cu \text{ silicate} \rightarrow CuS + Fe \text{ silicate}$$

In this manner each element in a homogeneous fluid body would tend to be concentrated into the liquid phase (metal, sulfide, or silicate) in which it was most stable chemically. The three phases, because of density differences, would then have become gravitationally concentrated, having the heavy metal phase in the core region, the lighter silicate phase in the mantle and crustal regions, and the sulfide phase possibly near the core-mantle boundary. The distribution of elements is thus not according to their weight but according to their chemical behavior.

A fourth group of elements would not have been able to condense into the liquid phase but would have remained as a gas phase. These elements were concentrated in the primative atmosphere of the earth at this stage.

There are, thus, four fundamental types of elements: those tending to concentrate in the metallic phase, those tending to concentrate in the sulfide phase, those tending to concentrate in the silicate phase, and those tending to concentrate in the gas phase. Goldschmidt has named them *siderophile*, *chalcophile*, *lithophile*, and *atmophile* elements, respectively. He assigned the elements to groups according to their chemical activity, and the results are listed in Table 1-7. We see that 12 of the most common elements are

Table 1-7. Goldschmidt's Geochemical Classification of the Elements*

Siderophile (metallic phase)	Chalcophile (sulfide phase)	Lithophile (silicate phase)	Atmophile (gas phase)
Fe Co Ni†	Cu Ag	Li Na K Rb Cs	H N (C) (O)
Ru Rh Pd	Zn Cd Hg	Be Mg Ca Sr Ba	Inert gases
Os Ir Pt	Ga In Tl	B Al Sc Y	
Au Re Mo	(Ge) (Sn) Pb	Rare earths	
Ge Sn	As Sb Bi	(C) Si Ti Zr	
		Hf Th	
C P	S Se Te	(P) V Nb Ta	
(Pb) (As) (W)	(Fe) (Mo) (Cr)	O Cr W U	
		(H) F Cl Br I	
		(Tl) (Ga) (Ge)	
		(Fe) Mn	

*From B. Mason, "Principles of Geochemistry," 2d ed., p. 55, Wiley, 1958.
†The elements in parenthesis occur in more than one group and are minor in groups where they are in parentheses. The 15 most abundant elements are underlined.

of major importance in the silicate spheres. Of these, phosphorus would tend to be more highly concentrated in the metallic core and would thus be relatively minor in the surficial rocks. Chromium occurs in nature dominantly with ultramafic igneous bodies and would be concentrated in the mantle and minor in the surficial rocks.

Composition of the Crust

Clarke and Washington (1924)* made the first approximation of the composition of the earth's crust. They assumed that igneous rocks make up 95 percent of the upper 10 km of the crust and based their average on a numerical average of 5,159 "superior" analyses of igneous rocks available at that time. Their computation was recalculated to 100 percent after eliminating water and minor constituents. The percentages are as follows:

SiO_2	60.18	CaO	5.17
Al_2O_3	15.61	Na_2O	3.91
Fe_2O_3	3.14	K_2O	3.19
FeO	3.88	TiO_2	1.06
MgO	3.56	P_2O_5	0.30

Several objections to their method have been raised. First, the analyses available were not of uniform worldwide distribution but were dominantly from the North American and European continents; vast oceanic areas were only sparsely represented. Thus their average is, at best, the average of the continental crust. The second objection is that the method emphasized rare and unusual rocks, since they were more frequently analyzed than common rock types. This was not considered a serious objection, because the compositional range of rare rocks is very wide, and they would probably average out. The third objection is that no allowance was made for the actual amount of each rock type represented by the analyses. Thus one analysis from a batholith carried as much weight as one analysis from a small dike.

The third objection has been met in two ways. Goldschmidt† pointed out that if an average sample of a large area of crust consisting dominantly of crystalline rocks could be obtained, it would be a reliable sample of the crust as a whole. He used glacial clay from southern Norway as such a sample, because it represented a rock flour scraped from the Fennoscandian Peninsula. His results are remarkably similar to those of Clarke and Washington:

SiO_2	59.19	Na_2O	2.05
Al_2O_3	15.82	K_2O	3.93
$Fe_2O_3 + FeO$	6.99	H_2O	3.02
MgO	3.30	TiO_2	0.79
CaO	3.07	P_2O_5	0.22

---

*F. W. Clarke and H. S. Washington, The composition of the earth's crust, *U.S. Geol. Survey Prof. Paper 127*, 1924.
†V. M. Goldschmidt, *op. cit.*, p. 54.

A second approach to the composition of the crust which meets this third objection and takes into account the oceanic segment of the crust has been made by Poldervaart (1955).* He divided the earth into four major geologic divisions: the oceanic basins, with an area of 268×10^6 km² and a thickness of 5 km; the continental sectors, with an area of 105×10^6 km² and a thickness of 35.75 km; the young folded mountain belts, with an area of 42×10^6 km² and a thickness of 38.25 km; and the suboceanic and continental shelf and slope areas, with an area of 93×10^6 km² and a thickness of 16.25 km. He designated a fifth division for volcanic islands, with an area of 2×10^6 km² and a thickness of 14.5 km. These subdivisions are indicated in Fig. 1-5.

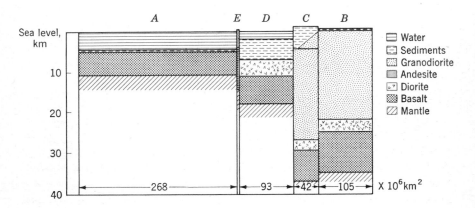

Fig. 1-5. Subdivision of the crust into major geologic divisions for the purpose of calculating the average composition of the crust. Segment A represents the oceanic basins; B, the continental sectors; C, the young folded mountain belts; D, the suboceanic and continental shelf and slope areas; and E, the volcanic islands. (*After A. Poldervaart, Geol. Soc. America Spec. Paper 62, 1955.*)

Poldervaart then calculated the best average values of each lithic type in each environment, recalculated the values on the basis of volume estimates to mass compositions, and added the values to give the estimated average composition of the crust above the Mohorovicic discontinuity. Values for the four major geologic environments and the total lithosphere are given in Table 1-8.

By comparing Poldervaart's results with those of Clarke and Washington and those of Goldschmidt, the effect of considering oceanic basins which have large areas of basaltic rocks is apparent: it is to lower the average content of silica and alkali and to raise the content of iron, magnesium, and

*A. Poldervaart, Chemistry of the earth's crust, in A. Poldervaart (ed.), Crust of the earth, *Geol. Soc. America Spec. Paper 62*, pp. 119–144, 1955.

Table 1-8.   Composition of the Lithosphere*

	SiO_2	TiO_2	Al_2O_3	Fe_2O_3	FeO	MnO	MgO	CaO	Na_2O	K_2O	P_2O_5
Oceanic areas	46.6	2.9	15.0	3.8	8.0	0.2	7.8	11.9	2.5	1.0	0.3
Suboceanic area	49.5	1.9	15.1	3.4	6.4	0.2	6.2	13.2	2.5	1.3	0.3
Young folded mountain regions	58.4	1.1	15.6	2.8	4.8	0.2	4.3	7.2	3.1	2.2	0.3
Continental shield regions	59.8	1.2	15.5	2.1	5.1	0.1	4.1	6.4	3.1	2.4	0.2
Lithosphere	55.2	1.6	15.3	2.8	5.8	0.2	5.2	8.8	2.9	1.9	0.3

*From A. Poldervaart, Crust of the earth, *Geol. Soc. America Spec. Paper 62*, p. 133, 1955.

calcium. Poldervaart's estimate is probably a more accurate average composition of the crust than either of the others.

Temperature within the Earth

Many measurements have been made in mines, wells, and tunnels which show that temperature increases with depth in the earth. From such measurements temperature gradients can be calculated, and they are found to vary widely from place to place. Measurements in a drill hole in the Norris Geyser basin of Yellowstone Park showed a temperature of 205°C at a depth of only 81 m; the surface temperature was 16°C. This indicates a gradient of 0.43 m/°C. However, Yellowstone Park is an area of very young igneous activity. In contrast, the Precambrian rocks of the Witwatersrand in South Africa have a gradient as low as 148 m/°C. Hundreds of measurements have been made, but the most representative gradient for the earth seems to be 35 m/°C.

In view of the earth's radius, the depths at which thermal measurements have been made are insignificant; therefore, to interpret temperature at great depth from the surface gradient is very risky. A straight-line extrapolation of the surface gradient gives the impossibly high temperature of nearly 200,000°C at the center of the earth. The interpretation of temperature at depth, then, must be based on other data and other assumptions. The main problems are the source of heat and the thermal conductivity of materials at depth. It seems likely that most of the heat now being lost at the earth's surface is generated by the decay of radioactive isotopes. These isotopes are distributed among the rocks in such a manner that the higher the silica content of an igneous rock, the higher the rate of radioactive-heat generation. In fact, according to average-radioactive-content data, a 35-km layer of granite would generate more heat than the earth is now losing at its surface. On the basis of the earth models, it can be said that temperatures rise much more slowly with depth and that the major portion of the heat now being lost is therefore generated near the surface.

The core-mantle boundary offers a situation which can be interpreted as being indicative of temperature at depth. Temperature above and below the Gutenberg discontinuity must be approximately the same and must be such that the base of the mantle is below its melting temperature, therefore solid, and that the top of the core is above its melting temperature, therefore liquid. Verhoogen (1960)* analyzed the situation with respect to the thermodynamic properties of the possible materials involved. He concluded that the most likely temperature would be less than 2700°K. Temperatures above this level are controlled by the mechanisms of heat transfer, that is, by radiation or convection. In the initial cooling of a molten earth, the temperature gradient at depth would have been controlled by convection. Thus a gradient would have been established so that the mantle temperature increased slowly with depth. If the gradient became too steep in the liquid mantle, the rate of convective overturn would have increased, and the temperature-depth curve would have become flatter. If the gradient became too low, the rate of convective overturn would have decreased so that the temperature-depth curve would have become steeper. Thus an optimum temperature gradient, the adiabatic gradient, would have been maintained throughout the crystallization of the mantle and would have been "frozen into" the solid mantle. Once crystallization had taken place, cooling would have been much slower because of very slow convection and conduction. The gradient would thus have remained very nearly the adiabatic gradient. On this basis Verhoogen calculates the temperature at a depth of 200 km to be very near 1500°C and at a depth of 100 km to be near 1100 to 1200°C. Therefore at depths of 100 to 200 km, temperatures are probably not far below the melting temperatures of rocks, and relatively minor disturbances could lead to melting and magma generation.

These speculations on temperatures within the earth are simply speculations and depend entirely on the assumptions made with respect to the distribution of heat in the early history of the earth and on radioactivity. It is instructive to compare some of these speculations. Birch, in a Geological Society of America Presidential Address (published February 1965)† on the earth's thermal history, concludes that in a cold essentially homogeneous earth, radioactively generated heat would have initiated melting of iron about 4,500 million years ago. The gravitational settling of this material to form the core would have released some 600 cal/g and would have resulted in fusion of a portion of the silicate phase. This would have risen to form a liquid upper mantle and crust containing all the radioactive elements. The process of core and upper-mantle separation would have taken place in the interval between 4,500 and 3,500 million years ago. Tempera-

*J. Verhoogen, Temperatures within the earth, *Am. Scientist*, vol. 48, pp. 134–159, 1960.
†F. Birch, Speculations on the earth's thermal history, *Geol. Soc. America Bull.*, vol. 76, pp. 133–159, 1965.

tures on the order of 4000 to 5000° would have been reached in the deep interior during the core segregation, and presumably surface temperatures would have exceeded 1000°.

In contrast, Donn, Donn, and Valentine (1965)* considered the early history of the earth through the age and character of the oldest known rocks and the history of solar-type stars and arrived at a very different interpretation. Again beginning with a cold homogeneous earth, they conclude that melting of the silicates in the mantle could not have occurred during the first billion years of the earth's history (prior to 4,000 million years ago). Further, they conclude that any water on the earth's surface was frozen prior to 4,000 million years ago. Thus two modern interpretations arrive at two very different pictures of temperature in the earth and at the surface in its early history. It is clear, however, that molten material can be generated at shallow depths in the mantle at the present time. Verhoogen's estimate of shallow-depth temperatures is thus a reasonable one.

*W. L. Donn, B. D. Donn, and W. G. Valentine, On the early history of the earth, *Geol. Soc. America Bull.*, vol. 76, pp. 287–306, 1965.

2 MAGMATIC PROCESSES

Basic Definition

In elementary geology the college student is taught that there are three major divisions of rocks: igneous, sedimentary, and metamorphic. At the elementary level, igneous rocks are relatively easy to define as those rocks resulting from the crystallization of minerals and the solidification of glass from siliceous melts at high temperatures or as those rocks resulting from the solidification of molten matter which originated within the earth. However, modern investigation of the physicochemical nature of silicate melts makes it apparent that many rocks which were previously called igneous were in all probability not developed by the direct crystallization of silicate melts. Moreover, many igneous-appearing rocks have been shown by field and petrographic evidence to have been formed in the solid state at temperatures well below the melting range. Such rocks are the result of metasomatic processes, i.e., the exchange of ions between minerals of solid rocks and migrating fluids over a wide range of temperatures and pressures.

In volcanic regions the formation of igneous rocks can be observed as the obvious product of high-temperature silicate melts which have reached the surface from the deeper part of the earth. A direct extrapolation from these observations would lead to the conclusion that such a melt could crystallize at depth, giving rise to igneous rocks. However, a magma mass in such a subterranean position has never been seen, and therefore the assumption that igneous rocks have evolved from such masses must be based on other evidence. The most valid evidence is the similarity of composition, mineralogy, and texture to clearly magmatic volcanic rocks; the effects of heat on the walls of such masses; and the apparent fluid nature of the material as indicated by contact relationships. From this evidence we must accept many bodies as being probably of true igneous origin. In the following discussion we shall accept all such bodies as igneous unless there is strong field or petrographic evidence of other origins.

GENERATION OF MAGMAS

Magma

The AGI Glossary defines magma as a "naturally occurring mobile rock material, generated within the earth and capable of intrusion and extrusion, from which igneous rocks are considered to have been derived by solidification. It consists 'in noteworthy part' of a liquid silicate-melt phase, which is liquid owing to the temperature attained, and a number of solid phases such as suspended crystals of olivine, pyroxene, plagioclase, etc. In certain instances a gas phase may also be present." This definition may be shortened to "a naturally occurring mobile silicate-rock melt which contains or does not contain suspended crystals and dissolved gases." Magmas extruded at the surface normally carry at least some suspended crystals, and from this it may be inferred that suspended crystals are also normally present at depth. The definition of magma includes these crystals and is not restricted to the melt phase. However, there is no clear definition of what percentage of a magma may be crystals. There is evidence that some materials may be mobile if they contain as little as 10 percent liquid. Whether this should be called magma or not is an open question. At any rate, magma involves a molten silicate material, and this involves melting of previously existing solid rock.

Melting Character of Rocks

In contrast to simple substances such as ice or metals, a crystalline aggregate consisting of several minerals does not have a specific melting point. Instead, such mixtures melt over a range of temperatures, and the composition of the melt changes as melting proceeds. Thus if a polymineralic rock is heated, a liquid phase will appear when a certain minimum temperature is reached. The liquid will contain all the constituent elements of the rock in some proportion. At the same time the remaining unmelted crystals will be depleted in some of the constituents. If after a portion of the rock has been melted no more heat is supplied, the melting will stop, and the proportions and composition of liquid and unmelted solid will stay the same. If heat is again supplied, the melting will continue, and the composition of both melt and remaining crystals will change progressively until the rock is entirely molten. Only then will the melt have the exact composition of the parent rock. The temperature of this ultimate complete melt must be markedly higher than the temperature of the first melt which appeared.

The experimentally determined melting relationships in the common rock-forming minerals indicate that the first liquid to form is relatively rich in silicon, potassium, sodium, and aluminum and relatively poor in iron, magnesium, and calcium. Thus, assuming that the parent rock being melted is basaltic in composition, the changing composition of both the melt and the remaining unmelted crystals can be followed. Assuming 100 g* of basalt

*Consider 100 g of rock containing 49.9 percent SiO_2 as containing 49.9 g SiO_2; 16 percent Al_2O_3 is equivalent to 16 g Al_2O_3; etc.

of the composition given in column A, Table 2-1, if 10 percent of the basalt is melted, the result is 10 g of liquid of the approximate composition given in column B-1. This 10 g contains $10 \times 0.739 = 7.39$ g of silica (column B-2), leaving $49.9 - 7.39 = 42.51$ g of silica in the solid (column B-3). But there are now only 90 g of solid, which contain $(42.51 \div 90) \times 100 = 47.2$ percent silica (column B-4). The same calculation has been made for each oxide. Comparison will show that each oxide present in a greater percentage in the liquid than in the parent rock has decreased in percentage in the remaining crystals; conversely, each oxide present in the liquid in a lesser percentage than in the parent rock has increased in percentage in the unmelted crystals. Proceeding on to 30-percent melting, the melt now is a liquid of the composition approximated in column C-1, Table 2-1. The liquid has become less silicic and the crystals even more impoverished in silica. Figure 2-1 illustrates the changing composition of the liquid as melting proceeds according to the calculations of Table 2-1. This is a purely illustrative example and is not intended to imply the precise compositions of the liquid for the melting of every rock.

Melting of a rock may thus be only partial, called *selective melting*, or complete, called *pure melting*. Selective melting always gives a liquid phase which is richer in silica and alkalies than the parent rock. Pure melting always gives a melt of the same composition as the parent rock. A rhyolitic magma or an andesitic magma can be developed from a basalt by selective melting, or a basaltic magma can be developed from a basalt by pure melting. A basaltic magma cannot be generated from any rock less mafic than basalt, such as rhyolite or andesite, but can be generated by selective melting of more mafic materials, such as some peridotites.

In an earth model grading downward from granite to basalt above the Mohorovicic discontinuity and then consisting of eclogite below the discontinuity in accordance with Kennedy's hypothesis, a granitic magma could be generated by pure melting of the upper crust or by selective melting of the lower crustal basalt or the eclogite. A basaltic magma, however, could be derived only by pure melting of either the basaltic portion of the crust or the eclogite. Moreover, if the upper mantle is some type of peridotite, as suggested by Kushiro and Kuno (1963),[*] or a mixture of peridotite and basalt, as suggested by Ringwood (1962)[†] or Engel, Engel, and Havens (1965),[‡] then basalts could be derived by selective melting of the upper mantle. In that case it is doubtful if a granite magma could be derived from the mantle, because the proportion of such a melt would be too small to separate from the unmelted materials.

[*]I. Kushiro and H. Kuno, Origin of primary basalt magmas and classification of basaltic rocks, *Jour. Petrology*, vol. 4, pp. 75–89, 1963.

[†]A. E. Ringwood, Mineralogical constitution of the deep mantle, *Jour. Geophys. Research*, vol. 67, pp. 4005–4010, 1962.

[‡]A. E. J. Engel, C. G. Engel, and R. G. Havens, Chemical characteristics of oceanic basalts and the upper mantle, *Geol. Soc. America Bull.*, vol. 76, pp. 719–734, 1965.

Table 2-1. The Changing Composition of Liquid and Crystals during Melting of a Basalt

Oxide	A. Basalt	B. After 10% melting				C. After 30% melting			
		Melt composition (rhyolite) (1)	Grams of each oxide removed in the liquid (2)	Grams of each oxide remaining in the solid (3)	Composition of the remaining solid, % (4)	Melt composition (dacite) (1)	Grams of each oxide removed in the liquid (2)	Grams of each oxide remaining in the solid (3)	Composition of the remaining solid, % (4)
SiO_2	49.9	73.9	7.39	42.51	47.2	66.8	20.04	29.86	42.7
TiO_2	1.4	0.3	0.03	1.37	1.5	0.6	0.18	1.22	1.7
Al_2O_3	16.0	13.8	1.38	14.62	16.3	16.5	4.95	11.05	15.8
Fe_2O_3	5.5	1.5	0.15	5.35	5.9	2.4	0.72	4.78	6.8
FeO	6.5	0.9	0.09	6.41	7.1	1.9	0.57	5.93	8.5
MgO	6.3	0.4	0.04	6.26	7.0	1.4	0.42	5.88	8.4
CaO	9.1	1.2	0.12	8.98	10.0	3.5	1.05	8.05	11.5
Na_2O	3.2	3.4	0.34	2.86	3.18	4.0	1.20	2.00	2.86
K_2O	1.6	4.5	0.45	1.15	1.28	2.7	0.81	0.79	1.13
P_2O_5	0.5	0.1	0.01	0.49	0.54	0.2	0.06	0.44	0.62
Totals	100.0 g solid		10.00 g liquid	90.00 g solid			30.00 g liquid	70.00 g solid	

Oxide	D. After 50% melting				E. After 75% melting				F. 100%
	Melt composition (andesite) (1)	Grams of each oxide removed in the liquid (2)	Grams of each oxide remaining in the solid (3)	Composition of the remaining solid, % (4)	Melt composition (1)	Grams of each oxide removed in the liquid (2)	Grams of each oxide remaining in the solid (3)	Composition of the remaining solid, % (4)	Melted liquid composition
SiO_2	60.4	30.2	19.7	39.4	54.4	40.80	9.10	36.40	49.9
TiO_2	0.8	0.4	1.0	2.0	1.1	0.82	0.58	2.30	1.4
Al_2O_3	17.6	8.8	7.2	14.4	17.3	12.98	3.02	12.10	16.0
Fe_2O_3	3.4	1.7	3.8	7.6	4.5	3.38	2.12	8.50	5.5
FeO	3.2	1.6	4.9	9.8	5.4	4.05	2.45	9.80	6.5
MgO	2.8	1.4	4.9	9.8	3.8	2.85	3.45	13.80	6.3
CaO	5.8	2.9	6.2	12.4	8.0	6.00	3.10	12.40	9.1
Na_2O	3.6	1.8	1.4	2.8	3.3	2.47	0.73	2.90	3.2
K_2O	2.1	1.05	0.55	1.1	1.8	1.35	0.25	1.00	1.6
P_2O_5	0.3	0.15	0.35	0.7	0.4	0.30	0.20	0.80	0.5
Totals		50.00 g liquid	50.00 g solid			75.00 g liquid	25.00 g solid		100.0 g liquid

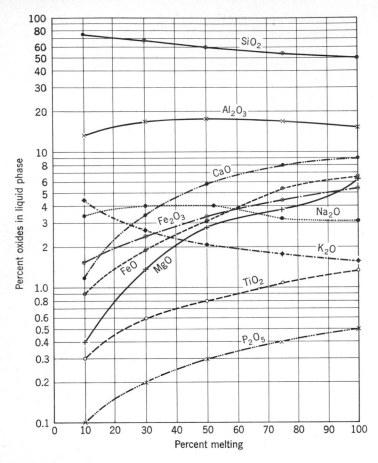

Fig. 2-1. Changing composition of the liquid as melting proceeds, according to the calculation of Table 2-1.

Sources of Heat

Sources of heat for the melting of rocks has been a problem of long standing to which no answer is as yet available. Early concepts considered magmas to be residual pockets of liquid which migrated slowly toward the surface after being formed by the original heat of the earth at great depths. Modern studies show no residual magmatic bodies in the mantle and indicate that most magmas probably come from a depth no greater than 100 km.

Before internal heat sources are considered, the heat received from the sun and its consequences should be discussed. The earth receives heat from the sun at the rate of 2 cal/(min)(cm²), which amounts to about 10^{24} cal/year over the total surface. This heat is largely radiated into space or converted into mechanical energy in the circulation of the atmosphere and hydrosphere. Some of the solar heat is absorbed into the earth, causing a temperature fluctuation in the rocks to a depth of a very few feet or tens of feet. It is obvious that this heat will not melt rock and create magma. However, solar energy is very important, because it maintains the surface temperature at 10 to 20°C and thus sets the lower limit of temperature of the geothermal gradient. Without solar energy the surface temperature would be at absolute zero or near to it, and the entire geothermal gradient would be lowered. Absolute zero is -273.1°C. Therefore, solar energy raises the temperature at shallow depths by nearly 300°C and requires near-surface rocks to be at temperatures much closer to their melting range than they would be if there were no sun.

Several processes may contribute to the heat required for the generation of magmas. The mechanisms which have been proposed can be grouped into the following classes: (1) radioactivity, (2) pressure changes, (3) depression of rock masses into higher-temperature zones, and (4) mechanisms transferring heat upward from the mantle by liquids or gases.

Radioactivity

As radioactive isotopes decay, heat is liberated at a determinable rate. Three major radioactive materials are present in the earth in sufficient quantity to cause an appreciable heating effect: uranium, thorium, and potassium 40. They liberate heat as follows:

U liberates 0.74 cal/(g)(year)
Th liberates 0.20 cal/(g)(year)
K40 liberates 0.000026 cal/(g)(year)

From the average percentage of these materials in the crust, it can be calculated that 1 g of average crustal material generates the following amounts of heat:

U in 1 g of crust develops 2.2×10^{-6} cal/year
Th in 1 g of crust develops 2.6×10^{-6} cal/year
K 40 in 1 g of crust develops 0.9×10^{-6} cal/year

In the uppermost 16 km of the crust there are 2.2×10^{25} g, each gram of which produces the heat given above. The total radiogenic heat of the crust thus amounts to 10^{20} cal/year. This enormous amount of heat is on the same order of magnitude as the heat lost at the surface through the present geothermal gradient. If the radioactive elements are uniformly distributed throughout the earth's surface, the heat being generated is being dissipated and cannot cause local melting. However, if it could in part be concentrated in one area, melting could be accomplished. No completely satisfactory mechanism for this localization has been proposed.

A possible importance of radioactive heat in early geologic history should be recognized. The radioactive isotope of potassium now makes up 0.12 percent of all potassium. However, it has a relatively short half-life of 8.4×10^8 years in comparison with the age of the earth, 4.5×10^9 years. This would indicate that at the beginning of geologic history the radioactive isotope would have made up 3 percent of all potassium and would thus have contributed about 25 times as much radioactive heat as at the present time. This may account for the tremendous igneous activity of the early geologic column. The other important radioactive elements would also have been slightly more abundant than at present, but their effect would not have been so great because of their very long half-lives.

Pressure Changes

Rocks, like most solids, expand when they melt, and therefore their melting range is dependent on pressure. A rock under high pressure can be heated above the temperature at which it would begin to melt if it were under a lower pressure. This phenomenon has been called on to explain the origin of magmas. If a rock were superheated under high pressure and the pressure were then reduced rapidly under such an environment that the heat could not be dissipated, the rock would melt. The development of deep-seated faults and the resultant reduction of pressure along the fault zone are the processes most likely to accomplish this pressure reduction.[*]

Barth (1962)[†] has reemphasized the importance of pressure reduction, particularly with respect to the development of extensive basaltic magmas. On the basis of new experimental data on the transmission of transverse (S) waves through liquids of high viscosity, Barth points out that silicate liquids at the pressures attained in the upper mantle would behave seismically as solids because of their high viscosity. Therefore, if fracturing extending into the mantle could reduce pressure a few thousand bars, the resultant marked decrease in viscosity would render the already liquid material highly mobile without requiring additional heat.

Orowan (1960)[‡] demonstrates that deep-seated faulting is not likely to involve slippage along an open fracture in solid rocks, because very high frictional resistance is required at depth. He suggests that faulting involves a plastic creep deformation, and this offers a source of local magma generation. As stress accumulates in a deep-seated region, it is progressively concentrated into narrow zones of plastic deformation in the crystalline rocks. Once flow is initiated in one zone, the stress is further concentrated in this zone until the rocks fail and displacement takes place. Shear melting may

[*]Yu. M. Scheinmann, A mechanism of formation of oceanic magma, in: The crust of the Pacific Basin, *Am. Geophys. Union Mono.* 6, pp. 181–186, 1962.
[†]T. F. W. Barth, "Theoretical Petrology," 2d ed., pp. 131–136, Wiley, 1962.
[‡]E. Orowan, Mechanism of seismic faulting, in: D. Griggs and J. Handin (eds.), Rock deformation, *Geol. Soc. America, Mem.* 79, 1960.

result, accompanying the high rate of flow, and sheetlike masses of magma will be generated. Thus a series of closely spaced parallel dikes can develop, and the magma will be essentially the same as the wall rock. This is a more likely process than generation of magma by release of pressure during open faulting.

Depression of Rock Masses into High-temperature Zones

Silicic igneous rocks melt at a lower temperature range than mafic igneous rocks. It would therefore appear possible to generate a silicic magma by the downwarping of crustal materials into a mafic substratum which is at a temperature below its melting range but above the melting range of the silicic rocks. Such a process may contribute to the generation of the great granite batholiths along the axes of geosynclines. According to the geosynclinal theory of mountain building, the collapse of geosynclines during the culmination of orogeny involves the downward buckling of the granitic crust into the underlying hotter mafic sima. The relationships are shown in Fig. 2-2. There is some question whether temperatures at depths to which the granitic crust would be downwarped are sufficiently high to cause melting. If the temperature were sufficiently high for melting, there would already be a minor fraction of granitic melt present as a result of the selective fusion process discussed earlier. Nevertheless, downwarping of any type of rock would result in an elevation of temperature, supplying some of the heat necessary for melting.

Many major structural features of the continental margins and adjacent oceanic floors have been interpreted in ways involving the depression of rock masses and subsequent magma generation. Magnetic data of the Pacific Ocean floor off the west coast of North America indicate a set of large-scale faults having tremendous horizontal displacement. Holmes (1965)[*] points out that the displacement is most consistent with gliding of the oceanic crust under the continents. The underridden portion of the continent is an area of voluminous Mesozoic to contemporary igneous activity. Similarly, Coats (1962)[†] concludes from a petrographic and structural study of the Aleutian arc that a portion of the magma involved in the volcanic activity was derived from sediments and volcanics of the ocean floor which were depressed to great depths. He invokes large-scale thrusting of the island arc over the oceanic crust as a mechanism of moving surficial rock types to great depths.

Mechanisms Transferring Heat Upward
from the Deep Mantle

Several processes can be visualized by which heat may be transferred from deep within the mantle to the subcrustal regions. One of these processes

[*] A. Holmes, "Principles of Physical Geology," 2d ed., pp. 939–943, Ronald Press, 1965.
[†] R. R. Coats, Magma type and crustal structure in the Aleutian arc, in: The crust of the Pacific Basin, *Am. Geophys. Union Mono.* 6, pp. 92–109, 1962.

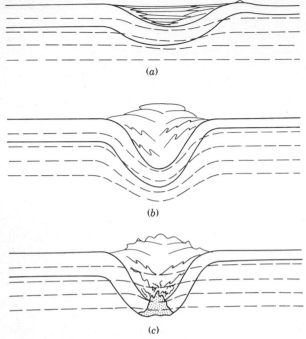

Fig. 2-2. Generation of magma by depression of the crust in a tectogene. The dashed lines represent four hypothetical isogeotherms. (*a*) The isogeotherms are depressed under the geosyncline due to the blanketing effect of sediments; (*b*) the situation immediately following geosynclinal collapse; (*c*) the situation after sufficient time has passed to allow the isogeotherms to rise from the base of the crust.

involves a convective overturn of at least a portion of the mantle. The physical nature of the material of the mantle must be quite different from that of surface rocks due to high temperatures and pressures. Geophysical evidence indicates that mantle material must behave as a plastic under long-term stresses but as a brittle material under short-term stresses down to 700 km, the maximum depth of earthquakes. Below this depth the mantle behaves as a plastic under all stresses. Such material would expand on heating and become less dense than the surrounding somewhat cooler material. As a result of density differences and plasticity, the hotter mass would rise. At the same time colder material would move in laterally to occupy the space, and a convection cell would be set up. Figure 2-3 illustrates this overturn. The result would be local elevation of temperature immediately below the crust. Mathematical analysis (Griggs, 1939)* of this system shows that it is physi-

*D. T. Griggs, A theory of mountain building, *Am. Jour. Sci.*, vol. 237, pp. 611–650, 1939.

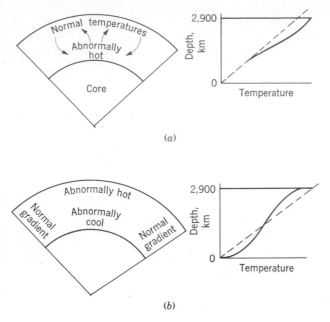

Fig. 2-3. Convective overturn in the mantle, causing elevation of temperature under the crust. (*a*) Before overturn;
(*b*) after overturn.

cally possible if the convective cell is large enough. However, the system
presents two problems: (1) the source of local heat at depth is unknown, and
(2) if the overturn had taken place repeatedly, the mantle would be homogeneous and contain no discontinuities.

The second objection, that of discontinuities in the mantle, may not be
a serious one if the seismic discontinuities are the product of inversion of
one crystal modification to another. Ringwood (1962)* proposes three
structural inversions in mantle material in the range of 400 to 1,000 km. He
suggests that at a pressure equivalent to a depth of 400 km, the inversion of
pyroxene to olivine plus stishovite (a high-density silica polymorph) would
occur; at a depth of 600 km, an inversion of normal olivine to olivine with a
spinel-type structure would occur; and at a depth of around 1,000 km, the
breakdown of spinel to periclase and a magnesium silicate with a corundum-
type structure would occur. These changes would result in a continuous
density increase in material of homogeneous composition and would coin-
cide with minor seismic discontinuities in the mantle.

A second mechanism for transfer of heat from the mantle has been
proposed: it involves the streaming of hot liquids or gases up the surface of

*Ringwood, *loc. cit.*

deep fractures and into the subcrustal zones. Three depth zones of earth-quakes extending from approximately 20 km to approximately 700 km are associated with modern young mountain systems. These earthquake foci are distributed in a pattern which indicates a major shear zone extending diagonally downward into the mantle to a depth of about 700 km. Reduction of pressure along this zone could conceivably cause the melting and ascent of magma from the high-temperature zones. However, if actual molten material could not be generated, the liberation of hot vapors and solutions from the depth and their subsequent ascent to subcrustal levels would at least occur. This concept has been illustrated by Eardley (1957)* for the Andean crustal structures (Fig. 2-4). It is, perhaps, the most attractive pro-posal for the source of magmatic heat. As yet, however, no one hypothesis of the origin of magmatic heat is entirely satisfactory.

ASCENT AND EMPLACEMENT OF MAGMAS

The ascent of the newly generated magma mass may be a relatively rapid process, as exemplified by the Hawaiian volcanoes, or a very slow process

*A. J. Eardley, The cause of mountain building—an engima, *Am. Scientist*, vol. 45, pp. 189–217, 1957.

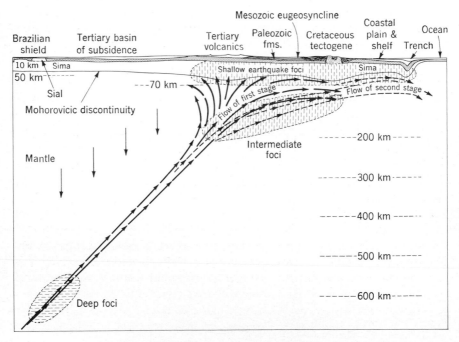

Fig. 2-4. Mantle fracture zones as a source of heat. Andean crustal structure accord-ing to A. J. Eardley (*Am. Scientist, vol. 45, pp. 89–217, 1957*).

similar to salt dome intrusions. Geophysical evidence from the 1959–1960 eruption of Kilauea on Hawaii indicates that magma began to rise from a depth of 55 km 2 years prior to the eruption and was stored in a subsummit chamber at a depth of about 4 km (Eaton, 1962).* In contrast, many authors now conceive of granitic intrusions as being a product of slow plastic deformation which involves not only the intrusion but also the surrounding wall rock (Read, 1957; Holmes, 1965).†‡ Such intrusive processes must of necessity be very slow. In either case there are four basic causes for the ascent of the intrusive body, each of which is discussed separately:

1. Gravitative pressure due to density differences
2. Volume change due to reduction of pressure
3. Horizontal pressure due to tectonic movements
4. Magmatic stoping

Gravitative Pressure due to Density Differences

Hydrostatic pressure, that is, nondirectional confining pressure, on a deepseated magma body is the result of the weight of the overlying mass of rock or magma. The weight of the overlying column is dependent on the height of the column and the density of the material in the column. A rock expands when it melts; therefore the density of the magma will be less than the density of the rock from which it formed. Table 2-2 gives some generalizations

Table 2-2. Rock and Magma Densities*

	Solid			Liquid	
	20°	*1000°*	*1100°*	*1000°*	*1100°*
Average sediment	2.5	2.42	2.41		
Average granite	2.75	2.63	2.62	2.40	2.39
Average gabbro	3.00	2.92	2.91	2.75	2.74

*From F. F. Grout, "Petrography and Petrology," p. 193, McGraw–Hill, 1932.

on density differences between rock and magma. If a basaltic-magma chamber underlies basalt rock and the irregular upper surface of the magma chamber ranges in depth from 60 to 70 km, the situation shown in Fig. 2-5

*J. P. Eaton, Crustal structure and volcanism in Hawaii, in: The crust of the Pacific Basin, *Am. Geophys. Union Mono.* 6, pp. 13–29, 1962.
†H. H. Read, "The Granite Controversy," Thomas Murby & Co., pp. 334–337 and 374–398, 1957.
‡Holmes, *op. cit.*, pp. 277–281.

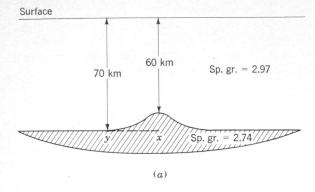

(a)

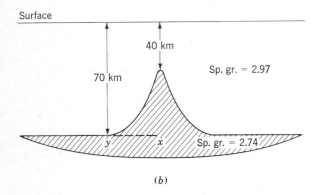

(b)

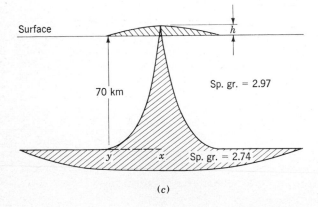

(c)

Fig. 2-5. Ascent of magma due to hydrostatic pressure differences at points X and Y. See text for discussion. Cross-hatched area represents the magma chamber. In (c) a volcanic cone is built on the surface to height h.

will develop. The hydrostatic pressure at point X will be the result of the weight of 60 km of material of specific gravity 2.97 plus 10 km of material of specific gravity 2.74, whereas at point Y the hydrostatic pressure will be the result of the weight of 70 km of material of specific gravity 2.97. The weight at Y will thus be greater than at X, and as a result there will be a pressure differential which will cause magma to move higher in the uplifted portion of the chamber. After the magma has risen so that its highest point is at a depth of 40 km, the pressure at point X will be $(40 \times 2.97) + (30 \times 2.74)$. An even greater pressure differential will exist when the magma column has risen to the surface. Then the pressure at X will be 70×2.74, and the pressure differential will be at a maximum. The magma will continue to rise, building up a cone on the surface to a height h such that the pressures at points X and Y are equal. Then

$$(70 + h)\, 2.74 = 70 \times 2.97$$
$$h = 5.84 \text{ km}$$

Hawaii is a basaltic volcano rising through a dominantly basaltic crust. Seismic evidence from earthquakes associated with eruptions indicates that the magma rises from a depth of approximately 70 km. The peaks of the active volcanoes are at an elevation of $\pm 28,600$ ft, or 8.7 km, above the surrounding sea floor. Thus a major portion of the rise of the magma can be accounted for by the density difference between the magma and the surrounding rock.

In contrast, the ascent of a light granitic magma into sediments at the base of a deformed geosynclinal prism would present a different problem. Here, because the sediments and the magma have very similar specific gravities, the magma will lose its hydrostatic head and will be unable to penetrate any great distance upward into the sedimentary prism. Similarly, Scheinmann (1962)* points out that a peridotite magma generated in the mantle could not rise by density contrast, since the magma and the surrounding rock have very similar densities. Such magmas can rise only by tectonic squeezing.

Volume Change due to Reduction of Pressure

A second cause of ascent is the change of pressure on the magma as a result of its rise to lower-pressure depths in the crust. If in the ascent of a basaltic magma, the liquid rises from a depth of 70 km to the surface (see above), a reduction of pressure on the mass from a maximum of 20,000 atm to 1 atm would take place. As a direct result of this pressure reduction the magma mass would expand, and this expansion would result in further upward movement of the magma.

A second direct result of the pressure drop is the evolution of gases

*Scheinmann, *op. cit.*, pp. 181–186 (especially p. 185).

from the liquid. Solubility of gas in a liquid is directly proportional to the pressure on the system. Thus, at high pressures more gas can be retained in solution than at low pressures. This is familiarly demonstrated in the behavior of carbonated drinks. When the cap is removed, gas is evolved. A similar process operates in a magma, in that the solubility of the gases is reduced with the reduction of pressure during ascent. The dissolved gases are evolved as bubbles, markedly increasing the volume of the magma, which in turn causes further upward migration. This evolution of gas as a result of pressure loss is called the *first boiling of the magma*, but it does not involve the complete loss of volatiles.

A clear demonstration of the effect of this gas expansion has been given by Woollard (1954).* A gravity study of the island of Oahu in the Hawaiian chain indicates that the best specific-gravity assumption for the mass of the island is 2.3, in spite of the fact that the islands are composed dominantly of basalt. The explanation of this low specific gravity is found in the highly porous nature of the basalt rock as a result of abundant gas cavities.

Horizontal Pressure due to Tectonic Movements

Magmas rising into a zone of tectonic activity may be squeezed and forced upward or laterally into the surrounding rocks. One product of such a force is the injection of magma along the crests of anticlines and the troughs of synclines, forming intrusive bodies called *phacoliths*. Similarly, a magma body intersected by a thrust fault may result in the injection of magma along the thrust surface, giving rise to a body called a *sole injection*. The development of concordant contacts between large intrusives and their walls may be in part due to squeezing and kneading of the magma into its wall rock. This may partly be the case in some concordant batholiths.

Magmatic Stoping

Magmas create room for upward and, to some extent, lateral movement by surrounding and engulfing various-sized fragments of their wall rock. This process is called *stoping*; it ranges from the engulfing of single crystals to the disruption of tremendous blocks of wall rock. Stoping processes are classified into three categories based on the abundance and size of the stoped blocks: (1) piecemeal stoping, (2) cauldron subsidence, and (3) roof foundering.

Piecemeal stoping consists in the incorporation of numerous relatively small blocks of wall and roof material by the magma. These blocks are frequently of greater density than the magma and tend to settle downward through the magma. Fragments lighter than the magma would float toward the top of the chamber. In either case the margin of the magma body becomes cluttered with scattered stoped blocks of wall rock, ranging in size

*G. P. Woollard, Crustal structures beneath oceanic islands, *Royal Soc. London Proc.*, vol. 222, A, pp. 361–387, 1954.

from single crystals to blocks measurable in hundreds of feet but still small relative to the size of the magma body (Fig. 2-6). The resulting included blocks are called *xenocrysts* if they are composed of single wall-rock crystals and *xenoliths* if they are rock fragments.

Piecemeal stoping of the roof is assumed to have several causes. First, because rocks are poor conductors of heat, the local thermal gradient around a magma body is quite steep in the early stages of heating; that is, rock at a relatively short distance from the magma contact is only little affected by its heat (Jaeger, 1957).* Consequently, rupturing stresses are set up in the wall rock as a result of differential thermal expansion. Second, high-pressure

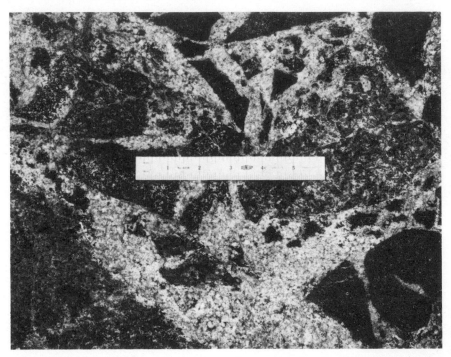

Fig. 2-6. Xenoliths of gabbro in granite at the margin of an intrusion. Mellen, Wisconsin.

zones can be built up in the rock as a result of trapped water or gases, such as water or carbon dioxide, liberated by reactions in the high-temperature zone. Third, the injection of thin stringers of magma (Fig. 2-7) (apophyses) into the rock under hydrostatic pressure develops tremendous shearing stresses at the ends of the stringers, thus ripping the rock apart. These processes combine to rupture the wall rock by shattering it into fragments

*J. C. Jaeger, The temperature in the neighborhood of a cooling intrusive sheet, *Am. Jour. Sci.*, vol. 255, pp. 306–318, 1957.

Fig. 2-7. Apophyses of granite cutting gabbro one-half mile from the contact. Mellen, Wisconsin.

which settle into the magma. Piecemeal stoping probably will not account for the major ascent of the magma from depth, but it will account for ascent of the last tens of meters or possibly the last kilometer. The most important features of piecemeal stoping, however, are its effects on the detailed nature of the contact and the development of numerous xenoliths in the magma body. A magma body emplaced by this process will commonly have a blocky discordant contact between the magma mass and its roof or wall. Frequently, the contact zone consists of a gigantic breccia of wall rock intruded by dikes of various size and stringers of magma.

The second stoping process is *cauldron subsidence*, which involves the subsidence of blocks of large size compared with the magma chamber. This is particularly effective in the ascent of stocklike bodies. When a cylindrical body of magma is rising in the crust, two situations can develop: (1) The pressure in the magma chamber may be greater than the strength of the roof rocks, and in such a case, a conical fracture will develop over the magma chamber and have its apex in the top of the chamber. The mass of roof rock so isolated may be lifted up bodily and magma injected along the fracture as a cone sheet (Fig. 2-8). (2) The magma pressure may be less than the weight of the overlying rocks; a steeply conical fracture may develop in the roof and have its base in the magma chamber. This block can then subside into the magma, resulting in cauldron subsidence. As the magma moves up into the fracture so formed, it will chill to form a ring dike. Repeated subsidence

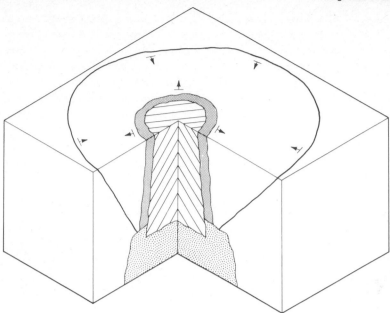

Fig. 2-8. Block diagram showing the relationships of a ring dike (solid) and cone sheet (stippled) to the subjacent intrusion (crosses). The block stoped by cauldron subsidence is cross-hatched.

of the sunken block of roof rock has produced a series of concentric ring dikes in several areas, and variations of magma pressures have developed related cone sheets.

Roof foundering is the largest category of stoping phenomena. Theoretically, a large magma body could rise so close to the surface that the strength of the roof rocks would be insufficient to support the thin roof over a broad area. As a result the roof could break up into blocks of relatively large dimensions and then be uplifted and tilted differentially to give a chaotic structure. Such roof structures are not uncommon over batholithic bodies. Under extreme conditions of roof foundering, the ruptured blocks could in part settle into the magma, leaving a portion of the magma exposed at the surface. Such an unroofing has been invoked by Hamilton and Myers (1967)* for the relationship between the Idaho batholith and its roof of silicic volcanics and for other similar situations.

Geophysical evidence accumulated over many large intrusive igneous bodies casts serious doubts on the effectiveness of large-scale stoping in magma emplacement. If magma rose by large-scale subsidence or foundering of denser crustal blocks through the liquid, then a given volume of

*W. Hamilton and W. B. Myers, The nature of batholiths, *U. S. Geol. Survey Prof. Paper 554-C*, 1967.

dense rock would have been replaced by an equal volume of less-dense rock. The fact that the more-dense rock would still be present at depth would result in only a small gravity deficiency over the intrusive body. In actuality there is consistently a large deficiency of gravity over batholithic intrusions, mitigating against the presence of sunken crustal blocks at depth.

In general, stoping can be considered important in controlling the nature of the contact between a magma and its wall and in contributing wall-rock fragments to the magmas. Stoping, however, is probably not an important process in contributing any large proportion of the vertical ascent of the magma. The major portion of the ascent is probably due to density differences between magma and solid rock and to horizontal pressures in the crust. The relative importance of these processes will vary, depending on the environment of the igneous activity (i.e., compressional versus tensional), the character of the magma, and the character of the rock through which it is ascending.

COMPOSITION OF MAGMAS
Average Composition of Igneous Rocks
The composition of a magma cannot normally be determined; only the composition of the resulting igneous rock can be determined. This is because a gas phase is normally present in the magma and is lost during crystallization or boils off from the magma. Even the analysis of igneous rocks is frequently somewhat misleading because of secondary modifications or weathering changes which have taken place or because of the difficulty of making exactly duplicatable analyses of silicate materials. The best approach to compositions, then, is through averages and compositional ranges.

The average composition of igneous materials cannot be approached through an arithmetical average of all analyses, because the available analyses emphasize the uncommon rock types. For example, Washington (*U. S. Geological Survey Prof. Paper* 99) lists 18 analyses of rare feldspathoidal rocks from Magnet Cove, Arkansas, while listing only eight analyses from the Keweenawan area of Michigan, of which only three are basalt or related rocks. The igneous rocks of Magnet Cove are confined to an area of 2.4 by 3 miles, whereas the basalts of the Keweenawan area are the dominant rocks in a belt 3 to 12 miles wide and 150 miles long in Michigan alone, with similar tremendous areas in Wisconsin, Minnesota, and Canada. A direct average would thus favor the small area with the interesting, but quantitatively unimportant, rocks. Therefore a weighted average must be used in which the area (or better, the volume) of the different rock types is taken into account.

R. A. Daly (1933)[*] presents reasonably accurate measurements of

[*]R. A. Daly, "Igneous Rocks and the Depths of the Earth," p. 35, McGraw-Hill, 1933.

outcrop areas of various igneous-rock types in the Appalachian and Cordilleran belts, as taken from 75 U. S. Geological Survey Folios published up to January 1912. These data are listed in Table 2-3. It is evident from this listing that the calc-alkalic rocks make up by far the greatest bulk of igneous rocks, the total percentage of all the alkalic rocks being less than 0.01 percent. On a continentwide basis this relationship seems to be maintained, and the proportion of alkalic rocks is probably less than 0.1 percent of the igneous rocks. Within the group of calc-alkalic rocks, Daly* makes the following generalization for the world as a whole:

Ratio of total areas, granite to diorite	20 : 1
Ratio of total volumes, rhyolite to andesite	1 : 10
Ratio of total areas, granite to gabbro	20 : 1
Ratio of total volumes, rhyolite to basalt	1 : 50

On the basis of Daly's measurements in the Cordillera and in the Appalachian system, Knopf calculated the average igneous-rock composition as follows:

SiO_2	61.64	Na_2O	3.40
Al_2O_3	15.71	K_2O	2.65
Fe_2O_3	2.91	H_2O	1.26
FeO	3.25	TiO_2	0.73
MgO	2.97	P_2O_5	0.26
CaO	5.06	MnO	0.16

An instructive calculation of the weight and volume percentages of the major elements of rocks has been made by Barth (1962) and is given in Table 2-4. This clearly shows the negligible volume effect of the small ions, particularly silicon and aluminum, and the extreme volume importance of the large oxygen ion. A silicate rock can thus be conceived of as a network of large oxygen ions, with all other ions filling the interstices. An appreciation of this relationship is particularly important when we come to consider metamorphic and metasomatic processes.

The distribution of the different oxides in igneous rocks is important. This is best approached through frequency diagrams. These diagrams are constructed by plotting the number of analyses which contain a given range of percentage of an oxide against that percentage. A large collection of analytical data is required. Three frequency diagrams which include the eight most important oxides are given in Figs. 2-9 to 2-11. The silica diagram is most interesting, in that it exhibits two marked peak frequencies, one including the range of composition of the basaltic and gabbroic rocks and the other including the granite and rhyolite compositions. On the basis of the ratios presented above, the peak at 52.5 percent SiO_2 corresponds to

*Daly, *ibid.*, p. 39.

Table 2-3.   Relative Areas of Igneous Rocks from U.S. Geological Survey Maps of the Appalachian and Cordilleran Areas*

	Pacific Cordillera, square miles	Appalachian belt, square miles	Total square miles
Plutonic rocks			
Calc-alkalic rocks			
Precambrian granite	2,089.0	1,151.0	3,240.0
Paleozoic and later granite	402.0	194.0	596.0
(Total granite)	(2,491.0)	(1,345.0)	(3,836.0)
Granodiorite	2,040.0		2,040.0
Quartz monzonite	11.0		11.0
Quartz diorite	45.3		45.3
Diorite	103.5	10.0	113.5
Gabbro diorite	98.5		98.5
Gabbro	226.4	47.5	273.9
Anorthosite	52.0		52.0
Alkalic rocks			
Syenite	24.4		24.4
Monzonite	17.5		17.5
Nepheline syenite	3.5	0.3	3.8
Shonkinite	8.7		8.7
Fergusite	< 1.0		< 1.0
Missourite	0.1		0.1
Theralite	6.3		6.3
Ultramafic rocks			
Peridotite	73.3		73.3
Pyroxenite	2.2		2.2
Totals	5,204.7	1,402.8	6,607.5
Hypabyssal rocks (intrusive)			
Calc-alkalic rocks			
Granite porphyry	17.9	2.0	19.9
Quartz porphyry and rhyolite	26.5	1.0	27.5
Dacite porphyry	7.8		7.8
Quartz hornblende porphyry	2.0		2.0
Quartz monzonite porphyry	4.6		4.6
Diorite porphyry	20.1	1.6	21.7
Hornblende porphyry	1.0		1.0
Quartz diabase	3.0		3.0
Diabase	150.0	118.0	268.0
Alkalic rocks			
Syenite porphyry	38.4	2.5	40.9
Monzonite porphyry	9.4		9.4
Nepheline syenite porphyry	< 0.1		< 0.1
Phonolite	2.7		2.7
Pseudoleucite porphyry	0.5		0.5
Totals	284.0	125.1	409.1

Table 2-3. Relative Areas of Igneous Rocks from U.S. Geological Survey Maps of the Appalachian and Cordilleran Areas (Continued)

	Pacific Cordillera, square miles	Appalachian belt, square miles	Total square miles
Extrusive rocks			
Calc-alkalic rocks			
Rhyolite	2,145.7	1.0	2,146.7
Dacite	82.1		82.1
Mica andesite	3.0		3.0
Hornblende andesite	21.6		21.6
Pyroxene andesite	3,966.0		3,966.0
Augite porphyry	255.0		255.0
Quartz basalt	8.0		8.0
Basalt	3,079.0	130.0	3,209.0
Alkalic rocks			
Trachyte	6.5		6.5
Latite	4.6		4.6
Phonolite	5.5		5.5
Trachydolerite	0.3		0.3
Teschenite	0.2		0.2
Nepheline basalt (Texas)	1.2		1.2
Nepheline melilite basalt (Texas)	2.8		2.8
Limburgite	2.5		2.5
Total	9,584.0	131.0	9,715.0
Total igneous-rock area mapped	15,072.7	1,658.9	16,731.6

*From R. A. Daly, "Igneous Rocks and the Depths of the Earth," pp. 35–36, McGraw-Hill, 1933.

Table 2-4. Weight and Volume Percentage of Elements in the Lithosphere*

Element	Weight %	Volume %	Atomic radii, Å
O	46.71	94.24	1.40
Si	27.69	0.51	0.36
Ti	0.62	0.03	0.56
Al	8.07	0.44	0.50
Fe	5.05	0.37	0.70
Mg	2.08	0.28	0.65
Ca	3.65	1.04	0.99
Na	2.75	1.21	0.95
K	2.58	1.88	1.33

*From T. F. W. Barth, "Theoretical Petrology," 2d ed., p. 17, Wiley, 1962.

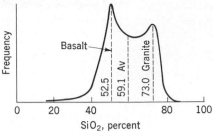

Fig. 2-9. Frequency distribution of silica in igneous rocks. Note that the average silica content is not the most frequent occurrence.

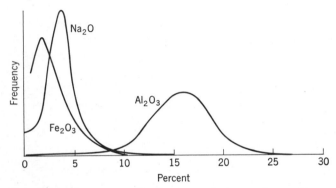

Fig. 2-10. Frequency distribution of the major oxides, except silica, in igneous rocks. (*After W. A. Richardson and G. Sneesby, Mineralog. Mag., vol. 19, pp. 303–313, 1922.*)

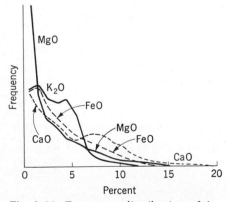

Fig. 2-11. Frequency distribution of the major oxides, except silica, in igneous rocks. (*After W. A. Richardson and G. Sneesby, Mineralog. Mag., vol. 19, pp. 303–313, 1922.*)

basalt, and the peak at 73 percent SiO_2 corresponds to granite. The intermediate rocks are much less frequent in the analyses; they are dominantly andesites. The average compositions of the important plutonic-rock types are given in Table 2-5, after Nockolds (1954).

Petrographic Provinces and Parental Magmas

In the preceding discussion of composition, reference was made to alkalic and calc-alkalic rocks. It has long been recognized that igneous rocks can be grouped compositionally and, further, that the igneous rocks of one area and time interval frequently have a common chemical characteristic. Thus the concept of the petrographic province has developed wherein it is recognized that the igneous rocks of a given geographic area and tectonic environment — and of a given geologic time — have common chemical or mineralogical similarities. Rocks of the same area but of a different time may have a different chemical characteristic and thus be distinguished from the first group. The fact that a group of rocks has a common peculiarity in spite of its representing a wide lithic range can best be explained by assuming that all components of the group are products of modification of a common parent magma and, further, that in a different petrographic province the rocks were derived from a slightly different parent magma.

The identification of the parent magmas now becomes an important question. What criteria should be applied to their identification? First, parent magmas should be abundant rock types, because it is reasonable to assume that the parent magma should be much more abundant than the rocks derived from it by its modification. Second, the important parent magmas should have a wide geographic distribution to account for the widespread igneous rocks. Third, parent magmas should have been important over a long interval of geologic time. Only two igneous-rock types meet these requirements: basalt and granite. These two may be considered parental, and most other igneous rocks can be developed from them by some modification process.

Basalt occupies a unique position in the igneous world: it has been extruded in vast quantities from the earliest geologic time to the present in almost every environment. The oldest rocks of the Canadian Shield contain abundant basaltic flows, and the current eruptions on Hawaii are of basaltic magma. Some of the major basalt eruptions and their ages are listed in Table 2-6.

These bodies are found on all continents and are distributed throughout the geologic column. Many of them involve volumes of magma ranging from 100,000 to 200,000 cubic miles as a series of many thin flows.

Basalt is also the dominant rock type of the volcanic oceanic islands. The Hawaiian Islands are dominantly basalt, as are the other noncoraline islands of the central Pacific. In the Atlantic Ocean, the Azores, Ascension

Table 2-5.   Average Chemical Compositions of Selected Plutonic Rocks*
(The numbers in parentheses indicate the number of analyses averaged)

	Calc-alkalic granites (72)	Alkalic granites (48)	Granodi-orites (137)	Quartz diorites (58)	Diorites (50)	Pyroxene gabbros (38)	Olivine gabbros (53)	Alkalic gabbros (42)
SiO_2	72.08	73.86	66.88	66.15	51.86	50.78	46.83	43.94
TiO_2	0.37	0.20	0.57	0.62	1.50	1.13	0.97	2.86
Al_2O_3	13.86	13.75	15.66	15.56	16.40	15.68	17.38	14.87
Fe_2O_3	0.86	0.78	1.33	1.36	2.73	2.26	1.91	4.35
FeO	1.67	1.13	2.59	3.42	6.97	7.41	8.20	7.80
MnO	0.06	0.05	0.07	0.08	0.18	0.18	0.14	0.16
MgO	0.52	0.26	1.57	1.94	6.12	8.35	10.03	9.31
CaO	1.33	0.72	3.56	4.65	8.40	10.85	11.36	12.37
Na_2O	3.08	3.51	3.84	3.90	3.36	2.14	2.03	2.32
K_2O	5.46	5.13	3.07	1.42	1.33	0.56	0.40	0.92
H_2O^+	0.53	0.47	0.65	0.69	0.80	0.48	0.63	0.66
P_2O_5	0.18	0.14	0.21	0.21	0.35	0.18	0.12	0.44

*Selected from S. R. Nockolds, Average chemical compositions of some igneous rocks, *Geol.*

Island, Saint Helena, Tristan da Cunha, and Gough and Bouvet Islands are all dominantly basalt. Kerguelen and other islands in the Indian Ocean are composed of basalt as well. Thus basalts are characteristic not only of the continents but also of the oceans. In addition, basalt of some type is present in almost all volcanic regions, even where other magma types are more abundant. Therefore basalt must be accepted as meeting the criteria of a parental magma.

Granites, likewise, have been developed in tremendous quantities throughout geologic time. However, they are more limited in geographic

Table 2-6.   Major Basalt Eruptions

Area	Age
Keweenawan of Lake Superior	Late Precambrian
Deccan traps, India	Cretaceous-Eocene
Columbia-Snake River Plateau, Washington, Oregon, Idaho	Miocene
Stormberg lavas, South Africa	Jurassic
Parana basalts, South America	Jurassic
New Jersey traps	Triassic
Thulean basalts, North Atlantic	Paleocene and Eocene
Karroo basalts, South Africa	Carboniferous and Permian

Pyroxenites (46)	Peridotites (23)	Dunites (9)	Alkalic syenites (25)	Calc-alkalic syenites (18)	Nepheline syenites (80)
50.50	43.54	40.16	61.86	59.41	55.38
0.53	0.81	0.20	0.58	0.83	0.66
4.10	3.99	0.84	16.91	17.12	21.30
2.44	2.51	1.88	2.32	2.19	2.42
7.37	9.84	11.87	2.63	2.83	2.00
0.13	0.21	0.21	0.11	0.08	0.19
21.71	34.02	43.16	0.96	2.02	0.57
12.00	3.46	0.75	2.54	4.06	1.98
0.45	0.56	0.31	5.46	3.92	8.84
0.21	0.25	0.14	5.91	6.53	5.34
0.47	0.76	0.44	0.53	0.63	0.96
0.09	0.05	0.04	0.19	0.38	0.19

Soc. America Bull., vol. 65, pp. 1007–1032, 1954.

range than basalts, in that they are confined to the continents and are lacking in the oceanic basins. The great granitic and granitic-gneiss terranes of the Precambrian shield areas are familiar features of all continents. Younger granite batholiths of the Paleozoic and Mesozoic orogenic belts are also well known. Whether these are truly igneous bodies crystallized from a magma or whether much of this material is a product of ultrametamorphism or metasomatism will be discussed later. At any rate many of them are of an igneous character and have had other igneous-rock types derived from them in secondary quantities. Thus granite can be considered parental, recognizing that it is confined to the continental areas.

The only other rock type which commonly occurs in batholithic-sized bodies is the anorthosites; but all the large anorthosite bodies are of Precambrian age, and therefore, this rock type fails to qualify as parental over the whole range of geologic time. In some occurrences peridotite may be considered a parental magma derived directly from the mantle. However, many peridotites and anorthosites are best explained as derivatives of a parental basalt. Andesite is the dominant extrusive of many young orogenic belts and has been considered parental. Andesite is usually associated with basalt, and many of its peculiar characteristics are best explained on the basis of a derivative nature rather than a parental nature. Therefore granite and basalt are the two common parent magmas, and most other magma types are derivatives of them.

CRYSTALLIZATION OF MAGMAS
AND THE DEVELOPMENT
OF DERIVATIVE ROCK TYPES

Three major processes and two minor processes have been recognized as important in the development of derivative magmas from a parent. These are

1. Crystal fractionation
2. Assimilation of wall rock material
3. Mixing of parental magmas
4. Liquid immiscibility
5. Gas fluxing

Of the five, the first is the most important. Crystal fractionation results in the continuous change of composition of the liquid phase as crystals form and are progressively removed from the liquid. Assimilation involves the change of composition of the liquid phase as a result of reaction between magmas and solid wall rock of contrasting composition. Mixing of two parent magmas may be important in the origin of andesites, since it results in a magma whose composition is intermediate between the compositions of the parents. Liquid immiscibility is the breakdown of one liquid into two liquids which separate physically, like oil and water. It is important in the separation of some sulfide liquids from basaltic magmas and, possibly, in the separation of carbonate-rich liquids. Gas fluxing is probably a minor process and involves the change in composition of a magma as volatiles escape from one portion of a magma chamber and are concentrated in another. The effect of gas under high pressure at depth may be important in controlling the direction in which crystal fractionation may proceed. Each of these processes will now be discussed in some detail.

CRYSTAL FRACTIONATION

Reaction Relationship

The concept of crystal fractionation, or crystal differentiation, was developed to a high degree by N. L. Bowen from experimental evidence of the physicochemical nature of crystallization of silicate melts. The basis of the concept is that magmas do not crystallize at a specific temperature but, rather, over a range of temperatures and, further, that the mineral species crystallizing and the composition of specific minerals crystallizing vary with temperature. The composition of the minerals forming should change continuously by reaction with the magma.

One line of evidence for this reaction relationship between crystals and magma is in the characteristic mineral associations and antipathies. Many minerals consistently occur together in rocks, and other minerals never or only rarely occur together. Some of these nonassociations—antipathies—may be explained on the basis of chemical incompatibility, but for many others there is no such basis. Some typical associations and antipathies are:

Associations
        Quartz-potash feldspars-sodic plagioclase-biotite
        Intermediate plagioclase-hornblende-biotite
        Calcic plagioclase-calcic pyroxene-olivine-magnesian pyroxene
Antipathies
    Chemically incompatible
        Quartz-olivine
        Quartz-feldspathoids
    Chemically compatible
        Quartz-calcic plagioclase
        Calcic plagioclase-potash feldspars
        Calcic pyroxene-potash feldspars
        Calcic plagioclase-abundant biotite
        Potash feldspars-olivine

These antipathic minerals can best be explained by assuming that the pairs of minerals either crystallize at markedly different temperatures or crystallize from magmas too widely divergent in composition to be associated. Moreover, the commonly associated minerals must crystallize from about the same composition of magma and at the same temperature. Bowen illustrates this by saying, "The controlling factors are thus analogous to those which determine that little girls ordinarily play 'London Bridge' with other little girls, occasionally with their mothers, seldom with their grandmothers and never with their great-grandmothers."[*] In the same manner the antipathic minerals belong to different generations of magma crystallization.

    A second, and very compelling, argument for reaction between crystals and liquid comes from laboratory study of the physicochemical relationships of some igneous minerals. The plagioclase feldspars are a prime example. The plagioclases form a complete solid-solution series from pure melts to all possible compositions, representing mixtures of the end members albite ($NaAlSi_3O_8$) and anorthite ($CaAl_2Si_2O_8$). Each composition of plagioclase crystallizes from a pure melt at only one temperature, and the relationship is such that the crystallization temperature drops with increasing amounts of albite. Thus, starting with a sample of given composition, the first crystals to form will be richer in anorthite than the sample. As the temperature falls, all previously crystallized plagioclase will react with the liquid, becoming progressively richer in albite until the crystals all have the composition of the original sample and the liquid is exhausted. If the process is stopped at some

[*]N. L. Bowen, "The Evolution of the Igneous Rocks," p. 21, Princeton University Press, 1928.

temperature above that of complete crystallization and this temperature is held constant, the crystals and liquid will remain unchanged in composition and proportion until the temperature is changed. This process is discussed in detail in the Appendix. The reaction between crystals and liquid with each increment of temperature change is called a *continuous-reaction series* and is common to all solid-solution series of minerals. Continuous-reaction series are recognized in the following igneous minerals:

Solid-solution series formed from pure melts of end members	High-temperature end member	Low-temperature end member
Plagioclase	Anorthite 1550°C	Albite 1118°C
Olivine	Forsterite 1890°C	Fayalite 1205°C
Nepheline	Potash nepheline 1750°C	Soda nepheline 1526°C
Alkali feldspars (sanidine) above ~ 700°C	$KAlSi_3O_8$ at 5 kbar 876°C	$NaAlSi_3O_8$ at 5 kbar 748°C

Temperature data from W. A. Deer, R. A. Howie, and J. Zussman, "Rock Forming Minerals," 5 vols., Longmans, 1962 and 1963.

In nephelines and alkali feldspars the relationship is not as simple as in plagioclases. Some more-complex minerals, such as pyroxenes and perhaps amphiboles, show similar variation of composition by reaction with the liquid. Many of these solid solutions undergo complex reorganization of crystal structures at lower temperatures, which complicates the details of mineralogy but does not radically affect the reaction relationship with liquids.

A second type of reaction relationship between crystals and liquid can be demonstrated experimentally in the relationship between olivine, pyroxene, and the silica of a liquid. Starting under equilibrium conditions and with certain compositions of a liquid consisting of forsterite (Mg_2SiO_4) and silica (SiO_2), forsterite will be the first mineral to appear as crystals. It will continue to crystallize out as the temperature falls until at a specific temperature it becomes unstable in association with the silica-enriched liquid. At this temperature the forsterite will react with a portion of the silica of the liquid to form clinoenstatite ($MgSiO_3$), according to the reaction

$$Mg_2SiO_4 + SiO_2 \rightarrow 2MgSiO_3$$

The reaction will continue at a fixed temperature until either the olivine or the liquid is exhausted. When one of the reactants has been depleted, the temperature can again fall. This relationship is discussed in detail in the Appendix. Such a situation is called a *discontinuous-reaction relationship*. It differs from a continuous reaction, in that as reaction occurs, one mineral is converted into another mineral, whereas in the continuous reaction, one composition of a mineral is converted into another composition of the same mineral.

Working from the associations and antipathies, from the experimental data, and from consideration of the relationship seen between minerals in rocks, Bowen* set up his all-important reaction series, as follows:

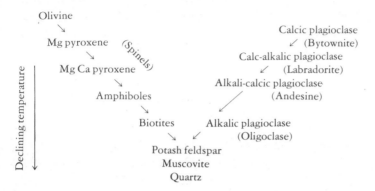

The plagioclases form a continuous-reaction series on the right, and the ferromagnesian minerals form a discontinuous-reaction series on the left. Potash feldspar, muscovite, and quartz are not formed as a product of reaction between the alkalic plagioclase and biotite but are merely the residuum of the magma remaining at the end of crystallization.

Bowen's reaction series implies that, given a magma of appropriate composition and given an appropriate cooling rate and environment, crystallization will begin with olivine and bytownite. The olivine will shortly react with the liquid to form pyroxene, and the plagioclase crystallizing out will become more sodic — a labradorite. It is then possible to go through the entire series, winding up with a magma from which alkalic plagioclase, potash feldspar, quartz, and micas may crystallize. Under other cooling and environmental conditions the same magma may be exhausted long before this granitic residue is reached. Similarly, starting with a magma whose composition and temperature are such that andesine and hornblende will be the first crystals to form, crystallization may produce a liquid from which oligoclase, orthoclase, quartz, and micas may form; or the liquid may be exhausted in the formation of andesine and hornblende. In this second case, however, the minerals higher in the reaction series than andesine and hornblende should not appear at all. The question now becomes, What determines whether the late liquid low in the reaction series will be formed or not?

The control of the composition of the late liquid depends on the extent

*Bowen, *op. cit.*, p. 60.

to which the possible reactions between crystals and liquid take place. If reaction is complete, the liquid will be exhausted before it reaches the composition from which the late minerals can form. If the reaction is poor or if no reaction takes place, the liquid will reach a composition from which the residual minerals will form. This can be understood by referring back to the earlier discussion of plagioclase crystallization. With each increment of temperature change as plagioclase crystallizes, *all* preexisting plagioclase crystals should change to that composition stable at the new temperature. This requires that the anorthite content of all preexisting crystals decrease and the albite content increase. Such a compositional change requires a diffusion of sodium and silicon into the crystal from the liquid and a simultaneous diffusion of calcium and aluminum out of the crystal into the liquid. If this reaction occurs, the anorthite content of the liquid decreases slowly and the albite content of the liquid increases slowly as the temperature falls. But if the reaction does *not* take place, then there is not a continual return of calcium and aluminum to the liquid nor is there the continual removal of sodium and silicon from the liquid to change the composition of the preexisting crystals. As a result the anorthite content of the liquid decreases rapidly and the albite content increases rapidly. When the temperature is reached at which the liquid should be gone and all the crystals should have the composition of the original sample throughout, there will still be liquid left over which will be abnormally rich in albite and poor in anorthite. Theoretically, at least, if there is absolutely no reaction, the last drops of liquid could be practically pure albite.

A similar relationship can be seen in the discontinuous-reaction series in the olivine + silica → pyroxene reaction. As olivine crystallizes from the forsterite-silica system discussed above, the silica content of the liquid gradually increases. When the reaction takes place to form pyroxene from the olivine, this excess silica is used up. If, however, this reaction is inhibited by some mechanism, the excess silica is not used up but remains in the liquid and is available for the formation of quartz. Thus quartz may be formed from an original liquid which should have given no free quartz according to its composition. The next question then becomes, What prevents reaction to make possible the formation of the low-temperature minerals?

There are three major processes which inhibit reaction between crystals and liquid, and each involves the physical separation of the crystals from the liquid. They are (1) crystal settling, (2) filter pressing, and (3) zoning and mantling. Crystal settling is particularly effective in the early phases of crystallization and involves the gravitative removal of crystals from liquid. Filter pressing is particularly effective late in the crystallization process and involves the removal of the liquid from a dominantly crystalline mush. Zoning and mantling are effective anytime and involve the coating of a crystal by its reaction product.

Crystal Settling

If a magma is intruded into a chamber and allowed to cool, crystallization will begin as widely scattered crystals "floating" in a liquid environment. The crystals and the liquid will usually have different specific gravities, and therefore if the liquid is not too viscous, the crystals will begin to separate under the influence of gravity. Those crystals heavier than the liquid will settle to the bottom of the chamber and accumulate, and those crystals lighter than the liquid will float to the top of the chamber. Thus in this highly liquid environment there will be a rain of crystals separating from the liquid. Usually the earlier crystals are dominantly heavier than the liquid and move to the base of the chamber. Inasmuch as the crystals forming will not have the composition of the original liquid, one result of crystallization is a change of composition of the remaining liquid. Later a temperature may be reached at which these early formed crystals should react with the liquid to form a new mineral or to modify their composition. Now, because the crystals have been removed from the bulk of the liquid, the necessary reaction between crystals and liquid is impossible, and exchange of material between crystals and liquid to develop a new composition is impossible. Thus the liquid is relatively enriched in those constituents which should have been removed later by reaction with the crystals.

A prime example of the effect of crystals settling is the early formation and gravitative removal of olivine from a basaltic magma. Many basaltic magmas have compositions and are at temperatures such that olivine must be the first mineral to form, but at a lower temperature all the early olivine should be used up by reaction with the magma to form pyroxene. As this early olivine forms, its constituents are subtracted from the liquid and the composition of the liquid is modified, becoming relatively richer in those constituents which the olivine does not remove. Thus, given a basalt liquid of the composition in column A, Table 2-7, if 10 percent by weight of olivine of composition in column B crystallizes out, the composition of the liquid would be changed to that shown in column E. In this table it will be noted that the percentage of each oxide except ferrous oxide and magnesia has increased in the liquid as a result of the crystallization of the olivine. Silica has increased in the liquid in spite of the fact that olivine removes silica. This is due to the silica content of the olivine being less than that of the original liquid.

If, in the above example, the olivine crystals were left scattered through the liquid, then at a lower temperature they would react with the liquid according to the equation

Olivine + silica $\rightarrow$ pyroxene
$(Mg, Fe)_2SiO_4 + SiO_2 \rightarrow 2(Mg, Fe)SiO_3$

In this reaction another 3.85 g of silica is removed from the liquid, leaving

Table 2-7.   Early Formation and Gravitative Removal of Olivine from a Basaltic Magma

Oxide	Basalt* liquid % (A)	Olivine† composition % (B)	10% olivine removes, g (C)	90% liquid, g (D)	Liquid composition, % (E)
SiO_2	49.9	38.5	3.85	46.05	51.22
Al_2O_3	16.0	.....	0	16.0	17.81
Fe_2O_3	5.5	.....	0	5.5	6.11
FeO	6.5	23.0	2.30	4.2	4.67
MgO	6.3	38.5	3.85	2.45	2.72
CaO	9.1	.....	0	9.1	10.11
Na_2O	3.2	.....	0	3.2	3.56
K_2O	1.6	.....	0	1.6	1.78
TiO_2	1.4	.....	0	1.4	1.56
P_2O_5	0.5	.....	0	0.5	0.56

*Average of 198 basalts, from R. A. Daly, "Igneous Rocks and the Depths of the Earth," p. 17, McGraw–Hill, 1933.
†Theoretical composition of olivine with forsterite-to-fayalite ratio of 3:1.

86.15 g of liquid containing 42.2 g of silica, or 49.0 percent silica. Gravitative removal of the olivine, however, will prevent this reaction and leave the 3.85 g of silica to crystallize later, possibly as free quartz.

There are many examples in nature in which the gravitative settling of early formed olivine has had exactly this effect. One of the best examples is in the Palisades basaltic sill of the Hudson River valley in New Jersey. Here, in a large sill whose original composition was such that no olivine should have survived the reaction to form pyroxene, there is an olivine-rich zone at the base and a quartz-bearing zone near the top. The early gravitative accumulation of olivine prevented the reaction to form pyroxene, and as a result the excess silica which should have been largely used up in the reaction was stored in the liquid to form a free-quartz phase.

Recent studies (Challis, 1965)* of the ultramafic bodies in New Zealand offer an excellent example of extreme gravitative accumulation of early crystals. The bodies are layered with alternating assemblages of olivine (dunite) and olivine-enstatite (harzburgite). The olivines occur as tabular euhedral crystals which are dominantly oriented so as to lie on their broad faces, the position in which settling crystals would come to rest. Enstatite in the harzburgite occurs as large anhedral crystals enclosing the olivine, as though representing a portion of the liquid trapped between the accumulating olivines. Challis interprets these rocks as having accumulated in a

*G. A. Challis, The origin of New Zealand ultramafic intrusions, *Jour. Petrology*, vol. 6, pp. 322–364, 1965.

subvolcanic chamber, such as that demonstrated under Kilauea (Eaton, 1962).* The related liquid phase in the area is represented by voluminous olivine-poor extrusive basalts and andesites of Permian age.

Gravitative separation of early crystals may be equally effective in minerals other than olivine. The early separation of very calcic plagioclase, for instance, will cause the late liquid to be enriched in the sodic plagioclase molecules. Similarly, the early separation of hornblende crystals from a granitic magma will result in the enrichment of the liquid in silica as a result of the very low silica content of hornblende. In addition, the liquid will be enriched in potash, because at lower temperatures hornblende should react with the liquid to form biotite, a potash-rich mineral. Thus crystal settling will invariably modify the ultimate composition of the liquid phase.

Filter Pressing

Later in the crystallization history, reactions between the crystals and the liquid may be prevented by the physical removal of the residual-liquid phase. As a result the liquid may form a separate rock mass of extreme composition, but of only minor volume, instead of being used up in the final adjustment of crystal compositions and minor reactions. The separation of liquid from crystals may be accomplished by the squeezing out of the liquid by the weight of the accumulating crystals, by the mechanical rupture of the magma chamber, causing pressure release, or by the trapping of the crystals in a magma mass moving along a restricted conduit.

As crystals accumulate on the floor of a magma chamber, they will develop a continuous mesh of crystals and interstitial liquid. The heavier crystals will tend to become compacted, squeezing the lighter liquid out of the interstices and upward in the magma chamber. This may be considered one type of filter pressing. In this process there is a moderately complete separation of the two phases through the downward movement of the crystals and the simultaneous upward movement of liquid. Challis (1965)† explains the dunite layers in the New Zealand ultramafic bodies as resulting from filter pressing of the liquid out from around accumulating olivine crystals. He suggests that periodic seismic vibrations associated with volcanism would assist in the expulsion of the residual liquid. Jackson (1961)‡ had earlier suggested the same mechanism as an aid to filter pressing in the Stillwater Complex of Montana.

Late in the crystallization stage, a separation of the liquid phase can be brought about by mechanical forces (Fig. 2-12). The forces may be compressional, tensional, or shearing stresses. Whenever rupture occurs, a

*J. P. Eaton, Crustal structure and volcanism in Hawaii, in: The crust of the Pacific Basin, *Am. Geophys. Union Mono.* 6, pp. 13–29, 1962.
†Challis, *op. cit.*, p. 361.
‡E. D. Jackson, Primary textures and mineral associations in the ultramafic zone of the Stillwater complex, Montana, *U.S. Geol. Survey Prof. Paper 358*, p. 65, 1961.

C 2 3 9 9

Fig. 2-12. Segregation of the late liquid from a mafic body by filter pressing. The dark zone in the middle of the specimen contains an abundant quartz-alkali feldspar intergrowth segregated from a gabbro by shearing. Transition zone in the Sudbury norite, Levak Station, Ontario.

pressure differential is set up and the most mobile phase, that is, the liquid, will move into the low-pressure areas. In a compressional environment with rupturing of the wall rock, the mobile liquid will be squeezed out and injected into fissures to form a new magma chamber. With tension in the roof rocks over a crystallizing body, fractures will develop which create low-pressure zones into which the mobile liquid will move. Shearing of a crystal mush will develop low-pressure zones into which the liquid will move as a consequence of dilatancy.

The alkalic rocks of Litchfield, Maine, offer an example (Barker, 1965).[*] Here a composite intrusive body developed from a nearly solid crystal mush lubricated by a nepheline-syenite magma. One portion of the body consists of a mechanically granulated accumulation of the suspended alkali feldspar crystals with minor sodic pyroxene and amphibole. No nepheline is present in this portion of the body, and apparently most of the liquid was filter-pressed out. This grades laterally into syenite and nepheline syenite in which the ferromagnesian minerals are biotite and magnetite without pyroxene or amphibole. A final small body from this portion of the complex is a nepheline-syenite pegmatite with 2-in. bundles of biotite. This small body

*D. S. Barker, Alkalic rocks at Litchfield, Maine, *Jour. Petrology*, vol. 6, pp. 1–27, 1965.

evidently represents the liquid phase which lubricated the initial mass which was drawn off into a separate 30-ft dike.

A third, but probably minor, process of filtering occurs where a liquid with suspended crystals is moving through a narrow channelway. At a constriction in the channel the crystals may hang up, forming a porous plug of crystals through which the liquid may move but which stops any subsequent crystals brought to the plug by the moving liquid. As a result, the liquid which has gone on past the filter plug will be unable to react with the crystals formed earlier from it and will thus have a different end product than it would have had. Such a situation has been observed by the author where a dike of porphyritic rock has terminated as a series of anastomosing stringers in the wall rock. The stringers were only a few times wider than the phenocrysts, and plugs were formed which allowed the liquid to go into the stringers but filtered out the crystals. In this example the filtered product was on a very small scale, but if the stringers had coalesced into a new dike, the resultant rock could have been quite different from the original.

Zoning and Mantling

At any time in the crystallization history, crystals may be effectively removed from the liquid by coating them with the product of reaction between the crystals and the liquid. In the minerals which show a continuous reaction with the liquid, this coating takes the form of a series of zones of contrasting composition of the same mineral. The process causing the development of these zones is called *zoning*. In the minerals which show a discontinuous reaction with the liquid, the early formed crystals may become coated with the mineral formed by reaction of the early crystals with the liquid, thereby removing the remnants of the crystal from the liquid. This process is called *mantling*. Each of these processes will have only a local effect on the composition of the liquid; however, if a major proportion of the growing crystals in a magma is zoned or mantled, the cumulative effect on the composition of the magma may be considerable.

Zoning can best be understood by considering the relationships in the plagioclase feldspars. As plagioclase crystallizes from the liquid, only one composition of plagioclase will form at each temperature, and the composition of all earlier formed crystals should change to the one forming. Modification of the composition of the solid crystals must take place by diffusion of ions through the crystal structure in both directions: sodium and silicon going into the crystal and calcium and aluminum coming out. Depending on the rate of cooling, solid diffusion may be a slow process compared with crystallization, and as a consequence, a layer of newly crystallized material may be deposited on the crystal before an appreciable adjustment in the bulk composition of the inner crystal can take place. This will make diffusion through the crystal even slower and under many environments will effectively prevent further compositional adjustment. The result is the same as it would be if the crystals were physically removed from the liquid. The effect on the liquid composition is shown in Table 2-8, in which it has been assumed that 20 percent plagioclase has crystallized from a basaltic liquid under two different environments. In one, crystallization has been too rapid for the modification of composition so that the plagioclases are zoned; in the other, crystallization has been slow enough so that there is a complete adjustment. The last temperature in each is such that the last plagioclase has the composition $An_{60}Ab_{40}$. The results clearly show that zoning of plagioclase enriches the liquid in silica and soda and impoverishes it in alumina and lime compared with the liquid product of crystallization with complete reaction.

Zoning is only occasionally visible by megascopic examination. In some plagioclase feldspars it can be seen as slight differences of color or as symmetrically distributed minute inclusions deposited on the surface of a crystal during a period of no growth (Fig. 2-13). Microscopically, zoning is very common in plagioclases, pyroxenes, and olivines and can be determined by optical properties (Fig. 2-14). It is very important to remember, however, that thin zones may represent an appreciable bulk of material. Starting with a 1-cm³ cube, for example, doubling its volume requires adding a uniform layer only 1.3 mm thick; tripling its volume, to 3 cm³, requires an additional layer 0.91 mm thick; 4 cm³ requires an additional layer 0.725 mm

Table 2-8.   Effect of Zoning of Plagioclase Feldspar on Liquid Composition

Assumed plagioclase compositions

	SiO_2	Al_2O_3	CaO	Na_2O
An_{86}	46.6	34.4	17.4	1.6
An_{80}	48.0	33.4	16.3	2.3
An_{75}	49.3	32.6	15.3	2.8
An_{67}	51.4	31.2	13.7	3.8
An_{60}	53.0	30.1	12.3	4.6
Assumed initial composition of basaltic magma	50.0	13.0	10.0	2.8

Calculated composition of liquids as plagioclase crystallizes

		SiO_2	Al_2O_3	CaO	Na_2O
Crystallize 20%	4% An_{86}	50.14	12.10	9.69	2.85
zoned plagioclase from	4% An_{80}	50.23	11.17	9.40	2.87
the magma, yielding	4% An_{75}	50.27	10.20	9.14	2.88
liquids of compositions	4% An_{67}	50.23	9.20	8.92	2.83
given in next column	4% An_{60}	<u>50.09</u>	<u>8.16</u>	<u>8.75</u>	<u>2.75</u>
Crystallize 20% An_{60} homogeneous plagioclase		<u>49.25</u>	<u>8.73</u>	<u>9.43</u>	<u>2.35</u>

Fig. 2-13. Zoning in feldspar. Zoning in the white plagioclase shows as dark inclusions. Zoning in the dark alkali feldspar shows as color bands. Jonesport, Maine.

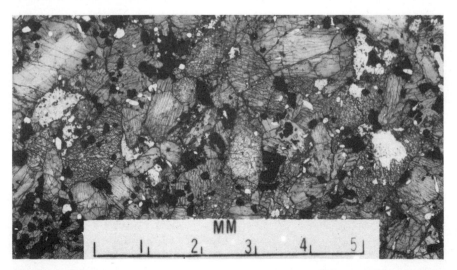

Fig. 2-14. Zoning in pyroxene. The light-colored cores of the crystals are augite, whereas the darker rim is low in calcium and high in titanium and iron. Projection of a thin section of ultramafic rock from Magnet Cove, Arkansas.

thick, etc. Stated another way, a layer 1 mm thick coating a 1-cm^3 crystal will have nearly three-quarters of the volume of the original crystal. Thus thin zones, which could be overlooked, are important.

Mantling is most commonly seen in rocks as a result of the reaction relationship between olivine and magma to form pyroxene. As discussed earlier, olivine will commonly crystallize from a basaltic magma at high temperatures, and at a lower temperature it will react with the magma to form pyroxene. This reaction involves removal of silica from the liquid. Such a reaction will, of course, take place most readily at the surface of the olivine crystal. Replacement of the interior of an olivine crystal by pyroxene involves a diffusion of silica into the crystal and simultaneous outward diffusion of excess magnesia. At the reaction temperature, additional crystals of pyroxene will be forming by direct crystallization. Again, under many environments the rate of reaction at the olivine surface and the rate of direct pyroxene crystallization may be much greater than the rate of diffusion through the crystal structure. The first nuclei of pyroxene will quite probably form on the surface of the olivine crystal, and there they will grow rapidly by reaction and by direct crystallization so that the olivine crystal will soon be coated with a continuous layer of pyroxene. At this point for any more olivine to be destroyed by the reaction, diffusion must take place through the zone of pyroxene as well as through the remnant olivine. This is a slow process; as a result, once a continuous coat of pyroxene has formed, the remnant olivine is as effectively removed from the liquid as it would be by physical gravitative separation. Thus silica which should have been used up in converting the olivine to pyroxene is stored in the liquid.

The effect of this type of mantling is shown in Table 2-9. Assuming an initial magma containing 50 percent silica from which 5 percent olivine of the composition shown crystallizes, the result will be enrichment of the magma in silica by 0.6 percent. Now, if 20 percent of the olivine (1 percent of the total bulk) is converted to pyroxene accompanied by some slight crystallization, giving a total of 4 percent solid olivine, 2 percent solid pyroxene, and 94 percent remaining liquid, the silica content will drop to 50.3 percent, but it is still above the original 50 percent. If the reaction of olivine to pyroxene had been complete, 5 percent olivine would have yielded approximately 8 percent pyroxene. The silica content of the remaining 92 percent liquid would drop to 49.3 percent. Thus the mantling clearly leaves the liquid richer in silica than it would have been had reaction been complete.

Mantling of olivine by pyroxene is quite commonly observable by megascopic means (Fig. 2-15). The olivine occurs as rounded and corroded granules embedded in the pyroxene. Mantling of the other minerals of the discontinuous-reaction series is generally not so distinct (Figs. 2-16 to 2-18), possibly owing to the greater complexity of the reactions involved. How-

Table 2-9.   Mantling of Olivine by Pyroxene

Given	Units	SiO_2	MgO	FeO
Olivine composition	wt %	39.8	44.3	15.9
Pyroxene composition	wt %	58.0	29.6	10.2
Basalt magma composition	wt %	50.0	5.5	9.9
Crystallize out 5% olivine				
Remove	g	1.99	2.21	0.80
Leaves 95 g liquid with	g	48.01	3.29	9.10
and 95 g liquid at	wt %	50.6	3.46	9.6
Remove 1% olivine by reaction				
Leaves 4% olivine solid at	g	1.59	1.77	0.64
Form 2% pyroxene solid at	g	1.16	0.59	0.24
Leaves 94 g liquid with	g	47.25	3.14	9.02
and 94 g liquid at	wt %	50.3	3.34	9.61
React 5% olivine with liquid to get				
8% pyroxene, removes	g	4.64	2.35	0.82
Leaves 92 g liquid with	g	45.36	3.14	9.02
and 92 g liquid at	wt %	49.3	3.42	9.8

ever, these minerals frequently occur in clusters which roughly follow Bowen's reaction series in their sequence within the clusters. Thus augite may form the center of a cluster and be surrounded by numerous smaller crystals of hornblende and finally by crystals of biotite.

Effects of Crystal Fractionation

The processes of crystal settling, filter pressing, zoning, and mantling operate to more or less continuously remove the crystals forming from the liquid. As a result, the composition of the liquid changes continuously. If the liquid is withdrawn from the chamber and isolated after partial crystallization, the rock produced by its crystallization will be markedly different from that which would have been produced from the original liquid. The general effects on the liquid are a continual enrichment in silica, soda, and potash and a continual depletion in magnesia, ferrous oxide, and lime.

The enrichment of the magma in silica as a result of the crystallization of plagioclase and the early removal of olivine has already been discussed. Hornblende crystallizing in the intermediate stages of the magma history and biotite crystallizing at the late stages both act to enrich the magma in silica, because both these minerals are relatively silica-poor. Hornblende averages approximately 40 percent silica and biotite averages approximately

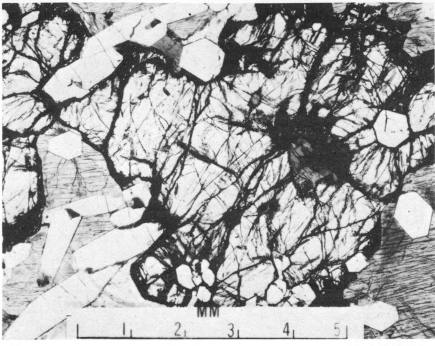

Fig. 2-15. Mantling of olivine by pyroxene. A large irregular granule of olivine (irregularly cracked) is surrounded by a larger single crystal of pyroxene (gray with straight-line cracks). The large white crystals are apatite. Projection of a thin section of olivine gabbro from Iron Mountain, Cripple Creek, Colorado. (*Ward's American Rock Collection.*)

Fig. 2-16. Mantling of alkali feldspar by plagioclase. The oval gray alkali feldspar is mantled by white plagioclase. This relationship is characteristic of granites called rapikivi. Jonesport, Maine.

Fig. 2-17. Mantling of olivine crystals by a cluster of small dark hornblende crystals. Lamprophyre dike near Benton, Arkansas.

Fig. 2-18. Mantling of rounded granules of olivine by pyroxene. The olivine has been altered to serpentine. Lamprophyre dike near Benton, Arkansas.

38 percent silica. The change of silica content of the liquid phase with crystallization can be seen in the following hypothetical example.

Start with an initial basaltic magma containing 50 percent SiO_2; 100 g magma contains 50 g SiO_2.

Step I

      Crystallize 10% olivine at 40% SiO_2 to remove 4 g SiO_2
           plus 10% plagioclase at 48% SiO_2 to remove 4.8 g SiO_2
           which combined remove 8.8 g SiO_2
      This leaves 80 g liquid with 41.2 g SiO_2
           $\therefore$ 51.5% SiO_2 [(41.2/80) × 100]

Step II

      Crystallize 20% pyroxene at 50% SiO_2 to remove 10 g SiO_2
           plus 20% plagioclase at 50% SiO_2 to remove 10 g SiO_2
           which combined remove 20 g SiO_2
      This leaves 40 g liquid with 21.2 g SiO_2
           $\therefore$ 53.0% SiO_2 [(21.2/40) × 100]

Step III

      Crystallize 10% hornblende at 40% SiO_2 to remove 4 g SiO_2
           plus 10% plagioclase at 54% SiO_2 to remove 5.4 g SiO_2
           which combined remove 9.4 g SiO_2
      This leaves 20 g liquid with 11.8 g SiO_2
           $\therefore$ 59% SiO_2 [(11.8/20) × 100]

Step IV

      Crystallize 5% biotite at 38% SiO_2 to remove 1.9 g SiO_2
           plus 5% plagioclase at 58% SiO_2 to remove 2.9 g SiO_2
           which combined remove 4.8 g SiO_2
      This leaves 10 g liquid with 7.0 g SiO_2
           $\therefore$ 70% SiO_2 [(7.0/10.0) × 100]

Sixty percent of the original liquid has crystallized out as an olivine gabbro and 30 percent as a diorite. The remaining 10 percent of the liquid contains 70 percent silica. Daly gives the average silica content of 546 granites as 70.8 percent. As far as the silica content is concerned, a mass of average basaltic magma can be differentiated so that the last 10 percent of the liquid has the composition of a granite.

The behavior of the alkalies sodium and potassium in the liquid is largely dependent on the feldspars. The early forming feldspar is calcic plagioclase, which initially is low in soda and very low in potassium. Therefore these cations are stored up in the liquid. As the plagioclase composition shifts with falling temperature, soda increases in the liquid at a decreasing rate. Eventually, in the majority of magmas, the soda content of the plagioclase reaches such a point that soda begins to decline in the liquid. No other soda-bearing mineral normally forms. In some alkali-rich and silica-poor magmas there is insufficient silica to tie up all the soda in plagioclase, and in such magmas nepheline and soda amphiboles and pyroxenes may appear in addition to the soda plagioclase. Only a very limited amount of potassium can be accommodated by the early forming plagioclase, and therefore potassium builds up in the liquid until a potassium-rich mineral becomes capable of crystallizing. In many magmas biotite is the first such mineral to become stable. It is usually followed by the potash feldspars or the potash-soda solid-solution feldspars or both. Normally, these minerals are among the very last to crystallize from a magma, and thus the potassium content increases throughout crystallization to the final residual liquid.

The lime content of the magma is controlled by both the feldspar continuous-reaction series and the ferromagnesian discontinuous-reaction series. In the early stages of crystallization, calcic plagioclase and olivine crystallize nearly simultaneously. Because olivine can accommodate practically no lime, the lime content of the liquid may actually increase. As soon as the monoclinic pyroxenes are stabilized, however, the lime content of the liquid begins to decrease rapidly, because both the major minerals forming (calcic plagioclase and augite) are lime-rich. When hornblende appears, it is also a lime-rich mineral so that lime continues to decline in the liquid. In the final stages of crystallization, the main lime-bearing mineral is sodic plagioclase, with or without hornblende.

Magnesium and iron are the main cations of the ferromagnesian discontinuous-reaction series, olivine-pyroxene-hornblende-biotite. Each of these minerals forms a continuous-reaction series; the magnesium end member is the high-temperature end and the iron end member is the low-temperature end. In the olivine solid-solution series, forsterite (Mg_2SiO_4)-fayalite (Fe_2SiO_4), the early olivine crystals are richer in forsterite, and the crystals become progressively richer in fayalite as crystallization proceeds. As a result of the crystallization of ferromagnesian minerals in large amounts in the early history of the magma, both iron and magnesium decrease in amount in the liquid. However, because of the solid-solution series, magnesium falls more rapidly than iron, and the ratio of iron to magnesium in the liquid increases.

The behavior of alumina during differentiation of the magma is variable, depending, apparently, on the rate of cooling of the magma body and the exact initial alumina content. Two different behaviors can be recognized and are represented by two rock assemblages. In many smaller bodies the alumina content of the magma decreases rapidly in the early stages of crystallization and then less rapidly as the silica content of the liquid increases. This behavior of alumina is characteristic of the association of gabbro and granophyre in many sills. However, in many volcanic associations the alumina content of the liquid increases slowly in the early stages of crystallization until it reaches a maximum in the mediosilicic ranges and then begins to decline gradually.

The change from increasing to decreasing alumina is apparently brought about by the appearance of hornblende as a solid phase in the mediosilicic range of magma compositions. Hornblende does not appear in the inter-mediate rocks of the gabbro-granophyre association in any important amount. The control of the behavior of alumina thus seems to be in the early stages of crystallization. At this stage the only important aluminum-bearing mineral forming is plagioclase feldspar. The effect of reaction or lack of reaction between plagioclase and liquid on the silica content of the liquid has already been considered. It will affect the alumina content in an analogous manner but with reversed effect. In the reaction of plagioclase with liquid to make a more sodic plagioclase, alumina is returned to the liquid. Therefore if the reaction between the plagioclase crystals and liquid is good, as it frequently is with slow cooling, the alumina content of the liquid may increase. However, if reaction between the plagioclase crystals with the liquid is poor, as it frequently is with rapid cooling, the alumina content of the liquid may decrease. Another factor which may influence the alumina content of the liquid in the early stages is the extent to which the orthorhombic pyroxenes form before the monoclinic pyroxenes appear. Hypersthene and enstatite are considerably poorer in alumina than is augite.

ASSIMILATION OF WALL-ROCK MATERIAL

General Statement

Assimilation is defined in the AGI Glossary as "the incorporation into a magma, of material originally present in the wall rock. The term does not specify the exact mechanisms or results; the 'assimilated' material may be present as crystals from the original wall rocks, newly formed crystals in-cluding wall-rock elements, or as a true solution in the liquid phase of the magma." Assimilation thus includes a number of processes whereby wall-rock material becomes a part of the magma. Inasmuch as the wall rock would rarely have the same composition as the magma, assimilation must result in a modification of the composition of the magma. Two major questions arise: How much material can be incorporated and by what processes? What effects will it have on rocks formed from the magma?

Sources of Heat for Assimilation

The problem of how much material may be incorporated involves the availability of thermal energy. The simplest situation is one in which a relatively cold monomineralic rock fragment is immersed in a magma whose temperature is above the melting point of the rock fragment. First, the solid fragment must be heated to its melting temperature. This involves supplying an amount of heat from the magma required by the heat capacity of the rock. Second, the rock must be melted. This involves supplying heat from the magma required by the latent heat of fusion of the rock. Third, the molten rock fragment must be heated to the magma temperature and must be dispersed in the magma. This requires further heat from the magma to accommodate the heat capacity of the molten rock fragment and the heat of solution of the new liquid in the magma. Each step of the process has removed heat from the magma. What heat, then, is available in the magma?

A magma may exist at a temperature above the temperature at which it will begin to crystallize. Such heat is called *superheat* and is immediately available for heating incorporated materials. However, the evidence from the rocks indicates that in most magmas there is no superheat. That is, most magmas as seen at the surface have already begun to crystallize. In our modern active volcanoes, such as those in Hawaii, even the most glassy specimens of lavas contain some phenocrysts — crystals already formed from the magma or crystals never melted because the magma did not get hot enough. Therefore, in most normal magmas superheat cannot account for much assimilation.

Once a magma has begun to crystallize, the heat necessary for assimilating foreign material must come from further crystallization of the magma. Crystallization liberates thermal energy, the latent heat of crystallization, which is numerically the same as the latent heat of fusion. The heat liberated by crystallization causes the temperature of the liquid to fall slowly as thermal energy is abstracted by heating wall rocks. Thus crystallization must accompany assimilation and limits the amount of material which can be assimilated. To visualize how this process works, assume that a liquid from which minerals low in Bowen's reaction series are crystallizing incorporates crystals higher in the series. The liquid cannot now melt the included crystals, because their melting temperature is above the temperature of the magma. However, the liquid will react with those high-temperature crystals and convert each into the related mineral which is crystallizing from the liquid. The reaction will involve an exchange of material between the included crystals and the liquid, modifying the composition of both. The liquid composition is therefore slightly changed, and the composition of the minerals crystallizing directly from it will also have to be slightly changed. All these reactions and exchanges of material between crystals and liquid require thermal energy to drive them. This energy can be supplied only by the simultaneous crystallization of more material from the liquid, thus releasing its latent heat of crystallization. This incorporation of high-temperature minerals in a low-temperature magma is thus an involved process, resulting in the conversion of the included crystals to a new lower-temperature mineral, the modification of the crystals forming by direct crystallization from the magma, and the simultaneous further crystallization of the magma.

The other situation, in which a high-temperature magma is incorporating lower-temperature minerals, is almost as complex. Here direct melting of the included crystals is possible. But the heat required for this melting can be obtained only by further crystallization of the magma. As a result of melting and solution of the included crystals, the composition of the magma will be changed, and therefore the composition of the crystals forming directly from it will also have to be changed. The incorporated crystals, the liquid, and the crystals forming are again all involved in the process, with the result that before any appreciable quantity is incorporated the liquid will be largely exhausted.

Effects of Assimilation of Igneous Rocks

The effects of assimilation of an igneous rock in a magma are relatively simple. If silicic rocks, granites, etc., are incorporated in a mafic magma, they may be melted and dissolved. However, only a limited amount of such xenoliths can be dissolved, because their solution will require simultaneous crystallization of the magma. These dissolved granitic constituents will be stored in the liquid phase and will increase the amount of late granitic differentiate which would normally develop by crystal fractionation.

Hall (1966)* describes an example in Ireland in which the partial assimilation of xenoliths through melting has resulted in a compositional change of the magma. Here stoped blocks of granodiorite have been selectively melted, adding the lowest-temperature-melting composition to a granitic magma. The result is a concentrically zoned body marked by increasing orthoclase and quartz content, decreasing anorthite content, and total plagioclase from the margin inward. The final product also involves the appearance of muscovite in place of biotite.

The assimilation of mafic igneous rocks in a silicic or granitic magma is more complex, because as was shown, these fragments cannot be melted but must undergo a reactive assimilation. This involves conversion of the higher-temperature minerals to the lower-temperature minerals with which the magma is saturated. Calcic plagioclase is converted to sodic-calcic plagioclase, and olivine and pyroxene are converted to hornblende and biotite. Usually, at the same time, new potash feldspars and quartz crystals, identical to those forming by direct crystallization of the magma, develop in the xenolith (Fig. 2-19). This reactive assimilation of mafic xenoliths in granitic magma is frequently seen in all stages of development in batholiths and stocks (Fig. 2-6). The different stages of the process are frozen into the rock by the required simultaneous crystallization.

Effects of Assimilation of Sedimentary Rocks

Assimilation of sedimentary rocks in a magma is more complex because of the marked differences possible between the compositions of sediments and igneous rocks. However, normal sediments add nothing to a magma which

*A. Hall, A petrogenetic study of the Rosses granite complex, Donegal, *Jour. Petrology*, vol. 7, pp. 202–220, 1966.

Fig. 2-19. Effect of assimilation of mafic xenoliths in silicic magma. The large feld-spars in the xenoliths are identical to those in the intrusion. The highly irregular portions of the borders and light-colored patches in the xenoliths are also the result of partial assimilation. From a porphyritic syenite dike near Wausau, Wisconsin.

is not already present; they merely add the elements in markedly different proportions. The effect of the addition of sedimentary materials is to cause modifications in the composition and proportions of the minerals crystallizing from the magma. An example of these effects is shown in Table 2-10, where the change of ideal minerals is calculated for the assimilation of 5, 10, and 15 percent average limestone in average basalt magma. Comparing the composition of the limestone with that of the basalt shows that the addition of limestone will enrich the magma markedly in lime and slightly in magnesia but will impoverish the magma in all other constituents. The calculated composition of the magma with 5, 10, and 15 percent dissolved limestone shows these changes. Mineralogical changes should show up as increased lime content in the lime-bearing minerals (pyroxene and plagioclase) and an increase in the silica-deficient minerals (olivine and nepheline). These changes can be followed in the lower half of the table, the calculated ideal minerals. Plagioclase becomes more calcic but decreases in amount owing to the impoverishment of the liquid in alumina and silica. Pyroxene becomes as calcic as possible but varies in amount in response to increasing lime and decreasing silica. The silica-deficient minerals olivine and nepheline increase in total amount. Thus the response to assimilation of sediments

Table 2-10.   Effects of Assimilation of Limestone in Average Basalt Magma

Oxide	Average limestone, %	Average basalt, %	Basalt + 5% limestone, %	Basalt + 10% limestone, %	Basalt + 15% limestone, %
SiO_2	5.2	51.0	50.0	48.9	47.8
TiO_2	0.1	1.4	1.37	1.34	1.31
Al_2O_3	0.8	15.6	15.27	14.86	14.51
Fe_2O_3	0.5	1.1	1.10	1.09	1.08
FeO	0.5	9.8	9.56	9.30	9.05
MgO	8.0	7.0	7.22	7.39	7.56
CaO	43.0	10.5	12.33	14.02	15.63
Na_2O	0.1	2.2	2.15	2.10	2.05
K_2O	0.3	1.0	0.99	0.98	0.97
CO_2	41.9				

Calculated ideal minerals (norm)

Orthoclase		6.12	6.12	6.12	6.12
Plagioclase:		48.09	47.25	42.75	34.05
Composition		An_{62}	An_{61}	An_{66}	An_{80}
Nepheline		. . . . .	. . . . .	1.70	5.68
Pyroxene:		39.99	33.61	33.94	41.12
$CaSiO_3$		23.51	40.05	51.21	51.07
$MgSiO_3$		41.51	33.62	27.70	28.70
$FeSiO_3$		38.98	26.33	21.39	20.23
Olivine:		2.28	8.90	11.40	8.98
Composition		Fo_{55}	Fo_{52}	Fo_{55}	Fo_{54}
Titaniferous magnetite		4.36	4.36	4.20	4.05

involves only an adjustment in composition and proportions of the minerals.

Assimilation of sandstone and shale by basaltic magmas involves less drastic modification than assimilation of limestone. The average shale is markedly richer in silica and potash and slightly richer in alumina than basalt. Assimilation of such material therefore results in the possible appearance of quartz and potash feldspar in amounts greater than would normally develop by differentiation. The composition and proportions of plagioclase and pyroxenes would be only slightly affected. Assimilation of average sandstone would prohibit the formation of any undersaturated mineral, such as olivine or nepheline, and would increase the quartz content of the late-differentiation products.

The assimilation of limestone by granite has long been considered important in the development of some syenites and feldspathoidal syenites. The process has been envisioned as involving the desilication of the magma by its reaction with lime from limestone or lime and magnesia from dolomites to form lime or lime-magnesia silicates, such as wollastonite, diopside, tremolite, and grossularite. The strongest evidence supporting this concept is the moderately common association of lime silicate rock and syenite at the contact between granite and limestone. However, from an analysis of the changing composition of a magma as limestone is added, it would appear that simple assimilation will not give rise to undersaturated rocks. The assimilation of 15 percent average limestone in average granite will only lower the quartz content from 28.5 to 18.6 percent. This, of course, is assuming that the product of assimilated limestone is uniformly distributed throughout the granite mass. Syenitic composition can be developed from a granite with the simultaneous development of a lime silicate-rich zone, a skarn, by processes involving diffusion of material across the limestone-granite boundary. The possible results of such a process are illustrated in Table 2-11. Here two weight units of average granite are reacted by diffusion with one weight unit of average limestone to give one

Table 2-11.   Assimilation of Limestone in Granite*

Oxide	200 g average granite gives oxides, g	100 g average limestone gives oxides, g	100 g average syenite	161.1 g residual lime silicate
SiO_2	142.8	5.3	61.6	86.5
Al_2O_3	29.4	0.8	16.7	13.5
Fe_2O_3	3.2	0.5	2.8	0.9
FeO	3.6	0.0	3.3	0.3
MgO	1.8	8.1	2.5	7.4
CaO	4.0	43.5	4.5	43.0
Na_2O	7.0	0.1	4.0	3.1
K_2O	8.2	0.3	4.6	3.9
CO_2		41.9		2.5

Calculated ideal minerals (norm)

Quartz	28.50		6.72	
Orthoclase	24.56		27.24	14.62
Plagioclase:	39.35		47.96	22.58
Composition	An_{26}		An_{29}	An_{30}
Diopside	. . . . .		6.77	24.60
Hypersthene	3.65		6.71	
Wollastonite	. . . . .		. . . . .	31.74
Calcite	. . . . .		. . . . .	5.55
Magnetite	3.18		4.18	0.91

*Composition of average granite and syenite from R. A. Daly, "Igneous Rocks and the Depths of the Earth," McGraw-Hill, 1933; composition of average limestone from F. W. Clarke, The data of geochemistry, *U.S. Geol. Survey Bull.* 770, 1924.

unit of average syenite plus about one and one-half units of lime silicate-rich rock. The weight change is accommodated by the loss of volatiles (CO_2) which would enhance the diffusion of material.

Assimilative reactions between granite and sandstone and granite and shale are much simpler than the reaction between granite and limestone. Average sandstone is richer in silica than is average granite, and therefore its assimilation enhances the amount of quartz formed from magma. The only other oxide more frequently abundant in sandstones than in granites is lime, which is usually present as a carbonate cement. The addition of lime may cause the plagioclase to become slightly enriched in anorthite, but because of the low alumina content of most sandstones, the total amount of feldspar will be diminished. The slight enrichment in lime may also cause the appearance of lime-bearing ferromagnesian minerals, such as hornblende or even diopside in place of some of the more normal biotite. The average shale is slightly enriched in alumina and the ferro-magnesian constituents and slightly impoverished in silica and alkalies as compared with the average granite. Consequently, there is very little change in the mineralogy of granite as a result of the assimilation of shale. The deficiency of alkalies and the slight excess of alumina may combine to cause a reduction in the amount of total feldspar and an increase in the amount of alumina-rich minerals, such as micas, or even the appearance of an aluminum silicate, such as sillimanite. The controlling factor here is not that most shales are abnormally rich in alumina but, rather, that they are too poor in alkalies to tie up all the alumina as feldspars.

In summary, assimilation normally results in modification of mineral compositions and proportions but rarely causes drastic changes in the direction of differentiation. These limited effects are due to the restricted amount of material which can be assimilated before crystallization is complete and the fundamental similarity in composition of all rocks. Nothing is added to the magma from the assimilated rock which was not already present in the magma. The amount and precise composition of a final liquid differentiate may be modified, depending on the composition of the incorporated wall rock. The most striking effect is seen at the contacts between granite and limestone, where an exchange of material across the contact develops syenite by desilication of the magma and lime-silicate skarn from the limestone.

MIXING OF PARENT MAGMAS

The Andesite Problem

One of the intriguing problems of petrology is the relatively great abundance of andesite flows and the comparative rarity of equivalent intrusive diorites. A related problem is the scarcity of rocks of intermediate composition in differentiated sills consisting of mafic rocks at the base and silicic rocks at the top. The evidence seems to indicate that the large volumes of andesite flows are due to the mixing of parental granitic and gabbroic magmas in a volcanic environment, giving rise to an intermediate between the two parents — andesite.

The great andesitic volcanoes are found in the circum-Pacific belt of young mountain ranges as shield volcanoes and composite cones built up of basalt, andesite, dacite or quartz latite, and rhyolite. In many, andesite is quantitatively dominant, whereas basalt and rhyolite are minor. The eruption of magma has not normally been a sequence of compositions which would indicate a simple process of differentiation. Some volcanoes, such as Newberry Volcano in Oregon, have been so extreme in their behavior as to erupt basalt and rhyolite alternately throughout their history. Andesitic volcanoes develop after the main orogenies in the areas, while the young mountain range is being actively uplifted. According to geophysical evidence there is a marked thickening of the crust during orogeny, possibly through the downward thrusting of crustal rocks to a depth much greater than normal. The thickened crust, with its relatively high radioactivity, could generate sufficient heat to melt not only the crustal granites but also the more mafic underlying materials. During the ascent of these magma masses, the silicic and mafic bodies could readily become mixed, at least in part, to give rise to an intermediate magma.

The mineralogical character of phenocrysts in many andesites indicates an assemblage of crystals derived from two contrasting magmas. Frequently, two compositions of plagioclase of the same apparent generation occur as phenocrysts in the same rock. It has been pointed out that only one composition may crystallize from a magma at any one time. The presence of sodic and calcic plagioclase of about the same size in one rock can be explained only by assuming the mixing of silicic and mafic magmas, both containing suspended plagioclase crystals. Subsequent crystallization occurred too rapidly for the phenocrysts to come to equilibrium with the mixed liquid. Similarly, complexly zoned plagioclase is frequently found in andesites. Plagioclase zoned as a result of differentiation normally has a core relatively rich in anorthite and a rim relatively rich in albite. However, in many andesites the zones alternate between sodic, calcic, and intermediate, in any order, from the core to the rim. Such alternation of zones is understandable if the crystals of two incompletely mixed magmas are moved from one liquid to another by gravitative settling or flotation by attached gas bubbles.

The phenocrysts of ferromagnesian minerals in andesites include the entire range of the discontinuous-reaction series — olivine, hypersthene, augite, hornblende, and biotite are all common. In a general way the ferromagnesian phenocrysts are associated with each other according to the reaction series, that is, olivine with pyroxene and biotite with hornblende. However, in many rocks they are associated without respect to position in the reaction series. In addition, hornblende and biotite are commonly surrounded by a reaction rim which includes pyroxene. This may be evidence of migration of the crystals to a higher-temperature zone by settling or by the process of extrusion. In some andesites, phenocrysts of olivine have been found in association with corroded phenocrysts of quartz and sanidine. Such an association also indicates a mixing of materials from two magmas.

A number of examples of mixed magmas are now known. Some of these are the result of a confluence of two flows of contrasting composition, such as the rhyolite-basalt flows of the Gardiner River of Yellowstone (Wilcox, 1944).* Others involve the intrusion of one magma into another, incompletely consolidated, magma body and result in net-veined complexes in which the younger, generally more mafic, magma has broken up into clots and pillows as it was chilled against the older cooler magma. The clots are separated by veins and stringers of the older magma (Blake, 1966).† The 1915 eruption of Lassen Peak, California, gives evidence of mixing of two magmas within the volcanic conduit (Macdonald and Katsura 1965).‡ The early stages of the eruption produced a flow of dacite glass which contains phenocrysts of plagioclase and minor phenocrysts of hypersthene, diopsidic augite, biotite and quartz, and rare phenocrysts of basaltic hornblende and magnesian olivine. Olivine and quartz are both rimmed by pyroxene, showing reaction with the magma. Biotite and hornblende were also attacked by the magma, the biotite phenocrysts having been reddened and the hornblende phenocrysts rounded and clouded peripherally by minute magnetite grains. Two types of plagioclase phenocrysts are present. One type is clear and complexly zoned and has compositions ranging from sodic labradorite to sodic andesine. The second type has cores identical with the first but shows broad clouded rims of intermediate labradorite containing tiny blebs of glass. Clearly, these crystals have been partially remelted. A late stage of the eruption produced a banded pumice with light-colored bands of dacite and dark bands of mafic andesite intimately mixed. Both magma types were clearly liquid at the time of extrusion. The mafic andesite contains phenocrysts from the dacite and is interpreted as representing a contaminated, more mafic (basaltic), parental magma. Macdonald and Katsura conclude that the magma from a composite body developed by the melting of two different source rocks, one mafic and the other silicic, at depth.

Thus there seems to be evidence that in the environment of deformed

*R. E. Wilcox, Rhyolite-basalt complex on Gardiner River, Yellowstone Park, Wyoming, *Geol. Soc. America Bull.*, vol. 55, pp. 1047–1080, 1944.
†D. H. Blake, The net-veined complex of the Austurhorn intrusion, southeastern Iceland, *Jour. Geology*, vol. 74, pp. 891–907, 1966.
‡G. A. Macdonald and T. Katsura, Eruption of Lassen Peak, Cascade Range, California, in 1915: Example of mixed magmas, *Geol. Soc. America Bull.*, vol. 76, pp. 475–482, 1965.

geosynclines, where the crust has been downwarped to considerable depth, two magmas can be simultaneously generated and, subsequently, partially mixed during the ascent to the surface. A granite magma is developed from the downwarped crust, and a basalt magma is generated from the base of the crust or the underlying material. The peculiarities of the phenocrysts can be accounted for by the mixing of contrasting magmas, each carrying suspended crystals but with insufficient time to restore equilibrium between crystals and resultant liquid, or by vertical movements of crystals through a magma chamber containing nonhomogeneous and incompletely mixed liquids of differing composition.

LIQUID IMMISCIBILITY
Magmatic Splitting

The early petrologists, predating the application of physicochemical methods to the study of magmas, considered that an intermediate magma could split into two complementary derivative magmas, one richer in silica and one poorer in silica than the parent. This concept is basically that of the physical separation of two immiscible liquids. Liquid immiscibility, as it has been applied to magmas, is explained as follows: at high temperatures a single homogeneous liquid of intermediate composition can exist; this liquid becomes unstable on cooling to temperatures still above the crystallization ranges; it then breaks down into two liquids of contrasting composition, one richer in certain chemical components than the original single liquid and the other poorer in the same components. The two resultant liquids, having different chemical composition, would consequently have different physical properties; they would therefore separate gravitatively and on crystallizing should give rise to contrasting rock types. Under other circumstances the resultant two liquids could be moved by intrusion or extrusion and be remixed in varying proportions to give rise to an entire series of intermediate compositions.

The process as just outlined is now known to be inapplicable to silicate melts in the range of igneous-rock composition. The physicochemical behavior of the immiscible liquids is well known, and the features of igneous rocks which have been pointed to as evidence of immiscibility do not satisfy the requirements. In addition, in experimental investigations of the silicate systems, immiscible liquids have been found which behave as predicted. The compositions of the liquids involved, however, fall far outside the range of composition of natural magmas. The composition of these liquids and of igneous materials is shown in Fig. 2-20. Thus the importance of such a process can be ruled out in the differentiation of the silicate phase of magmas.

However, liquid immiscibility does exist in the igneous environment between the silicate phase and a sulfide phase. In blast furnaces and smelters

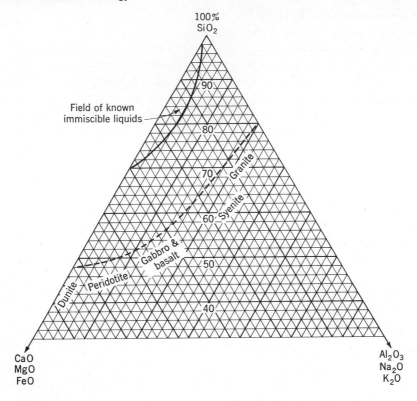

Fig. 2-20. The field of known immiscible liquids involving the rock-forming oxides compared with the composition of the common igneous rocks.

a familiar immiscible relationship exists between the silicate slag, the metallic sulfides, and the molten metals; each liquid separates gravitatively to a different level in the furnace. In a similar manner a sulfide melt has separated from the silicate magma in some mafic intrusive bodies. At elevated temperatures, mafic magmas may contain as much as 6 to 8 percent sulfide in solution, but as the temperature falls, the solubility of the sulfide decreases, forming droplets disseminated through the silicate magma. Because of their higher density the droplets accumulate at the bottom of the magma body to give a zone enriched in sulfides. As cooling continues, the silicates crystallize first, leaving an interstitial sulfide melt which subsequently crystallizes to a pyrrhotite-chalcopyrite-pentlandite mixture, the chief ore of nickel. This is the only situation in igneous rocks in which liquid immiscibility is clearly indicated. An immiscible-liquid relationship may exist between a carbonate-rich liquid (carbonatite magma), discussed below, and alkali-rich magmas.

EFFECTS OF VOLATILES

General Statement

The role of gases in the evolution of igneous rocks has been the subject of much debate, because it is so hard to evaluate. That gases are present in considerable volumes is clearly evident from volcanic activity, the presence of gas cavities in extrusive and even intrusive rocks, the presence of minerals containing essential volatile components, the obvious effects of the escape of gases on the wall rocks surrounding igneous bodies, and the apparent relationship between igneous activity and hydrothermal mineral deposits. Still, the amount of gas present, the way in which it escapes from the magma, and its possible effects on the magma composition are unclear. Five possible effects are discussed in the following pages: (1) modification of magma compositions in a large magma body by gas transfer; (2) modification of the crystallization sequence by water vapor under high pressure; (3) effects of local concentrations of gases in the late magmatic stage and the development of some pegmatites; (4) effects of disseminated late volatiles on the primary igneous minerals; and (5) carbonatites.

Gas Transfer

In a magma body of considerable vertical height, the streaming of gas from the lower portions of the magma may selectively remove certain constituents of the magma and release them in the upper portion of the magma because of reduction of pressure and temperature. In this way the lower portion of the magma may be impoverished in certain constituents and the upper portion enriched in these same constituents. Such a process could be effective only where convection in the magma is impeded by the shape of the body or the viscosity of the magma.

Thermodynamics requires that in a system consisting of several phases, all the constituents be present in all the phases. The distribution of the constituents, however, will not be the same in all phases. Thus in a crystallizing magma from which a gas is separating, there will be a liquid phase, a gas phase, and one or more solid-crystal phases. Each will contain all the elements in the system but in differing proportions. The gas phase will tend to be enriched in the more readily volatilized material, dominantly water and gases such as CO_2, H_2, HCl, HF, etc., but also appreciable amounts of metals, possibly as the volatile chlorides and fluorides. Under elevated temperatures and pressures at the base of the magma column, the metal ions, particularly the alkalies, will be taken up in the gas phase. As the gas rises into somewhat lower temperatures and markedly lower pressures, the solubility of these constituents in the gas will decrease and they will be returned to the liquid phase. Thus, in a vertical column the base may become somewhat impoverished in the alkalies and the top somewhat enriched. The extent to which such a volatile transfer will modify a magma cannot be assessed with the present knowledge.

Effect of Water Vapor at High Pressure

Studies by Yoder and Tilley (1962)* on the crystallization of basalts under high water-vapor pressure indicate that the sequence of crystallization is different from that at low pressure. Above a water-vapor pressure somewhat more than 10,000 bars, amphibole is the first mineral to appear and is followed by olivine and pyroxene. At lower pressures olivine and pyroxene appear before amphibole. Therefore amphibole phenocrysts form at depth and, as the magma ascends to lower-pressure zones, are resorbed by the magma. Inasmuch as amphibole is markedly different in composition from the magma, any relative movement of crystals and liquid will result in compositional changes. Many hornblendes are relatively rich in alkalies and poor in silica. Therefore their separation from a basaltic magma will enrich the liquid in silica and impoverish it in alkalies. Resorption of the same hornblende during ascent of a magma into which hornblende has been gravitatively accumulated will result in relative enrichment of the liquid in alkalies and some relative depletion of the magma in silica. Thus the two major types of basalts, the undersaturated alkali basalts and the saturated tholeiitic basalts, may be derived from a common parent through the effect of water vapor. Crystal differentiation of the alkali magma yields the sequence olivine basalt-trachybasalt-trachyte-phonolite, whereas the crystal differentiation of the tholeiitic magma yields the sequence basalt-andesite-dacite-rhyolite.

Late Volatile Concentrations
and Resulting Pegmatites

Pegmatites, to judge by their occurrence and character, probably form in several ways. Clearly, however, some of them have formed as a result of local late concentration of volatiles in the residual liquids of magmas. These appear as local bodies within an intrusive mass in which the texture changes, gradually or abruptly, from the normal magmatic texture to a pegmatitic texture. Such bodies usually measure a few feet to a few tens of feet in dimensions. Larger pegmatite bodies, which are usually intrusive into the margins of intrusive bodies or into the surrounding metamorphic rocks, also show evidence of high volatile content. The effects of volatiles are seen in the large size of the crystals, the abundance of minor minerals in which volatiles are essential, and the presence of vugs and gas cavities.

The accumulation of the volatiles into the late phase of the magma is brought about by the crystallization of the dominantly anhydrous primary magmatic silicates. As crystallization proceeds, the volatiles, and along with them the minor elements which cannot be accommodated in the structures of the primary minerals, are concentrated in the liquid. As a result the gas

*H. S. Yoder, Jr., and C. E. Tilley, Origin of basalt magma: an experimental study of natural and synthetic systems, *Jour. Petrology*, vol. 3, pp. 342–532, 1962.

pressure in the magma builds up until it is equal to the confining pressure on the magma system. At this point the magma "boils," but the vapor, because of the high temperature and pressure, has a high density, approaching that of an aqueous solution rather than that of steam. Rock materials are much more soluble in such a vapor than in steam. There thus result two fluids: one a silicate melt saturated with volatiles; the other a dense gas phase containing considerable quantities of dissolved rock material. Together these fluids crystallize slowly to give pegmatite bodies. The zonal nature of many pegmatites indicates that crystallization proceeds from the walls of the body inward, and the last fluid produces a central core.* The magma is highly fluid, and the drop of temperature must be very slow for the large crystals characteristic of pegmatites to develop. The volatiles are gradually lost through the walls of the body, and frequently, as a result, the outer zones are replaced by later minerals characteristic of the inner zones.

Disseminated Late Volatiles and Deuteric Alteration

A completely fresh specimen of igneous rock, particularly plutonic rock, which shows no evidence of alteration is very difficult to find. At least in part, the reason for this is that after crystallization is complete, the rock is still hot and gases are still present. These gases react to a greater or lesser degree to alter the primary igneous minerals to lower-temperature minerals, most of which are hydrous. The process is called *deuteric alteration*.

Several deuteric-alteration effects are commonly visible. In the silicic igneous rocks, the alteration of feldspars to clay minerals, such as sericite and kaolinite, is quite common. This process is similar to weathering, and it is usually impossible to distinguish between them. In deuteric alteration, however, normally transparent crystals, such as sanidine, become clouded, and other nontransparent crystals lose the normal bright luster of their cleavage faces. Deuteric alteration of biotite to chlorite is a common effect. Again the bright lustrous cleavage faces become duller, and flakes of chlorite may even be recognized. In the extrusive rocks, gas cavities and shrinkage cracks are often partially filled by crystals of materials transported by these late gases. Thus tridymite plates and complex spherical crystalline balls of cristobalite occur in many rhyolites along with less common minerals, such as topaz, apatite, sphene, and fluorite.

In mafic igneous rocks, deuteric alteration is even more common. Basalts have their gas cavities and cracks filled by beautifully crystalline zeolites, epidote, quartz, calcite, and feldspars. The coarse-grained rocks frequently show incipient alteration of pyroxenes to amphibole, giving material called uralite. Calcic plagioclase may alter to clay minerals in part

*E. N. Cameron, R. H. Johns, A. H. McNair, and L. R. Page, Internal structure of granitic pegmatites, *Econ. Geology* Mono. 2, pp. 98–105, 1949.

or, in some cases, give a very fine-grained greenish material called saussurite, a mixture of epidote and albite. Olivine is quite susceptible to deuteric alteration, reacting with water vapor to form serpentine, iddingsite, and talc. In ultramafic rocks, this alteration of olivine or magnesian pyroxenes to serpentine is frequently almost complete, giving rise to serpentine rock and some types of soapstone.

Carbonatites

Carbonatites are igneous-appearing rocks composed dominantly of carbonates. Pecora (1956)* reviewed the known occurrences of these rocks and concluded that they were not magmatic but were the result of the action of carbonate-rich solutions associated with igneous processes. In 1960, the volcano Oldoinyo Lengai in Tanganyika erupted a lava flow of soda-rich carbonate melt with the composition 12.74 percent CaO, 29.53 percent Na_2O, 7.58 percent K_2O, 8.59 percent H_2O, and 31.75 percent CO_2.† The possibility of a carbonate magma is therefore unquestionable. Most carbonatites, however, are composed dominantly of calcite; dolomite is usually less abundant, and the other carbonates are even less abundant. More than 50 other minerals have been reported from these rocks, including the rock-forming silicates and many ore minerals. Thus the Tanganyika flow does not approximate the average carbonatite in composition. However, all known carbonatites are associated with highly alkaline igneous rocks, and many are surrounded by a zone in which there has been extensive introduction of alkali in the wall rock. In the subvolcanic environment it would appear that the major fraction of alkali of the carbonatite magma is lost to associated silicate magma or to the walls, thus leaving a residue of calcite and dolomite.

The source of carbonatite magmas remains a mystery. A limestone source has been suggested as the origin of the material, but data on the ratio of carbon and oxygen isotopes in carbonatite indicate that they are very different from sedimentary carbonates (Holmes, 1965).‡ Furthermore, the extensive area of carbonatite plugs associated with alkali volcanoes in Africa is in a Precambrian shield area of little known carbonate sediments or metamorphic equivalents. It would seem that the source of the volatile material would have to be the lower crust or the upper mantle, and these rocks represent a phase of degasing of the earth.

---

*W. T. Pecora, Carbonatites: a review, *Geol. Soc. America Bull.*, vol. 67, pp. 1537–1556, 1956.
†J. B. Dawson, Sodium carbonate lavas from Oldoinyo Lengai, Tanganyika, *Nature*, vol. 195, pp. 1075–1076, 1962.
‡A. Holmes, "Principles of Physical Geology," 2d ed., p. 1072, Ronald Press, 1965.

3 SEDIMENTARY PROCESSES

Introduction

Sedimentary rocks are defined in the AGI Glossary as "Rocks formed by the accumulation of sediment in water . . . or from air. . . . The sediment may consist of rock fragments or particles of various sizes . . . ; of the remains or products of animals or plants . . . ; of the product of chemical action or evaporation . . . ; or of mixture of these materials. Some sedimentary deposits . . . are composed of fragments blown from volcanoes and deposited on land or in water." Sedimentary rocks thus involve a wide variety of materials and a variety of processes. They can, however, be grouped into two main categories: clastic materials and chemical materials. Most materials in either category have undergone related processes. Clastic materials are those which were developed as discrete solid particles in the weathering environment and have remained as such throughout their sedimentary history. This does not imply that they have not changed in their physical, chemical, or mineralogical character during the processes but, simply, that they have remained as solid particles. Chemical materials are those which have been transported in aqueous solutions or colloidal dispersion and have been precipitated by chemical or biochemical processes or evaporation. The short intervals of transportation involved in the destruction of old minerals and the development of new minerals during weathering are not included. Most sedimentary rocks contain materials of both types, but generally one of them dominates the rock.

Sedimentary rocks are of particular interest, because it is through them that the geologic history of a region is primarily reconstructed. The textures and structures in the rock, along with the fossils it may contain, give clues to the environment in which it was deposited. In addition, the clastic rocks may allow the reconstruction of the environment from which the sediment was derived and give some idea of the methods by which it was transported. Thus clastic sedimentary rocks are of particular interest.

The reconstruction of geologic history from clastic sedimentary rocks depends on the individual particles having remained solid throughout their history and, consequently, having been affected by each of the environments through which they have passed. There are five major factors which influence such particles. (1) The source rock from which the particles were derived controls, to some extent, the size, shape, and composition of clastic materials. A quartz-sand grain cannot be derived from a rock which contains no sand-sized or larger quartz grains. (2) The character of the weathering which produced the clastic grains from the parent rock, wherein the controlling factor is the balance between chemical and mechanical weathering and their modification of the size and composition of derived materials, is an influence. (3) The various agents and distances of transportation which are involved in sorting clastic materials and modifying their size and shape influence the particles. There are often three stages of transportation involved: the initial erosion of the weathered material by rain splash, sheet runoff, deflation, etc.; the continental transportation by streams, wind, or ice; and the subaqueous transportation and distribution in the area of accumulation. (4) The environment of deposition influences the size distribution of clastics, bedding characteristics, organic content, color of sediment, and other factors. The contrast between quiet and turbulent water and the chemical milieu with respect to pH and oxidation-reduction potential are dominant. (5) The postdepositional processes which convert unconsolidated sediment into sedimentary rock and which are of prime importance in the resultant rock are the final influence. Lithification and authigenic development of new minerals in the rock modify the texture, mineralogy, and physical properties of the sediment. It is not always easy to differentiate between processes of lithification and diagenesis and processes of metamorphism. Each of these five major influences on sedimentary rocks is considered individually and an attempt is made to see how each is recorded in the final product — a sedimentary rock.

INFLUENCE OF SOURCE ROCK ON SEDIMENTS

General Statement

The source rock has a direct influence on three characteristics of clastic material derived from it: the mineral composition, the size of the grain, and the shape of grain. These are particularly important in particles derived directly from the parent rock without development of new minerals. Thus size, shape, and composition of sand grains and coarser fragments are in part inherited features. This is true not only in single-crystal grains but also in polygranular and polymineralic particles.

INHERITED COMPOSITIONAL FEATURES

Mineral Stability

Minerals are stable in the environment in which they are originally formed. In other environments they tend to react to form other minerals, to go into solution, or to be modified by further growth. Igneous-rock-forming minerals are chemically stable; that is, they will not undergo change as long as they are in the temperature and pressure range of magmatic processes and in contact with a melt of restricted composition. When the temperature changes or the composition of the fluids in contact with crystals changes, a mineral may become unstable and subject to change, as exemplified by reaction relationships and deuteric alteration. The weathering environment is dramatically different from the magmatic environment, in that minerals are exposed to oxygen, carbon dioxide, dilute aqueous solutions, and a variety of acids at low temperature and pressure. As a consequence many of the minerals are unstable and are subject to change. However, the rate at which igneous minerals will react with the reagents of the weathering environment is highly variable, depending on the mineral species. Some minerals react at sufficiently rapid rates so that within a geologically short interval of time they are completely destroyed. Such minerals can be classed as *unstable* in the weathering environment. Other minerals react, but do so very slowly so that they may persist for a long time without complete destruction. These minerals can be considered *metastable*, that is, chemically unstable but with very low reaction rates in the weathering environment. Still a third group of minerals is essentially unaffected in the weathering environment, even with geologically long exposure. Such minerals can be considered *stable* in the weathering environment, and many of them are in fact chemically stable in their environment.

Almost all rocks have been developed or have been modified in environments quite different from that of weathering. Even sedimentary rocks have accumulated or have been lithified in contact with solutions quite different from those of percolating groundwater and have been exposed to reducing rather than to oxidizing conditions. Therefore, in the great bulk of source rocks there will be a mixture of stable, metastable, and unstable minerals, each of which can be contributed to new sediments under proper conditions. Igneous rocks, however, play a unique role as a source rock, because all sediments have ultimately been derived from igneous rocks. In each of the Precambrian shield areas the oldest rocks exposed are igneous, and all other rocks have been ultimately derived from them or from younger igneous material. It therefore becomes important, first, to consider the relative stability of igneous-rock-forming minerals in the weathering environment.

In an igneous terrane undergoing prolonged intense weathering and having a residual blanket of weathered material (a regolith) developed on top, the relative stability of the minerals can be determined by examining the mineral composition of different levels in the regolith. This is because the material at the top of the residual blanket has undergone the longest time of weathering, whereas that at the base has just begun to weather. Assuming no removal of clastic particles from the surface as weathering proceeds, the ground level is lowered through removal of material by solution, but at the same time, the surface of the unweathered rock is lowered at a greater rate by the weathering process at the rock-regolith boundary. This is illustrated in Fig. 3-1. The top of the blanket contains only the most stable of the original minerals, whereas the bottom of the blanket contains relicts, if nothing more, of even the least-stable minerals of the source rock. The process is illustrated in Fig. 3-2, which shows an igneous-rock assemblage containing six minerals of differing stability such that mineral 1 (m_1) is the most stable and mineral 6 (m_6) is the least stable. The surface of the regolith will contain only m_1 plus the decomposition products of m_2 through m_6. Below that will be a zone in which m_1 and relicts of m_2 (a mineral slightly less stable than m_1) will both be present along with the decomposition products of the other source minerals. In the third zone there will be m_1, plus partially decomposed m_2, plus relicts of m_3, plus decomposition products of all the others, and so on to the base of the regolith. At the very base the most-stable minerals will be unaffected, or nearly so, by weathering (m_1,m_2), the minerals of intermediate stability will be partially altered (m_3,m_4), and the least-stable minerals will be largely gone (m_5,m_6), but all will occur set in a matrix of decomposition products. Such an ideal situation, containing all the important rock-forming minerals, is not actually available. However, the relative stability of minerals can be pieced together from a number of different soil profiles. Goldich* has developed a stability sequence in this manner, as follows:

Least stable	Olivine	
		Calcic plagioclase
	Augite	
		Calci-alkalic plagioclase
	Hornblende	Alkali-calcic plagioclase
		Alkalic plagioclase
	Biotite	
	Potash feldspar	
	Muscovite	
Most stable	Quartz	

The similarity between this sequence and Bowen's reaction series is immediately noticeable. This stability sequence does not imply that one mineral in the sequence will weather to a lower mineral in the sequence (olivine does not weather to augite) but, rather, that under given weathering conditions, olivine will be completely destroyed by weathering before augite. The parallelism between Goldich's stability sequence and Bowen's

*S. S. Goldich, A study of rock weathering, *Jour. Geology*, vol. 46, pp. 17–58, 1938.

Original surface level

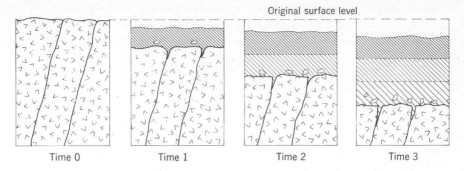

Time 0   Time 1   Time 2   Time 3

Fig. 3-1. Development of a regolith with time. With no surface removal of clastic material the ground level will be lowered, but the rock-regolith boundary will be lowered at a more rapid rate. The checked area is the parent igneous rock cut by joints. The cross-hatched area is regolith; the density of cross hatching decreases with the time interval in which the weathering took place.

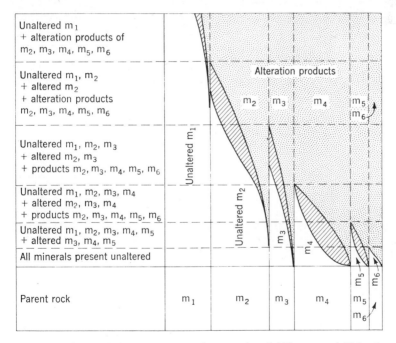

Fig. 3-2. Idealized disappearance of minerals of different stabilities in the weathered zone. The stippled area represents the weathering products of each mineral; the cross-hatched area represents the presence of partially decomposed minerals; and the blank area represents the presence of unweathered relicts of each mineral.

reaction series points out that for the quantitatively important silicate minerals, the greater the difference between the environment of formation and the weathering environment, the less stable the mineral will be in the weathering environment.

The same concepts can be applied to rock fragments, in that the stability of rock fragments is controlled in part by the stability of the important constituent minerals. The texture of the rocks also plays an important role. Given two rocks with about the same textural characteristics, the rock more dominantly composed of stable minerals, such as quartz and alkali feldspars, will be less susceptible to weathering than the rock more dominantly composed of unstable minerals, such as pyroxene and calcic plagioclase. Thus, medium-grained granite should be more stable than a medium-grained gabbro. Texture may reverse this relationship, in that a fine-grained dense rock is often less permeable to weathering solutions than is a coarse-grained dense rock.

Under appropriate weathering conditions, any of the constituents of the source rock may be contributed to a sediment as clastic grains. If a sediment contains important amounts of unstable minerals, it can be assumed that they are present either because the intensity of chemical weathering was not severe enough to destroy them or because the length of time that they were exposed to weathering was too short to destroy them. In either case the presence of unstable minerals as clastic grains in a sediment indicates that the source rock contained that mineral, and thus we can make interpretations as to the character of the source rock. However, if the sediment contains only stable clastic particles, two possible interpretations can be made: first, the source rock contained only stable minerals and, as a result, any intensity or duration of weathering could produce only stable minerals; or, second, the source rock contained unstable minerals and the intensity and duration of weathering were sufficient to destroy all minerals except the stable ones. Thus a pure-quartz sand could be derived from a previously existing pure-quartz sandstone by any type or duration of weathering, but a pure-quartz sand could be derived from a granite only if weathering were sufficiently severe and prolonged to destroy all feldspar, mica, and ferromagnesian minerals.

Contributions of Igneous Rocks to Sediments

A sediment derived directly from an igneous rock is termed a *first-cycle sediment.* Most first-cycle sediments, in addition to stable rock and mineral fragments, contain pebbles of igneous rocks or sand grains of metastable, or even unstable, igneous minerals. Probably only rarely in geologic time has weathering been prolonged enough and severe enough to give rise to a first-cycle clastic rock composed of only stable minerals. One such case may be the pure-quartz sandstones developed extensively at the base of the Cambrian around the Canadian Shield. Here the tremendous duration of erosion after the final Precambrian orogeny developed a pure-quartz sand from the Precambrian granites and gneisses. Much more commonly the relationship is like that described by Wahlstrom (1948)* between the granite of Precambrian age and the Fountain formation of Pennsylvanian age of Colorado. The Fountain arkose rests on a pre-Fountain soil profile 80 ft thick, exhibiting the relationships shown in Fig. 3-3. Wahlstrom determined the mineral

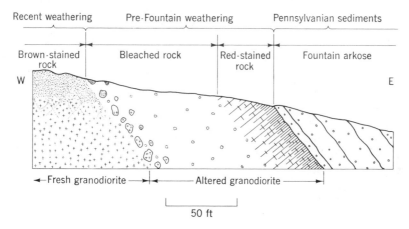

Fig. 3-3. Cross section at right angles to the Fountain granodiorite contact, Flagstaff Mountain, Arizona. (*After E. E. Wahlstrom, Geol. Soc. America Bull., vol. 59, 1948.*)

composition for different levels in the old soil profile, as shown in Fig. 3-4. The arkose was derived from similar materials higher up on the flank of a postweathering arch.

The presence of metastable minerals such as feldspar and micas is evidence of first-cycle sedimentation, but it is not proof, because these minerals may also be derived from metamorphosed sedimentary rocks. Of course, if they are present in polymineralic igneous-rock fragments, then this is clear proof. The best evidence of first-cycle origin, however, comes from the study of the minor accessory minerals of sands. These are the minor minerals present in most igneous rocks and relatively stable under weathering. They tend to be concentrated in the sand fraction developed from the soil. Frequently, these minerals can be separated easily from the lighter major constituents of a sand or sandstone and can be identified under a binocular microscope. When this is done, it is found that there is a relatively small

*E. E. Wahlstrom, Pre-Fountain and Recent weathering on Flagstaff Mountain near Boulder, Colorado, *Geol. Soc. America Bull.*, vol. 59, pp. 1173–1189, 1948.

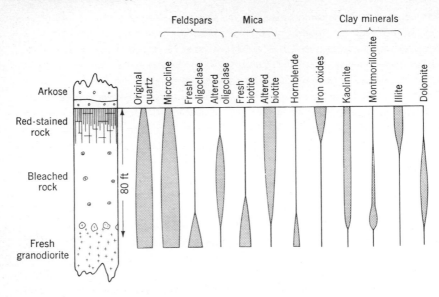

Fig. 3-4. Persistence with depth of important minerals in pre-Fountain mantle. (*After E. E. Wahlstrom, Geol. Soc. America Bull., vol. 59, 1948.*)

number of characteristic minerals derived from any one simple igneous terrane and, further, that the accessory minerals from silicic and mafic igneous rocks are characteristically different. Furthermore, for any one mineral species, such as zircon or tourmaline, all the accessory grains will have the same general character, that is, the same color, crystal habit, or character of inclusions. Some of the more common accessory minerals and the igneous rocks from which they are derived are listed below:

Silicic igneous rocks
   Apatite   Sphene
   Biotite   Zircon
   Hornblende   Magnetite
   Muscovite   Tourmaline
Mafic igneous rocks
   Augite   Leucoxene
   Apatite   Olivine
   Hypersthene   Rutile
   Ilmenite   Serpentine
   Magnetite   Chromite
Pegmatite
   Cassiterite   Tourmaline
   Fluorite   Muscovite
   Topaz

Contributions of Preexisting Sedimentary Rocks to Sediments

Many sedimentary materials have been derived directly from preexisting sediments or sedimentary rocks. In fact the older sediments may themselves have been derived from still older sediments. This introduces a concept of multiple cycles of sedimentation, in that once a clastic grain has been deposited, it is not forever fixed in that position but may be reexhumed and started on the weathering-transportation-deposition-lithification cycle again. Theoretically, at least, some of the earliest clastic particles developed in remote Precambrian time may still exist as clastic particles. What evidence indicates multiple cycles of sedimentation?

The best and most positive evidence of multiple-cycle sedimentation is the presence of clastic particles of rocks or minerals of strictly sedimentary origin. The most commonly encountered material of this type is sand and pebbles of chert. Most chert is of primary sedimentary or diagenetic origin, and it is almost always present in the coarser-grained multiple-cycle sediments. Similarly, pebbles of sandstone or quartzite are clear evidence of a sedimentary source rock. Pebbles and sand-sized grains of shale and limestone are also derived from preexisting sediments; however, some caution must be used in considering them as indicators of source, because they may be derived from sediments only very slightly older than the sediment that encloses them and may be from the same depositional site. Thus intraformational conglomerates, edgewise conglomerates, and clay flakes frequently develop in one short time interval and in the area of sedimentation as a result of desiccation. If, however, the limestone or shale pebbles are definitely from a different environment or age, they indicate primary sedimentary source rocks. Furthermore, because of their chemical and physical instability, the presence of such clastic fragments gives important evidence as to the weathering and transportation environment. This aspect will be discussed later.

A second evidence of multiple-cycle sedimentation is the absence of unstable or metastable minerals. This is not positive evidence, because complete chemical weathering of igneous material yields a first-cycle sediment composed of only stable minerals. However, pure-quartz sandstone is most frequently the product of multiple-cycle sedimentation. The presence of some metastable minerals does not prove a single cycle of sedimentation. Normally, during the first cycle most of the unstable and much of the metastable material is destroyed. During a second cycle the remaining unstable and most of the metastable material is destroyed. During subsequent cycles the last of the metastable material will go, leaving only stable mineral grains.

The thorough rounding of resistant minerals, especially quartz, indicates a tremendous amount of abrasion. As a result of extensive experimentation with the abrasion of clastic grains, Kuenen (1959)* concludes:

*Ph. H. Kuenen, Experimental abrasion, Part 3. Fluviatile action in sand, *Am. Jour. Sci.*, vol. 257, pp. 172–190 (especially p. 189), 1959.

"After a succession of ten sedimentary cycles with major stream transport each time, a quartz grain would still remain quite angular, provided no other factors intervened. To reduce a cube of 0.4 mm to a sphere, for instance, would require a river transport of several million kilometres." In contrast, Kuenen (1960)* found that eolian abrasion of quartz sand is much more rapid. A 1.7-mm cube was reduced to a sphere, losing 48.3 percent of its weight, over a distance of 327 km. Thus, if wind transportation can be ruled out of its immediate past history, a well-rounded quartz sand generally indicates multiple cycles of sedimentation.

The minor accessory minerals of sands present evidence of many cycles, in that they are composed dominantly of well-rounded exceptionally stable minerals. Tourmaline and zircon are particularly characteristic minerals. In addition, these minerals will be represented by many varieties based on color, shape, inclusions, euhedralism, and similar features. The explanation for the complexity of such suites can be visualized through a complex system of sediment parentage, such as presented in Fig. 3-5, where a sand under study has derived its material from three preexisting sands. Each of these parent sands has, in turn, drawn its material from preexisting rocks of different parentages and histories. The initial grains of resistant minerals were developed by crystallization from magmas, by recrystallization in metamorphic environments, or by authigenic growth in sedimentary rocks. It is not surprising that different grains of a single mineral derived from all these different sources should be different in color, habit, roundness, euhedralism,

*Ph. H. Kuenen, Experimental abrasion, Part 4. Eolian action, *Jour. Geology*, vol. 68, pp. 427–449, 1960.

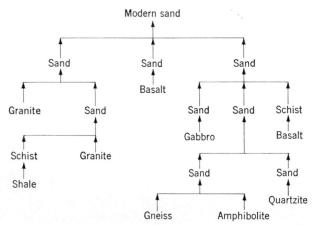

Fig. 3-5. The complex history which can be involved in a multiple-cycle sand. Accessory minerals derived from each type of parental rock will be different, yielding a very complex suite of very stable minerals.

etc. In contrast, a sediment developed directly from one igneous-rock suite will have a small number of varieties of these minerals.

A final evidence of multiple-cycle sedimentation is the presence of abraded grains which have been secondarily enlarged during a previous sedimentary history. During lithification of a sediment involving cementation by material of the same composition as the clastic grains, the cement is often deposited as crystalline overgrowths on the grains. These secondary overgrowths are in crystallographic continuity with the original grain, and the process is called *secondary enlargement*. For example, a quartz-sand grain may have silica added to it, converting the rounded grain to a euhedral or subhedral crystal. If such grains are subsequently put into transportation, the secondary enlargement will become abraded, and the original sand grain will still be recognizable. Such a single grain clearly shows two cycles of transportation and deposition and is clear evidence of multiple cycles. Abraded secondary overgrowths can be seen in thin sections.

Contributions of Metamorphic Rocks to Sediments

With the exception of the relatively minor characteristic metamorphic minerals, the contribution of metamorphic rocks to sediments is similar to that of igneous rocks. Quartz, feldspars, micas, amphiboles, and pyroxenes are the most abundant minerals in both igneous and metamorphic rocks, although they are present in different proportions. These minerals behave in the same manner in the weathering environment regardless of their parent rock type. Therefore the main contribution to sediments is the same regardless of the source rock.

The presence of rock fragments of clearly metamorphic origin is, of course, evidence of a metamorphic source. Schist, gneiss, and quartzite are common materials to be preserved as pebbles. Sand-sized grains of metamorphic rocks such as slate and phyllite are very common in some sands. The best direct evidence of a metamorphic source rock in many sediments comes from the presence of characteristic metamorphic accessory minerals. When the following minerals are common as accessory minerals in sands, it is indicative of metamorphic origin:

Actinolite	Kyanite
Andalusite	Tremolite
Epidote	Sillimanite
Garnet	Staurolite
Glaucophane	

Inherited Size and Shape Characteristics

Size and shape of clastic grains, particularly in sand and coarser materials, are in part controlled by the parent rock. These features may be modified during the subsequent transportation but are not generally completely destroyed. In pebble-sized particles, both size and shape are controlled by directional fabrics or structural elements of the source rock. Close-spaced bedding-plane parting in shale requires that clastic particles derived from it be flaky in shape. The closer the spacing of the partings, the smaller the resultant particles must be. Similarly, the schistosity of many metamorphic rocks results in the development of lenticular pebbles. In more-massive source rocks, joint systems and joints in conjunction with bedding control both size and shape of the original polygonal blocks from which pebbles are derived. Subsequent abrasion modifies the shape and gradually obliterates the original outlines. Figure 3-6 shows a series of pebbles derived from quartz crystals.

The sizes and shapes of sand particles are frequently more markedly controlled by the parent rock than are the sizes and shapes of pebbles. This is especially true in the case of the more mechanically resistant minerals such as quartz. It is apparent that a single-crystal sand grain must be smaller than that mineral in the parent rock; thus sand-sized quartz grains develop from coarsely crystalline igneous or metamorphic rocks but not from

Fig. 3-6. Front and top views of three quartz pebbles and a quartz crystal similar to the crystals from which the pebbles were derived. The first pebble clearly retains the hexagonal prismatic shape of the original crystal. The third pebble was broken prior to rounding. Saline River gravel, Sheridan, Arkansas.

aphanitic rocks. The shapes of quartz grains also seem to be determined by the parent. Quartz in metamorphic rocks tends to develop lenticular-shaped grains, whereas the grains of igneous rocks tend to be elongated along the c crystallographic axis. Abrasion-mill tests show that as sand-sized particles, quartz abrades very slowly and requires transportation over a long distance to become well rounded. Thus the original grain shape is preserved without much modification. Minerals with well-developed cleavages usually have their shapes controlled by the cleavage.

Under some extreme environments the shape of larger fragments can be completely obliterated relatively rapidly. Sandblasting in desert regions will cut and polish faces on pebbles which the wind cannot move, yielding faceted pebbles with one or more facets cut (Fig. 3-7). Similarly, larger fragments carried along at the base of a glacier will have flat surfaces ground on them. Such pebbles are usually striated and flattened on one or more sides (Fig. 3-8). The shapes are typically described as "flat irons."

NEW MATERIALS PRODUCED DURING WEATHERING

Up to this point only particles derived unchanged from the parent rock have been considered. A wide variety of new materials is produced during chemical weathering which contribute importantly to sediments, especially to the

Fig. 3-7. A ventifact, a wind-faceted pebble. The pebble is composed of vein quartz and is in place in a disintegrated granite rubble. Driftless area near Wausau, Wisconsin.

Fig. 3-8. Glacially faceted and striated cobble.

finer clastic materials and to the precipitated sediments. In general, the character of these materials is not as dependent on the character of the source rock as it is on the environment of weathering. The chemical composition of the parent rock does, however, control the relative proportions of these products. The materials produced by weathering may be grouped as clay minerals, oxides and hydroxides of iron, and materials removed in solution.

The clay minerals are basically hydrous aluminum silicates which may have other cations, particularly potassium, calcium, and magnesium, entering into some of them. There is an additional group of aluminum hydroxides and hydrous aluminum oxides which, though not strictly clay minerals, are developed under weathering environments and represent extreme leaching conditions. In general, all these minerals occur in crystals of such minute size that they cannot be distinguished, except by special techniques outside the scope of usual petrographic study. However, these minerals are volumetrically very important and have a great influence on the chemical composition of sediments. The character of the clay mineral present in a sediment may also be indicative of the weathering environment in which it formed. Six groups of clay minerals are of great importance.

The *gibbsite group* is composed of aluminum hydroxide and hydrous aluminum oxides without any other essential cations present. As relatively pure concentrates they are the minerals of the mineraloid *bauxite* and are the dominant aluminum minerals of the soil type known as *laterite*. Gibbsite and its relatives are the product of complete humid tropical weathering in environments of thorough leaching and removal of material in solution by circulating groundwater. Under these conditions all cations, including silicon, are removed by the groundwater, and the remaining alumina crystallizes into several structurally distinct minerals.

The *kaolinite group* is composed of hydrous aluminosilicates without essential additional elements. Again this represents the products of thorough leaching of the metals, alkali metals, and alkaline earth elements but differs from the gibbsite group, in that silicon is not totally removed. Structurally, the kaolinites consist of a sheet of silicon-oxygen tetrahedra coordinated with a sheet of aluminum-oxygen-hydroxide octahedra. The resultant double sheet is electrostatically neutral so that no additional ions are required. The various minerals of the group differ in the stacking patterns of these double sheets. Kaolinite, the most important mineral of the group, is the commonest clay mineral produced in normal humid climates.

The *montmorillonite group* is composed of hydrous aluminosilicates with important amounts of additional cations, particularly iron, magnesium, calcium, and sodium. It therefore forms in an environment of incomplete leaching of the elements liberated from the parent minerals by chemical weathering. The basic structure of all the minerals of this group is a three-layer "sandwich" made up of an aluminum-oxygen-hydroxide octahedrally coordinated layer between two silicon-oxygen tetrahedrally coordinated layers. Considerable compositional variation is possible through substitution. Aluminum substitutes in part for silicon in the tetrahedral layers in some species, and magnesium and iron substitute in part or entirely for aluminum in the octahedral layer in some species. As a result of these substitutions the three-layer unit is never electrostatically neutral but always has a residual negative charge. This is balanced by weakly bound cations between adjacent three-layer sheets. Calcium and sodium usually occupy this position along with a large and variable amount of water. The cations in this interlayer position are readily exchanged for other cations in solution in the groundwater contacting the clay minerals, and therefore the montmorillonites are said to have a *high cation-exchange capacity*. In addition, the interlayer water content is highly variable, and the crystals swell or contract as water is added or removed by desiccation. Thus the montmorillonites are spoken of as *expandable* clays. The environment of montmorillonite development is one of incomplete leaching and abundant cations in solution. Therefore the group is typical of soils developed in semiarid to arid regions. Montmorillonites also develop through subaqueous alteration of volcanic ash, giving rise to the bentonite clays. A high magnesium content in the source rock seems to be the dominant requirement for the development of this group.

The *illite group* is composed of hydrous aluminosilicates with essential potassium. They again represent incomplete leaching but require an abundance of potassium either in the source rock or in the solutions with which they are in contact. The basic structure is a three-layer sheet like that of the montmorillonites—an octahedrally coordinated aluminum-oxygen-hydroxide sheet bound to two tetrahedrally coordinated silicon-oxygen sheets. Aluminum substitutes in variable amounts for silicon in the tetrahedral sites, creating a valence deficiency in the triple layer; this is accommodated by variable amounts of potassium fixed in interlayer sites. Where the ratio of silicon to aluminum in the tetrahedral sheets is $6:2$ (the maximum possible), there will be two potassium ions in the unit formula, and the mineral is muscovite. Thus the illites are silica-enriched potassium-depleted micas. The presence of bound potassium between the sheets results in a low cation-exchange capacity and a nonexpanding structure, in contrast with the montmorillonites. Illite is the dominant clay mineral of many marine shales, and the question then arises as to whether the illite is a primary product of weathering or the result of diagenetic alteration of other clay minerals by reaction with potassium in sea water. If illite is the direct product of chemical weathering, its development would require alkaline soil conditions such as those developed in arid or subarid environments.

The *chlorite group* contains some members which occur as clay-sized particles in weathering products and sediments. These are derived by alteration of the primary ferromagnesian minerals and by diagenetic alteration of montmorillonitic clays. The chlorite clays are hydrous aluminosilicates with appreciable essential iron and magnesium. Structurally, they consist of a three-layer sheet like the montmorillonites, but between these sheets there are regular octahedrally coordinated sheets of composition $(Fe,Mg)_3(OH)_6$. Variable substitutions of aluminum for tetrahedrally coordinated silicon and of iron or magnesium or both for octahedrally coordinated aluminum are possible, giving a wide range of compositions. Some of the chloritic clays have a two-layer structure like that of kaolinite, but the compositional variation of these minerals is the same as that of the true chlorites. Some chlorite may be the direct product of chemical weathering, but chloritic clay is most common in marine shale, where it is probably a diagenetic product. The red clays of the deep ocean floor, for instance, contain an abundance of these materials.

The *mixed-layer clays* are the final group of clay minerals. In these, single-crystal units are made up of two or three distinct structural types of clay minerals. The interlayering may be regular, having a fixed number of layers of one clay type alternating with a fixed number of another type, or it may be completely random. Common mixed-layer clays include the pairs illite-montmorillonite, illite-chlorite, chlorite-montmorillonite, muscovite-illite, and biotite-montmorillonite, and the triple mixed-layer clay illite-chlorite-montmorillonite. The mixed-layer clays are in most cases derived by the modification of preexisting clay minerals resulting from weathering or diagenesis. Thus, if potassium is being leached from an illitic clay and there is available calcium in the weathering solution, a mixed-layer illite-montmorillonite will develop as an intermediate step between the two end minerals. Similarly, replacement of the calcium of a montmorillonite by magnesium in sea water will give an intermediate step of a montmorillonite-chlorite mixed-layer clay.

Iron oxides and hydrous oxides are a second major group of materials produced by weathering. Hematite and the mineraloid limonite are the common forms resulting from oxidation or oxidation plus hydration of ferromagnesian minerals. The minerals generally form as clay-sized particles so that again the identification of mineral species is difficult. Hematite, the anhydrous ferric oxide, is more characteristic of weathering in dry climates, whereas limonite or a mixture of hematite and limonite is more characteristic of humid climates. The term limonite is a catchall term to represent mixtures of several different minerals or mixtures of amorphous material. Two structurally different $FeO \cdot OH$ minerals occur, goethite and lepidocrocite. In addition to these and the amorphous materials, limonite generally contains variable amounts of absorbed water, colloidal silica, clay minerals, and organic decomposition products. Under exceptional conditions of thorough leaching, a residue to essentially pure iron oxides can be developed which is analogous to the aluminum residue bauxite.

Table 3-1 summarizes the clay minerals and related residual minerals.

The materials removed in solution during weathering consist dominantly of the alkali elements sodium and potassium and the alkaline earth metals calcium and magnesium. In general, sodium is removed to a greater extent than is potassium owing to the selective retention of potassium in clay minerals and the relative stability of the potash feldspars. Similarly, in general, calcium is removed to a greater extent than is magnesium as a consequence of the high magnesium content of some clay minerals. Iron in the oxidized state is essentially insoluble, and appreciable iron is removed in solution only under strong reducing conditions. Most of the aluminum liberated in weathering is tied up in the clay minerals of the weathering zone. However, in the presence of sulfate ions, aluminum reacts to form the soluble alums. Thus if sulfides such as pyrite or marcasite are undergoing oxidation, the sulfuric acid produced will react with available aluminum-bearing minerals, rendering aluminum soluble. Quartz is essentially insoluble, but during the breakdown of the silicate minerals, much silica is liberated and removed, as evidenced, for instance, by the development of the gibbsite minerals from feldspars. The removal of silica must be either as a hydrated colloid or as soluble alkali silicates. The relative solubilities of iron and aluminum hydroxides and silica are given in Fig. 3-9 (Pickering, 1962).*

*R. J. Pickering, Some leaching experiments on three quartz-free silicate rocks and their contribution to an understanding of laterization, *Econ. Geology*, vol. 57, pp. 1185–1206, 1962.

Table 3-1. Summary of Clay Minerals and Related Residual Minerals

Property	Gibbsites	Kaolinites	Montmorillonites	Illites	Chlorites	Iron minerals
Structure	Al–O–OH octahedra	Two-layer sheet	Three-layer sheet	Three-layer sheet	Three-layer sheet	Fe–O–OH octahedra
Interlayer cations	None	None	Ca, Na, H_2O	K	$(Fe, Mg)_3(OH)_6$ sheets	None
Cation substitution	None	Slight	Great	Moderate	Moderate	None
Expandability	None	None	Great	Slight	None	None
Occurrence in soils	Acid excessively leached soils	Acid leached soils	Alkaline soils from mafic and intermediate rocks	Alkaline soils from silicic and alkalic rocks	Variable distribution	Variable; present in most soils. Dominant in completely leached mafic soil
Occurrence in modern sediments	Rare	Common	Common	Abundant	Common	Minor
Occurrence in ancient sediments	Rare	Common	Common	Abundant to dominant	Common	Minor to dominant in ferruginous sediments

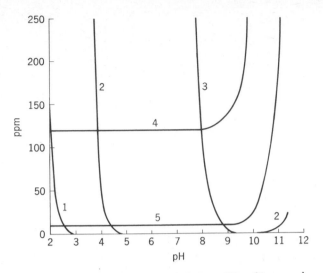

Fig. 3-9. Relative solubilities and the pH's of iron and aluminum hydroxides and silica. Curve 1–ferric hydroxide; curve 2–aluminum hydroxide; curve 3–ferrous hydroxide; curve 4–amorphous silica; and curve 5–quartz. (*After R. J. Pickering, Econ. Geology, vol. 57, 1962.*)

The relative removal of cations by weathering can be seen in the composition of the dissolved load of streams. Table 3-2 presents the compositions of dissolved solids, expressed in parts per million (ppm), for a number of rivers. The rivers were selected to represent a wide range of climatic conditions, but major trunk rivers draining a variety of environments were avoided. The average Ca:Mg ratio for these rivers is 3.3:1, and they range from 1.7:1 to 5.7:1. This ratio for the total crust, calculated from the data of Table 1-8, is 2.03:1. The ratio Na:K has a much higher range of values, 1.9:1 to 150:1. Omitting the three highest values in which leaching of salt deposits is probable, the ratios vary from 1.9:1 to 10:1, averaging 5:1. This ratio for the total crust, calculated from the data in Table 1-8, is 1.36:1. Clearly, then, there is selective removal of sodium and calcium in solution and selective retention of potassium and magnesium in the insoluble residues. The amount of dissolved iron in all samples where it was determined is low. It is interesting to note that those rivers with the highest iron content are in areas with abundant Pleistocene till and abundant lakes and swamps. Reducing conditions in swamps would liberate soluble iron from the incompletely weathered glacial materials.

The relative importance of igneous-, metamorphic-, and sedimentary-rock types as sources for modern sediments has been estimated for sand-

Table 3-2. Composition of Dissolved Solids in Rivers*

River	Station	Chemical analysis, ppm									Total ppm	Mean Discharge, cfs	Drainage area, square miles
		SiO_2	Fe	Ca	Mg	Na	K	HCO_3	SO_4	Cl			
Sacramento	Freeport, California	20	0.04	12	5.8	9.6	1.2	67	9.0	7.3	102	17,930	
San Joaquin	Vernalis, California	20	0.05	28	13	59	2.9	101	50	81	321	2,054	14,010
Columbia	Northport, Washington	5.1	....	21	4.1	1.7	0.9	74	13	0.5	86	95,700	59,700
Humboldt	Rye Patch, Nevada	43	0.02	50	14	105	14.2	325	71	61	530	300	16,100
Gila	Gilespie Dam, Arizona	25†	....	397	164	1,500	10†	29.9	1,500†	2,100†	6,130	11.8	49,650
Missouri	Wolf Point, Montana	7.2	0.01	53	21	41	3.8	193	131	9.0	370		82,290
Kansas	Bonner Spr., Kansas	12†	....	68	14	35	6.5†	205	69	42	361	11,740	59,890
White	Clarendon, Arkansas	9.5	0.06	22	8.5	2.9	1.5	107	3.9	4.1	117	23,540	25,497
Pecos	Girvin, Texas	10	....	736	435	3,540	50†	117	3,300	5,520	13,600	28.1	29,560
Brazos	Bryan, Texas	12	....	80	14	131	....	152	134	196	669	3,538	38,400
Neches	Evadale, Texas	13	....	7.9	3.3	20	....	20	21	26	102	5,174	7,952
Minnesota	Montevideo, Minnesota	14	0.23	69	35	18	7.6	223	155	4.5	443	1,507	6,180
Mississippi	Anoka, Minnesota	12	0.15	38	12	4.6	2.0	169	12	0.9	190	7,584	19,100
Ohio	Warsaw, Kentucky	4.8	....	38	9.9	21	2.5	61	86	30	237		83,200
Quinebaug	Jewett City, Connecticut	3.7	0.33	5.5	1.3	8.9	1.5	22	11	7.4	59	1,690	711
Lehigh	Easton, Pennsylvania	7.3	0.02	24	9.6	8.2	4.1	61	45	9.3	157		1,364
Schuylkill	Philadelphia, Pennsylvania	9.7	0.01	37	15	18	3.4	77	92	20	253	1,879	1,893
Roanoke	Jamesville, North Carolina	9.5	0.07	7.5	2.4	7.4	2.1	36	6.9	5.4	69	3,000	9,247
Edisto	Jacksonboro, South Carolina	5.7	0.15	5.0	1.4	8.0	0.8	15	4.3	13	60	961	2,870
Ohoopee	Reidsville, Georgia	11	0.12	3.6	0.7	3.5	0.7	13	1.9	5.5	45	961	
Flint	Culloden, Georgia	13	0.04	3.4	1.1	7.3	1.9	29	2.3	3.2	52	2,600	
Peace	Arcadia, Florida	11	0.09	28	7.9	15	1.7	44	57	19	203	910	
Kissimmee	Okeechobee, Florida	3.6	0.06	10	3.8	10	1.9	31	13	16	89	615	

*Data drawn from Quality of surface waters in the United States, U.S. Geol. Survey Water-Supply Papers 1941, 1942, 1943, 1944 and 1945, 1962.

†Data averaged from incomplete data for the water year 1962.

sized materials. The character of sand grains, their inclusions, and the character of the minor accessory minerals are the basis for this estimate. The average modern sand, according to Krynine, consists of 30 percent material reworked from older sediments, 45 percent material derived from all types of metamorphic sources, and 25 percent new material of direct igneous origin.

INFLUENCE OF WEATHERING ON SEDIMENTS

General Statement

The great bulk of rocks develop or are subsequently modified under environments markedly different from surface conditions. The rocks are therefore susceptible to physical or chemical modifications when exposed to the atmosphere, hydrosphere, or biological activity. Such changes are called *weathering* and can be divided into two main types, chemical and mechanical. *Chemical* weathering involves the production of new minerals from old minerals, or portions of them, and the removal of materials in solution or as colloids. *Mechanical* weathering involves reduction of grain size without the destruction of the minerals composing the grains. In most weathering, both chemical and mechanical processes are involved; however, the dominance of one or the other can usually be recognized in the character of the weathering products. Weathering is basically determined by climate, in that climate controls the temperature, the availability of water, oxygen, carbon dioxide, and natural acids, and the type and amount of plant cover. Bedrock character and topography are other important controls.

PROCESSES OF MECHANICAL WEATHERING

General Statement

Several processes operate to disintegrate rocks, thus reducing their grain size. Four main processes are discussed here: crystal growth, temperature changes, unloading, and organic action. These are by no means all the possible causes but are those which are probably most important.

Crystal Growth

The growth of any type of crystal in fractures or small openings in the rock can exert pressures on the walls of the opening sufficient to rupture the rock. The action of frost in temperate regions is probably the most familiar example, but the same result may be obtained by desiccation of brines and resultant growth of salt crystals, by growth of clay-mineral crystals on the surfaces of feldspar due to incipient chemical weathering, or by recrystallization during hydration of alteration products. In each case, rupturing of the rock is the result of a growing crystal exerting a force, up to the crystal's crushing strength, on its container to make room for itself.

Rock rupture by the freezing of water is usually explained as the result of the 9-percent expansion which takes place when water freezes. In the majority of situations in which freezing takes place, very little work in breaking rock is actually done by this volume expansion, and for several reasons. First, in most fractures the shape is such that expansion is accommodated by bulging up of the ice at a free air-water boundary. This is frequently seen in the freezing of milk in bottles left outside as the formation of a column of frozen milk above the top of the bottle. Second, if the ice plug cannot move upward, the increased pressure below the plug lowers the freezing point. Rock and soil are rather effective insulators of heat, and therefore a prolonged temperature well below 0°C is required for further freezing. Third, if rupture did take place as a result of ice expansion, the remaining water in the fracture would escape, and there could be no further effect of freezing until the fracture was sealed by some mechanism so that more water could accumulate (Fig. 3-10).

Fig. 3-10. Frost spalling of jointed and laminated metaargillite. Frost crystals growing in capillary films along closed joints have wedged off chips like those below. Gowganda Formation, Desbarats, Ontario.

The freezing of water proceeds by the growth of masses of ice crystals from water supplied along capillary openings (Reiche, 1945).* Rock openings available to moisture from the surface are of two types, capillary and supercapillary. Water in capillary-sized openings is very difficult to freeze, requiring considerably more supercooling to initiate freezing than does water in supercapillary openings. Roedder (1962),† in studying the freezing

*P. Reiche, "A Survey of Weathering Processes and Products," pp. 11–14, University of New Mexico Press, 1945.
†E. Roedder, Studies in fluid inclusions, *Econ. Geology*, vol. 57, pp. 1045–1061, 1962.

of solutions in microscopic cavities in crystals, found that due to supercooling, $-35°C$ is inadequate to freeze most inclusions. A temperature of $-78.5°C$ maintained for 30 min is generally adequate, although some samples did not freeze completely at $-196°C$. When small macrobodies of water in a rock begin to freeze, the first effect of the volume expansion is the escape of excess water into adjacent capillary films. Once freezing is complete in such a larger opening, the growth of the ice crystals can continue by the addition of water which is drawn to the growing crystals by the surface tension of the water film along the capillary-sized fissures. Such crystals can exert a pressure up to their crushing strength on the walls. The force opposing the crystals is the tensile strength of the rock, which at the shallow depths of weathering is low in comparison. Thus the small super-capillary openings are wedged apart, generating larger cavities in which crystals continue to grow. Crystal growth is stopped only by the exhaustion of water in the films. In climates of alternate freeze and thaw, the capillary water is renewed with each thaw, even when all the ice masses in the rock do not melt. The effectiveness of this process can be seen in the breakdown of fissile shales into small flakes. In freshly exposed outcrops of water-saturated shales, the destruction of single homogeneous blocks can be completed in one winter of alternate freeze and thaw. The effectiveness of desiccation of brines is equally good. In many deep-well samples, single cores of brine-saturated shale will break down into a series of thin disks in a matter of a few years storage in a core lab. The "poker chips" so formed are seen to be coated with a film of tiny salt crystals, the growth of which ruptured the shale along bedding planes (Fig. 3-11). Similarly, the growth of caliche in rocks of arid regions or of clay-mineral grains on the surface of feldspars in coarse-grained igneous rocks results in granulation of the rock.

Temperature Changes and Rock Disintegration

In desert regions, and particularly in high mountain regions, the surface of the rocks may be covered by a grus of granular disintegrated material and show very little evidence of chemical weathering. Such materials are particularly common on coarse-grained granites (Fig. 3-8). The production of the grus has been explained as resulting from heating during the day and rapid cooling at night, creating a temperature range on the order of $100°F$; the different thermal expansion of the minerals creates stresses sufficient to eventually break the rock down into individual grains. The stresses so developed can be calculated from the thermal-expansion characteristics of the minerals and are found to be significantly less than the strength of the rock (Tarr, 1915).* Experiments have been performed in which alternate heating and cooling of specimens over a temperature range

*R. S. Tarr, A study of some heating tests, and the light they throw on the cause of disaggregation of granite, *Econ. Geology,* vol. 10, pp. 348–367, 1915.

Fig. 3-11. Wedging of shale by the growth of thin gypsum crystals on bedding planes. Mecca shale of Pennsylvanian age, Mecca, Indiana.

much greater than that encountered in desert regions produced no granulation effects, even with temperature alternations equivalent to many years of exposure (Griggs, 1936).* In all probability this type of granular disintegration is the result of crystal growth between the component mineral grains of the rock. The incipient alteration of feldspar to clay on the surface of feldspar crystals may start the process, and the growth of frost crystals probably completes it.

Thermal expansion as a result of solar heat may cause rupture of large sheets of rock if a large unbroken surface is exposed to heating. The buckling of pavement on hot days illustrates what might happen. However, in a large outcrop the effect may be different from that of a pavement because of the third dimension of thickness. Pavements, since they are relatively thin

*D. Griggs, The factor of fatigue in rock exfoliation, *Jour. Geology*, vol. 44, pp. 783–796, 1936.

and resting on soil or rock fill, may buckle, whereas an outcrop which is a great thickness of identical material may not buckle. However, this is a possible contributing factor to exfoliation, where other agents, such as unloading, have developed fractures parallel to the surface.

Catastrophic temperature changes, such as those encountered in forest fires, can cause considerable granular disintegration of the surface of outcrops, particularly where the rock is coarse-grained. After a severe forest fire on Mt. Desert Island, Maine, piles of granulated granite were present at the foot of the smaller outcrops, where the fire could affect the rock. Similar disintegration of granite has been seen in fires in buildings, where granite pillars were exposed to severe heating. In time past, forest fires were probably much more frequent and widespread than now, and resulting rock disintegration was probably important. Recent study of forest-plant ecology indicates that such fires must have been frequent to develop the open virgin forests of giant trees (Cooper, 1961).* Such fires were restricted to ground level and thus to burned-off "blow-downs," fallen limbs, ground litter, and saplings and probably drastically reduced ticks and chiggers.

Unloading

Rocks, particularly intrusive igneous rocks and metamorphic rocks, are developed at great depth under high confining pressure. They are therefore compressed. Uplift and erosion reduce the confining pressures, and the rocks expand as a result. The expansion finds relief in the development of cracks and joints. In many plutonic igneous rocks these fractures are parallel to the present surface, giving a series of sheeting joints which are characteristically closely spaced at the surface and become more widely spaced with depth (Johns, 1943; Chapman and Rioux, 1958).† In many massive granites the sheeting very closely follows the present topography, raising these unanswered questions: Can the sheeting be controlled so closely by the topography? Is the topography the result of the sheeting? In either case unloading appears to be important and opens the rock to access by weathering agents to considerable depth (Fig. 3-12).

Mechanical Effects of Organic Activity

Plants and animals, particularly man, are active agents of rock breaking. The wedging apart of blocks of rock by the growth of tree roots is a familiar example. However, a more important agent is the work done by burrowing animals and insects. By continually working through the weathering material, these animals keep it porous so that the chemical weathering agents have a ready access to the unweathered material.

*C. F. Cooper, The ecology of fire, *Sci. American*, vol. 204, pp. 150–160, 1961.
†R. H. Johns, Sheet structure in granites: Its origin and use as a measure of glacial erosion in New England, *Jour. Geology*, vol. 51, pp. 71–98, 1943; C. A. Chapman and R. L. Rioux, Statistical study of topography, sheeting and jointing in granite, Acadia National Park, Maine, *Am. Jour. Sci.*, vol. 256, pp. 111–127, 1958.

Fig. 3-12. The effect of sheeting joints on access of weathering agents. Decomposed granite (from water edge to hammer head) underlying less severely weathered granite. Near Wausau, Wisconsin.

Many other agents produce fractures in rocks and therefore may be considered as contributing to mechanical weathering. Diastrophism, with its resultant joints and cleavage, is quite important in opening rocks to weathering. The mechanical effects of crushing along fault planes or along the soles of glaciers produce much mechanically reduced material. Glaciers, in this respect, may be thought of as thrust sheets and the resultant till as a fault breccia, or mylonite. Finally, the impact of grain on grain during transportation by running water, waves, or wind will mechanically reduce rock materials in size.

PROCESSES OF CHEMICAL WEATHERING
General Statement
Chemical weathering involves the reaction of rock materials developed under different chemical and physical environments with the chemical agents at or near the surface of the earth. The agents of chemical weathering are water, oxygen, carbon dioxide, natural mineral acids, and organic acids. In the weathering process two or more of these agents usually cooperate in the destruction of the rock. The processes involve the development of new minerals, particularly clay minerals and iron oxides or hydroxides, and the removal of material in solution or by colloidal suspension. In the process residual unweathered or incompletely weathered detrital grains are produced which may appear only slightly different from similar mechanically produced grains.

Weathering Effects of Water

Water is the most important agent in chemical weathering, but some of the processes in which water is involved require the presence of one or more of the other agents. Thus, hydrolysis involves the exchange of hydrogen ions from water and cations of a mineral structure, but its effectiveness is increased by the presence of readily ionizable acids such as the natural mineral acids or organic acids. Similarly, solution of material is primarily aqueous, but the solubility of many cations is controlled by their oxidation state, the presence of carbon dioxide, or the hydrogen-ion concentration. Thus water does not usually work alone but in combination with other active chemical-weathering agents.

Hydrolysis is the exchange of hydrogen ions from the water for cations from the mineral structure. In the process portions of the original structure are broken up into subcrystalline clumps of colloidal-sized materials which subsequently recrystallize to clay minerals. Hydrolysis is effective, because layers of water a few molecules thick attach to the surface of a crystal in very nearly perfect crystal coordination (Fig. 3-13). In such a layer a portion of the hydrogen ions of water will be coordinated with the oxygen of a water molecule and an oxygen of the crystal. Because of thermal oscillation of the ions, such coordinated hydrogen ions may be detached from the water film and move into the crystal. This will cause an excess positive charge in the crystal which will be neutralized by the expulsion of a positively charged cation. The metallic cation will be expelled, because its large size, relative to the hydrogen ion, requires a weaker bonding force in the structure (Frederickson, 1951).* Such an exchange of hydrogen and alkali ions can be readily demonstrated by covering crushed feldspar with distilled water. After an interval of 1 hr, the distilled water will be found to be alkaline, demonstrating a removal of hydrogen ions from the water and a consequent build up of hydroxyl ions in the water. In the case of the hydrolysis of feldspars, the removal of the cations (sodium, potassium, or calcium) leaves a partially coordinated hydrated aluminosilicate structure, probably in clots of colloidal dimensions, which will subsequently recrystallize to clay minerals.

It is apparent from the nature of this hydrolysis process that the presence of free hydrogen ions from acid sources will aid materially in the processes. The solution of carbon dioxide from the atmosphere or from decaying organic material in the soil zone gives carbonic acid, which is weakly ionizing, supplying hydrogen ions for the hydrolysis process. It is also important in stabilizing the cations released by hydrolysis in solution as soluble carbonates or bicarbonates or by removing them as insoluble precipitated

*A. F. Frederickson, Mechanism of weathering, *Geol. Soc. America Bull.*, vol. 62, pp. 221–232, 1951.

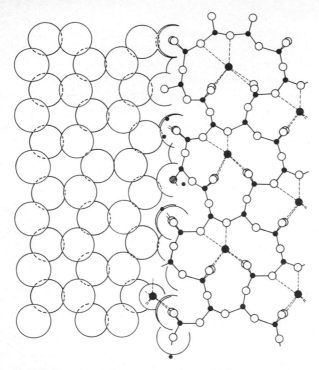

Fig. 3-13. Diagrammatic drawing indicating the structural relationships between an albite crystal and an adsorbed water film. The structure on the left is that of ice. Small dots represent hydrogen ions. (*After A. F. Frederickson, Geol. Soc. America Bull., vol. 62, 1951.*)

carbonates. The strong mineral acids such as hydrochloric and hydrofluoric acids may be locally effective in weathering in the vicinity of volcanic activity. The various acids of sulfur are produced locally in the weathering zone through the oxidation of sulfides such as pyrite or marcasite and, where present, will cause much weathering. The nitrogen acids are produced sparingly in the atmosphere through the action of lightning and may be carried down into the weathering zone where they will supply hydrogen ions.

Plants are important sources of hydrogen ions in the weathering zone, both during life and as a result of decay. Live plants obtain their supply of required cations from the soil by a hydrogen-exchange process. The surface of living rootlets carries a high negative electric charge which attracts a cloud of hydrogen ions around it (Figs. 3-14 and 3-15). Exchange of these hydrogen ions and the cations absorbed or in exchangeable positions on adjacent clay grains, or in some cases, exchange directly with unweathered minerals, liberates the cations to the groundwater from which they can be

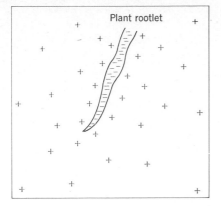

Fig. 3-14. Diagram of a plant rootlet which carries negative charges and is surrounded by an "atmosphere" of cations, most of which are H ions. (*From W. D. Keller, "The Principles of Chemical Weathering," Lucas Brothers Publishers, 1957.*)

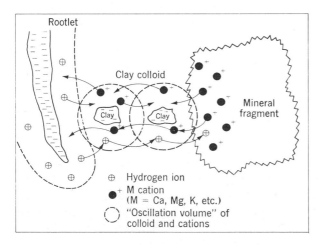

Fig. 3-15. Diagram illustrating the movement of cations by contact exchange between adjacent colloid particles. (*From W. D. Keller, "The Principles of Chemical Weathering," Lucas Brothers Publishers, 1957.*)

taken up by the plant. The decay of plants produces complex organic acids in the humus which are particularly effective in weathering, because their colloidal nature makes it possible for them to come into intimate contact with the unweathered materials. Such acids, if periodically renewed, can completely decompose feldspars in a matter of a very few years.

The silicate-rock-forming minerals are relatively insoluble in water so that direct simple solution is a minor process of weathering. The cations liberated by hydrolysis and other chemical weathering are, of course, removed in solution. Most commonly these cations are stabilized in solution by the presence of carbon dioxide. The chemically precipitated sediments are more susceptible to solution than any other rock types. Limestone, dolomite, and rock gypsum are comparatively soluble, particularly in the presence of acids and carbon dioxide. As a result the weathering behavior of these materials depends on the availability of water. In humid regions these rocks are extensively removed by solution, yielding subdued rounded outcrops largely buried by the insoluble residues from the rocks and by transported detritus. In contrast, limestone in arid regions frequently stands up as resistant beds. The shattering effects of mechanical weathering are healed by the minor solution and redeposition as a result of evaporation of the scarce water.

Hydration of minerals is a third process involving water in the weathering environment. The hydration of materials such as iron oxides to form limonite and anhydrite to form gypsum involves marked volume increases and therefore has a considerable mechanical effect as a result of chemical processes. Similarly, the clay minerals can have various states of hydration and can undergo volume changes associated with a change of the availability of water.

Oxidation-Reduction Processes in Weathering

A major proportion of the iron in primary igneous and metamorphic silicates is in the reduced ferrous state. The availability of oxygen in the weathering zone results in the early oxidation of the iron to the ferric state. This is immediately visible in the iron staining of rocks rich in ferromagnesian minerals. Oxidation and removal of iron from the silicates result in the partial destruction of the silicate structure, making the mineral more susceptible to other weathering agents.

Oxidation or reduction of the organic material in the soil zone has important effects on chemical weathering and the mobility of materials in the subsoil zone. Decomposition of organic debris, which is normally an oxidation process, liberates large quantities of carbon dioxide which aids hydrolysis and renders the cations soluble. In this zone of decay a reducing environment may be developed as a result of the availability of more carbon than the soil oxygen can accommodate. Therefore iron will be reduced to the relatively soluble ferrous bicarbonate or hydroxide and will be transported downward to be precipitated as the ferric oxide or hydroxide in a lower zone. Comparison of the solubility curves for ferrous and ferric hydroxides in Fig. 3-9 shows this relationship clearly. In some strongly reducing environments, such as areas of accumulation of organic-rich sediments, the iron may be removed entirely in solution or may be reprecipitated as the insoluble iron sulfides pyrite, marcasite, and melnikovite.

An economically important effect of oxidation in weathering is the supergene enrichment of sulfides. Most sulfides are oxidized to sulfates, which are more or less soluble. Particularly important is the oxidation of the iron sulfides, because they hydrolyze to form the various acids of sulfur and insoluble ferric hydroxide. The sulfur acids so formed will, in turn, attack other minerals, rendering them mobile. The same processes are important in the weathering of pyrite- or marcasite-bearing sediments. It is in this environment that aluminum is made soluble as one of the alums. This is the only way in which aluminum can be removed from the weathering products in important amounts.

CONTROL OF THE BALANCE BETWEEN CHEMICAL AND MECHANICAL WEATHERING

General Statement

The character of the sediment available in a region for transportation and subsequent deposition is controlled primarily by the character of the source rock and the balance between chemical and mechanical processes of weathering. The influence of the source rock is readily apparent. It is, for example, impossible to get feldspar grains from a rock containing no feldspar, regardless of the character of the weathering. Similarly, clay flakes cannot be produced from a pure-quartz sandstone. However, if the source rock contains a wide variety of minerals, some of which are stable and others unstable, the balance between chemical and mechanical processes will have an important effect on the sedimentary product. The dominance of chemical weathering will produce a sediment composed of the resistant minerals of the source rock plus the decomposition products of the unstable minerals of the source rock. However, a dominance of mechanical weathering will produce a sediment rich in the unstable minerals of the source rock.

The primary control of weathering character is climate. Total annual rainfall and mean annual temperature are the most important features of climate here, because they control the character of vegetation. These relationships and the resultant vegetation types are shown in Fig. 3-16.

The seasonal distribution of rainfall is important because of its influence on moisture loss by evaporation. If rain falls dominantly in the summer, the lines dividing desert, steppe, prairie, and humid forests will shift to the right because of greater evaporation loss. In contrast, if rain falls dominantly in the winter, these dividing lines will shift to the left, and steppe vegetation will persist with a lower total rainfall. These shifts may be as great as 3 in. of total rainfall per year. The steppe environment is the transition zone between arid and humid types of weathering and erosion. The grasses are typically bunch grasses, which do not develop an interlocking sod. In the

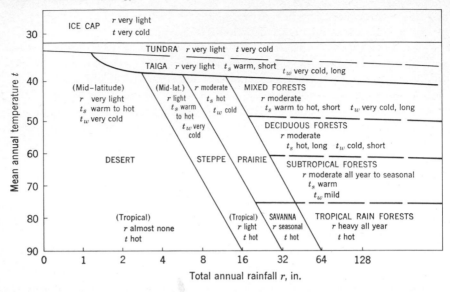

Fig. 3-16. The relationship between climate and vegetation. Symbols: r–character of rainfall; t–temperature; t_s–summer temperature; t_w–winter temperature.

dryer parts of the steppe, bunches are separated, leaving much of the soil bare. Additional effects of temperature can be identified, in that the rate of chemical reaction is temperature-dependent, and the effectiveness of frost action in mechanical weathering depends on the frequency of freezing temperatures.

Rainfall

Rain which falls on the earth can run off, infiltrate, evaporate, or be transpired by the plant cover. Over a broad area the only portion of the rain which will accomplish chemical weathering is that portion which infiltrates. The mineral character and porosity of the soil and the plant cover are the main controls of infiltration. Quantitative studies show that during a storm the rate of infiltration is quite rapid at first and then slows down markedly (Sherman and Musgrove, 1942).* This behavior is the result of changes in the character of soil materials as they are wetted. In the dry state the soil has numerous capillary and small supercapillary openings due to the activity of animals, root systems, and shrinkage of mineral-soil material in the dry state. During the early stages of rainfall these openings fill rapidly with water, but as the rain continues, they become plugged by solid particles washed into them or by swelling of organic and clay colloidal materials as they adsorb water. As a result the rate of infiltration decreases markedly (Fig. 3-17). Plant cover has

*L. K. Sherman and G. W. Musgrove, Infiltration, in: O. E. Meinzer (ed.), "Hydrology," chap. VII, pp. 244–258, McGraw-Hill, 1942.

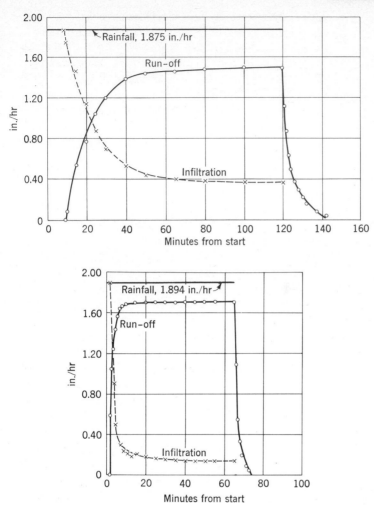

Fig. 3-17. Rainfall, surface runoff, and infiltration on forested land (55-year-old pine) and on bare and abandoned land in the Talla-hatchie River Basin, Mississippi. [*After L. K. Sherman and G. W. Musgrave, in O. E. Meinzer (ed.), "Hydrology," McGraw-Hill, 1942.*]

a very marked effect on this process. In a region of sparse or no plant cover, as soon as the surface layer of soil is wetted, sealing the pores, the remaining rainfall largely runs off down slope and is lost to all the slope area. There-fore only that very limited portion of the rain which accomplishes the initial surface wetting will be capable of further infiltration and consequent chemi-cal weathering. In contrast, in well-vegetated regions the runoff of rain is markedly impeded by the presence of a ground litter composed of matted root material, dead grass, leaves, or other debris. This material forms a mat, sometimes several inches thick, and acts as a dam to break up the flow of surface runoff, allowing much more time for infiltration. As a result a much greater percentage of the water is available for chemical weathering. The character of the vegetation makes little difference in this behavior. Forest, grassland, or tropical jungle all result in a thick ground-litter mat of some type. This behavior is familiar to anyone who has compared the time of runoff from a bare backyard with that of a nicely grassed front yard after a heavy rain.

As shown in Fig. 3-16, only the dry portion of the steppe and the desert regions do not support an effective plant cover of some type on most soils. Of course, if the soil is completely sterile, containing no nutrient materials, then no plants can grow, regardless of the amount of rain. Considering the long-term processes of geology, however, such a situation would be ephemeral, as will be pointed out later. Total annual rainfall and mean temperature both control the type of plant in the humid regions, but any type of cover will be effective in increasing infiltration and decreasing runoff. As rainfall increases, loss due to evaporation and transpiration usually increases, and the amount of infiltration probably does not increase greatly. Thus all humid regions will behave similarly as far as weathering is concerned, and the weathering will be dominantly chemical as a result of the availability of water in the weathering zone.

In contrast, the dry steppe and desert can support only a scanty plant cover, and much of the surface may be bare ground. Rainfall in such a region occurs typically as cloud bursts, a large amount of rain falling over a limited area in a short period of time. Such a situation is not conducive to any appreciable amount of infiltration. As a result the main weathering effects in such a region will be mechanical rather than chemical. The small amount of water and the resultant limited chemical weathering have their greatest effect in the mechanical breaking of the rock through the growth of crystals of new clay minerals or as the result of soluble salts being deposited by later evaporation of the water. Swelling and shrinking of colloids and clay minerals may also have considerable mechanical effect.

Temperature Controls

Temperature has several important influences on the balance between chemical and mechanical weathering. First, it influences the rate of chemical reactions. The warmer the mean temperature, the more rapidly chemical reactions proceed *if* the reactants are present. This "if" is very important, because many arid regions have high mean temperatures, but because of lack of moisture, there is little chemical action apparent. However, given the same bedrock character and the same amount of available moisture, there will be more chemical weathering accomplished per unit time in a warm climate than in a cool one.

A second, and frequently a more direct, influence of temperature is its control of the availability of water. Ice and snow cannot infiltrate into the ground and cannot produce chemical effects. In a region of prolonged annual subfreezing temperatures, that portion of the precipitation which falls as snow will be effectively lost to weathering processes. Generally, in such regions the ground freezes before the major portion of the snow falls and then thaws after most of the spring meltwater and slush have run off down hill. As long as the ground is frozen and all pore spaces are plugged with ice, no water can infiltrate downward. In contrast, in regions of alternate freezes and thaws the situation is ideal for the maximum amount of mechanical action to break rock as a result of ice-crystal growth. Ice crystals and masses can grow in fractures in the rock during freezes, exhausting the available moisture in the capillary openings. In a subsequent short period of thaw, the capillary openings will be recharged with water which will then be added to ice masses during the next freeze. In this manner large masses of ice may be built up in the soil during the winter and can reduce a fractured rock mass to a pile of rubble during one winter. This process can be observed in the rapid destruction of freshly blasted rock cuts along roads in regions of alternate freeze and thaw during winter. Blasting of the rock cut will develop the openings necessary for migration of water into the rock.

Topographic and Relief Controls

Over most of the continental area topography has a relatively small influence on the character of weathering. This is true because most of the land area has sufficient slope so that surface runoff is a possibility, and only a small portion of the land has slopes too steep for a weathered regolith to accumulate. Furthermore, many topographic features themselves are the direct result of the climate-controlled erosion of tectonically uplifted areas. There is a normal tendency to exaggerate slopes when looking at them and when drawing profiles. The visual exaggeration of slopes is the result of man's having a much greater perception of vertical angles than of horizontal distances. Thus a hill will appear higher than it actually is and much closer than it actually is, and this results in an exaggerated idea of the slope. One of the steepest sustained slopes in the United States is the east face of the Sierra Nevada Mountains, which, when one stands at the base, appears to be almost straight up. Actually, the slope averages a little more than 1,500 ft/mile, or no more than a 17° slope. The north slope of the Andes in Venezuela rises 10,000 ft in 6 miles from the coast. Discounting 1 mile of coastal plane, this gives a slope of 24°. This slope is in a humid region and is covered with tropical rain forest or bamboo and ferns in a dense growth almost to the top (Garner, 1959)* (Fig. 3-18). Thus, below the maximum elevation at which a dense plant cover can grow, the accumulation of a residual soil profile

*H. F. Garner, Stratigraphic-sedimentary significance of contemporary climate and relief in four regions of the Andes Mountains, *Geol. Soc. America Bull.*, pp. 1327–1368 (especially pp. 1334, 1335); 1959.

Fig. 3-18. Mountain valley on the north slope of Cordillera de la Costa, Venezuela. The channel is inclined away from the observer about 15° at this point. Vegetation and plant cover extend down the steep valley sides to the channel margins. The channel floor is bare bedrock with clastic sediments in pockets. (*Courtesy Dr. H. F. Garner.*)

depends primarily on whether or not the climate is such that a plant cover can grow to tie down the weathered material. In humid regions an effective plant cover will develop; rainwater will be slowed up in its runoff and therefore infiltrate. The result is a dominance of chemical weathering. In contrast, in arid regions inadequate plant cover will develop, material will be stripped off the slopes as soon as it is loosened by weathering, and no appreciable amount of water will infiltrate. The result will be a dominance of mechanical weathering.

Bedrock Control

The mineral composition and the permeability of the bedrock influence the character of the weathering quite markedly. Two different rock types in the same area frequently behave quite differently. In igneous rocks, composition and texture are both quite important. The presence of highly unstable primary igneous minerals will increase the susceptibility of a rock to weathering. Minerals such as nepheline or olivine will break down rapidly, causing the rock to become more permeable and therefore more readily weathered. Olivine-bearing zones in mafic igneous rocks frequently weather more rapidly than olivine-free zones in the same rock mass. Texture also plays an important role, in that some weathering environments seem to cause coarse-grained igneous rocks to be more susceptible than fine-grained dense rocks. Frequently, coarse-grained granites will weather more readily than associated fine-grained diabase, in spite of the higher relative stability of the granite minerals in Goldich's series. Apparently, the coarse-grained rocks have more and larger intergranular channelways along which the weathering agents can penetrate than do the fine-grained dense rocks. Joint and fracture systems in the rocks have a similar effect (Fig. 3-19).

The weathering behavior of sedimentary rocks and schistose rocks is influenced by a number of variables. Bedding thickness and character, compositional differences between beds, and proportions of soluble constituents such as carbonates are probably the most important. Thin-bedded sediments are much more susceptible to the mechanical action of crystal growth, due to either desiccation or freezing, than are thick-bedded units. As a result thinly laminated shales are often completely buried in outcrops by a gentle slope composed of shale flakes. In contrast, siliceous shales, which are thicker-bedded, break down into polygonal blocks and do so much less readily under average humid climates. The amount and distribution of soluble materials have a marked control. The intercalation of limestone with insoluble materials such as chert and quartz sandstone results in a much more rapid

Fig. 3-19. Weathering controlled by joints, leaving unweathered residual boulders. The weathering here has been dominantly mechanical, probably due to growth of clay and frost crystals. Road-material quarry in porphyritic syenite dike near Wausau, Wisconsin.

mechanical rupture of the insoluble materials than they would usually undergo themselves. Thick cherty limestone will probably yield more fragmented chert per unit of time than will massive cherts. Similarly, the character and amount of cement in sandstone control to a large degree its weathering characteristics. The amount of cement, as indicated by the porosity, is probably equally as important as the character of the cement. A thoroughly carbonate cemented sandstone would probably be less susceptible to weathering than a sand incompletely cemented by secondary enlargement of quartz grains. Again, climate has a marked influence on any of these features. Limestone in an arid region frequently weathers less rapidly than other less soluble materials because of its ability to heal mechanically developed fractures through the small amount of solution and redeposition which goes on. Conversely, in humid regions limestones frequently are the most susceptible of the common rock types on weathering, because the abundance of available water sweeps material out of the rock as soon as it is dissolved.

Summary

The character and products of weathering are controlled primarily by climate and parent-rock types. A humid temperate or humid tropical climate results in a marked dominance of chemical weathering, and the residual materials so formed are primarily clays, iron oxides and hydroxides, and sand-sized particles of the chemically resistant minerals of the parent rock. The metastable and unstable minerals will be largely destroyed in such an environment. The resulting sediment is termed a *mature sediment*. Arid climates, arctic climates, and high altitudes, above the level at which an effective plant cover can be maintained, result in a dominance of mechanical weathering. The sediment formed in such areas will be composed of silt, sand, and coarser-sized particles, all mechanically reduced from the parent rock. If the source rock contains metastable and unstable materials, these will not be destroyed or will be only partially destroyed, and the resulting sediment is termed an *immature sediment*. However, if the source rocks contain no unstable or metastable material, the product may be indistinguishable from that produced in a humid climate, except for grain size. Thus a mature sediment does not indicate the climatic environment from which it was immediately derived. Clay minerals may indicate either complete chemical weathering of preexisting aluminosilicate minerals, therefore humid climates, or mechanical breakdown of preexisting shales. Conversely, immature sediments must indicate a dominance of mechanical weathering of rock containing unstable and metastable minerals and therefore will usually indicate arid, arctic, or high-elevation weathering environments. Exceptions can, of course, be found, such as beach deposits by wave action on immediately adjacent sea cliffs, but these will generally be of limited distribution and will show clear relationships to the source rock.

INFLUENCE OF TRANSPORTATION ON CLASTIC SEDIMENTS

General Statement

Transportation of sediments is closely tied in with weathering processes, because climate is the prime factor in controlling both the type of material and the way in which the material is put into motion in the source area. Climate also controls the behavior of streams, which are the most important continental transportation systems. Transportation is also related to weathering, in that the removal of weathered materials makes possible the continuation of some weathering processes. Similarly, transportation is closely related to deposition, because deposition is the end point of transportation, and no clastic materials can be deposited which have not first been transported to the depositional site. Therefore much that was discussed earlier is applicable here, and much that is said here will be applicable to the next section. For convenience, transportation will be divided into three areas: (1) processes which put material into transportation in the source-rock area; (2) major processes involved in the movement of materials across the land surface; and (3) agents of transportation in marine and other standing bodies of water.

TRANSPORTATION IN THE SOURCE AREA

Rain Splash and Sheet Wash

The greatest bulk of sediments is transported to the depositional site by running water. The streams and rivers, in turn, derive most of their load from material carried to them through surface runoff during and immediately following rainfall or through springs and seeps of ground water. It is therefore again apparent that rainfall, in the form of its components infiltrated water and surface runoff, is a most important factor.

The first influence of rainfall is through the impact of falling raindrops on the ground. As raindrops strike the ground their kinetic energy is expended on the ground and loosens the smaller particles of the surface. Raindrops with a diameter of 5 mm falling through still air attain a terminal velocity of approximately 30 ft/sec. The kinetic energy of 1 in. of precipitation of such drops is approximately 65 ft-lb/ft². Assuming a soil density of 2, this is sufficient energy to lift a 6-in. layer 1 ft into the air. As drop size decreases, the terminal velocity decreases, and therefore kinetic energy decreases, as shown in Fig. 3-20. In areas of complete plant cover, this energy is expended on the foliage of the trees and grasses, and rain reaches the ground as slow drips and trickles which have little energy to loosen the surface of the weathered material. In contrast, in areas of inadequate or incomplete plant cover, this kinetic energy is expended on the ground and much work can be accomplished. Thus there is a difference in the effects of rainfall in humid and arid climates. In arid regions the rain immediately

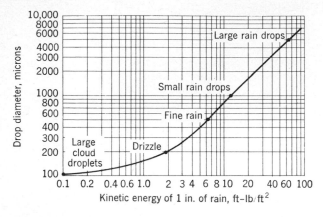

Fig. 3-20. Relation of the kinetic energy of 1 in. of rain to raindrop size, calculated from the terminal velocities of falling drops. (*Terminal velocities from J. A. Day, "The Source of Weather," Addison-Wesley, 1966.*)

starts to pick up a load of clastic materials on impact, whereas in humid regions nothing even approaching this load is picked up.

The sheet runoff of rainfall is equally as important and distinctive in the two fundamental climates. In arid regions the major portion of the rainfall runs off downslope as sheet wash and is relatively unimpeded by plant roots and ground litter. As a result the runoff velocity and volume are high, and large amounts of relatively coarse material may be carried downslope to stream channels or gullies. In humid regions, however, because of the intervention of the plant cover the proportion of runoff is smaller, and the velocity of the sheet wash is much slower. Because of the reduced velocity and because the mantle material is tied down by the plant-root systems, sheet runoff carries relatively little material, and what it does carry is fine-grained silt- and clay-sized mineral particles and organic debris. Extensive experimentation has been carried out on the effect of plant cover on erosion by sheet wash.* Figure 3-21 illustrates the results of such an experiment in a mountainous region of Utah and shows a clear relation between percent of vegetation and eroded soil. Table 3-3 presents the results of another such experiment, carried out in northern Mississippi, and points out the effect of type of plant cover on runoff and erosion.

Combining the character of weathering with the effects of rain splash and sheet runoff, a picture of the character of clastic materials supplied to

*Several papers in: Proceedings of the Federal Inter-Agency Sedimentation Conference, U.S. Dept. Agr., Misc. Pub. 970, 1965, present data on such experiments. See particularly: H. W. Lull and K. G. Reinhart, pp. 43–47; S. J. Ursic and F. E. Dendy, pp. 47–52; R. L. Fredericksen, pp. 56–59; O. L. Copeland, pp. 72–84; and E. L. Noble, pp. 144–153.

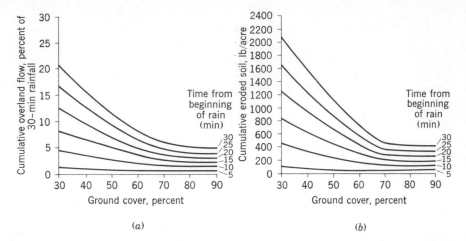

Fig. 3-21. Cumulative runoff (*a*) and cumulative soil eroded (*b*) by a simulated rainfall of 1.87 in. for 30 min by 5-min periods in relation to ground-cover density. This study was conducted on wheatgrass range sites in Utah on soil derived from granite. (*After O. L. Copeland, in Misc. Pub. 970, Agr. Research Service, U. S. Dept. of Agr., 1963, pp. 72–84.*)

Table 3-3. Sediment and Surface-water Yields from 75 Experimental Plots in Northern Mississippi*

Land use or cover type	Average annual rainfall, in.	Average annual runoff, in.	Annual sediment yield, tons per acre	
			Mean	Range
Open land				
Cultivated (corn)	52	16	21.75	3.28–43.06
Pasture	51	15	1.61	1.19–2.03
Forest land				
Abandoned fields	51	7	0.13	0.01–0.54
Depleted hardwoods	51	5	0.10	0.02–0.32
Pine plantations	54	1	0.02	0.00–0.08
Mature pine-hardwoods	51	9	0.02	0.01–0.04
Gullies	53	.....	182	84.3–399.3

*From S. J. Ursic and F. E. Dendy, Sediment yields from small watersheds under various land uses and forest covers, *U.S. Dept. Agr. Misc. Pub. 970*, 1965.

streams can be seen. In humid regions weathering is dominantly chemical, producing clays and residual resistant sands. Because of vegetation, rain splash will not act on the ground to loosen these materials, and sheet runoff will carry relatively minor amounts of the clay- and silt-sized materials to the streams. This behavior will be independent of slope as long as the slope does not exceed the angle of repose of the weathered material. In contrast, in an arid climate, weathering is primarily mechanical, producing sand- and gravel-sized material as well as silt and clay. Rain splash is effective directly on the ground, loosening material and putting all sizes up to coarse sand and fine gravels into motion. Sheet runoff on slopes is capable of moving all the sand and finer-sized materials, and even the pebbles and smaller cobbles may be swept along. Thus climate, through its control of weathering, determines to a large degree the type of material available and, through its influence on plant cover, determines the type of material brought to streams.

Wind Erosion

Wind deflation puts materials which have been reduced in size and loosened by other processes into transportation in the source area. However, for wind to be effective it must be able to operate directly on the ground. Therefore, again, plant cover as a reflection of climate is important, because an effective plant cover reduces the wind velocity at ground level to the point where no material can be picked up. This is independent of the type of plant cover, whether jungle, forest, grasslands, or even moss covers. Wind is therefore most effective in arid regions having sparse or no plant cover. It can be effective in humid regions, however, where plant cover has been destroyed by the development of a sterile soil. There are such local "deserts" in humid regions. The effects of man and his poor agricultural methods must be discounted in dealing with the geologic record, so there are many examples of modern deflation processes which are not applicable to earlier situations.

Wind picks up and transports out of desert regions the silt- and clay-sized materials produced by weathering and made available by sheet wash and streams. It also picks up and moves over the surface the sand-sized particles which are, in turn, concentrated in dunes. The coarser-sized materials are thus left behind by deflation processes as a desert pavement of pebbles and boulders. Once such a pavement has been developed, however, the action of the wind is not stopped but only slowed down. This is because the pebbles themselves will be subject to reduction in size through weathering and can even be removed periodically by heavy rainfall and sheet wash. Thus in arid regions much fine material can be put into transportation, leaving behind the coarser material in the source area.

Landslides

Landsliding and the other agents of mass wasting are important agents for transporting materials to streams for further transportation. In humid regions of high relief, landsliding may be the most important mechanism for bringing materials to the streams (Garner, 1959).* In such regions a thick residual regolith is stabilized on the slopes by the anchoring effects of vegetation. Periodically, however, thorough saturation causes the weight of the regolith plus water to become great enough to overcome the strength of the plant-cover–root system, and masses of regolith and partially decomposed bedrock may slide downslope to the stream floors (Fig. 3-22). Minor events on the slope, such as wind falls or brush fires, trigger such slides. The result is the rapid contribution of a wide range of materials, including weathered, partially weathered, and fresh materials, of a wide range of size to the streams. Such mechanisms can supply the constituents of the graywacke suite of sediments if the area of high relief is situated close enough to the depositional site so that the incompletely weathered materials are not destroyed during their subsequent transportation.

CONTINENTAL TRANSPORTATION

Behavior of Streams in Humid Regions

All streams and river systems in humid regions have one common characteristic; that is, they must all flow through to the sea. Internal drainage basins, such as lakes and swamps, are ephemeral features and eventually fill

*Garner, *op. cit.*, pp. 1336–1339.

Fig. 3-22. Landslide scars of Mount Naguata, Cordillera de la Costa, Venezuela. Steep heavily vegetated slopes which show landslide scars in a deep residual regolith. (*Courtesy Dr. H. F. Garner.*)

with sediment. Subsequently, the sediment fill is removed by further down-cutting. Thus all the sediment introduced into the streams will eventually reach the sea. Inland continental deposits of streams and lakes are preserved only as a result of drastic drainage changes.

Recent studies of the mechanics of stream action show that the volume, velocity, and carrying capacity of streams increase progressively from head to mouth and are independent of slope (Leopold and Maddock, 1953).* Humid-region rivers must increase continually in volume because of the addition of water from tributaries, from springs, and from groundwater seeps along their courses. Therefore all the clastic materials brought into the stream will be transported through to the sea. In humid regions the major fraction of the contributed clastics is fine-grained, and the bulk of this material is carried in suspension. In many areas there is very little of even this material (Lull and Reinhart, 1965)†. The clear forest streams have long been eulogized by the poets and sought by the trout fisherman. Such streams, with their small loads of suspended material, have small bed loads of coarser materials and, therefore, incise downward into their channels, stripping bedrock by slow corrosion and removing channel fills brought in during other climatic regimens.

Behavior of Streams in Arid Regions

The streams of arid regions show a marked contrast in behavior to humid region streams. In arid regions the groundwater table is very low, and there-fore only very large rivers originating in humid regions can cut deeply enough to flow perennially. The streams indigenous to arid regions are inter-mittent, flowing only during and immediately after a heavy rainfall. For the most part these streams have the character of flash floods and drain into internal basins. Therefore unless an arid region is immediately adjacent to a marine depositional site, very little stream-transported material will reach the sea.

The behavior of streams in deserts is quite familiar. Rain comes pri-marily as cloudbursts, dumping several inches of rain over a limited area in a short period of time. One probable reason for this is that unless a squall can produce a large volume of rain, the rain will evaporate in the hot dry desert air before reaching the ground. In the rain area, sheet wash accumu-lates into the shallow troughs, carrying large quantities of sands and gravels with it, and rushes as a highly turbulent muddy torrent downslope. When the stream gets beyond the immediate rain area, it begins to lose water rapidly by evaporation and infiltration into the parched regolith. As water is lost, the transporting power of the stream is lost (Quinn, 1957)‡ so that the

*L. B. Leopold and T. Maddock, Jr., The hydraulic geometry of stream channels and some physiographic implications, *U.S. Geol. Survey Prof. Paper 252*, 1953.
†H. W. Lull and K. G. Reinhart, Logging and erosion on rough terrain in the east, U.S. Dept. Agr. Misc. Pub. 970, pp. 43–47 (especially pp. 43, 44), 1965.
‡J. H. Quinn, Paired river terraces and Pleistocene glaciation, *Jour. Geology*, vol. 65, pp. 149–166 (especially p. 155), 1957.

channel becomes filled with sands and gravels and the stream ceases to exist. Rain may not return to the same area for months or years to further transport the coarse clastics, but in the meantime it will be operating on other discontinuous patches over the desert. Thus in an arid environment of low relief, a blanket of coarse clastics is built up and spread out as a thin layer over the entire region. In mountainous regions, valleys, depressions, and intermontane basins are filled with a deep accumulation of sands and gravels. In neither case does any appreciable volume of sediment reach the sea.

The contrast between the loads of arid- and humid-region streams is shown in Figs. 3-23 and 3-24. The two streams represented drain about the same area. The Pecos River (Fig. 2-23) in New Mexico is flowing through an arid region but is perennial, because it arises in the subhumid southern end of the Sangre de Cristo Mountains. It falls 7,162 ft from its source to the sampling station at Santa Rosa, New Mexico, but 4,800 ft of that is in

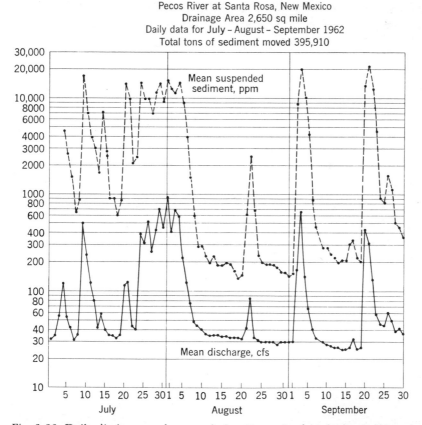

Fig. 3-23. Daily discharge and suspended-sediment load in the Pecos River at Santa Rosa, New Mexico, for the months of July, August, and September of 1962. (*Data from U.S. Geol. Survey Water-Supply Paper 1944, p. 562*)

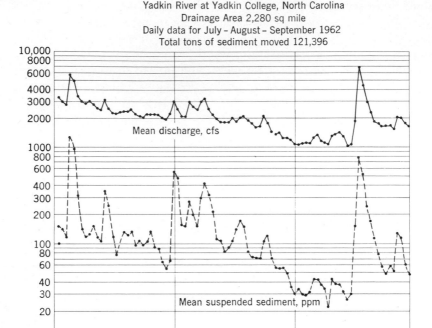

Yadkin River at Yadkin College, North Carolina
Drainage Area 2,280 sq mile
Daily data for July – August – September 1962
Total tons of sediment moved 121,396

Fig. 3-24. Daily discharge and suspended-sediment load in the Yadkin River at Yadkin College, North Carolina, for the months of July, August, and September of 1962. (*Data from U. S. Geol. Survey Water-Supply Paper 1941, p. 298.*)

the first 34 miles; for the 100 miles from Pecos to Santa Rosa, New Mexico, the gradient is only 23 ft/mile. In contrast, the Yadkin River in North Carolina (Fig. 3-24) is in the humid Appalachians, rising in the east slope of the Blue Ridge at 3,500 ft. In the first 3 miles it falls 1,500 ft, and the gradient for the next 110 miles to Yadkin College, North Carolina, averages 12 ft/mile. The charts depict the mean daily discharge and suspended sediment load for the months of July, August, and September of 1962. The humid-region Yadkin River, with a discharge of 1,000 to 3,000 ft³/sec, carries a very light load of sediment and during this interval moved a total of 121,396 tons of sediment. In contrast, the Pecos River has an extremely variable discharge and consistently carries a much heavier load of suspended sediment, moving a total of 359,910 tons during the same interval. The difference in gradient of the two streams through the 100 miles above the gaging station can hardly account for the tremendous difference in sediment transportation.

Eolian Transportation

The fine silts and clays picked up in desert regions by the wind are transported at high altitudes for long distances. If they settle out again in an arid region of poor plant cover, they remain only temporarily, to be picked up again by a later windstorm. Windblown dust accumulates only where it is protected from further wind transportation. Such sites of accumulation either are vegetated regions or the sea, where particles will settle to the bottom. Preliminary investigations (Chepil, 1965)* at Manhattan, Kansas, indicate that from 5 to 10 tons of dust accumulate per acre each year. In vegetated regions the dust that settles on plants is washed down by rainfall and may be added to the residual regolith or to sheet runoff from the regions so that its identity is lost in the larger amount of residual materials. If abnormally large amounts of dust accumulate so that its identity is retained, the product is a loess. Dust that settles in the sea accumulates as bedded deposits composed dominantly of silt-sized angular quartz grains.

Sand-sized materials transported by the wind do not migrate far from the arid source area and reach the sea only if the dune area is immediately to the windward of the sea. Because of the high porosity and permeability of dune sands, they are not subject to any appreciable amount of sheet-wash erosion. Therefore they build up permanent continental deposits until such time as there is drastic change in the geologic environment due to land- and sea-level changes or marked drainage changes.

Sterile-soil situations under humid climates are another area of eolean transportation. Sterile soils develop under prolonged humid climates, where all the nutrient materials of the soil are leached out (Quinn, 1962).† As a result vegetation is destroyed and wind action becomes important. The wind removes the finer materials in the same manner as in arid regions and develops sand dunes of the coarser material, which is dominantly quartz. The dunes migrate downwind until they are blown into permanent streams and rivers of the humid drainage. As the sterile silt and sand are removed, the incompletely weathered B and C horizons are exposed, and new vegetation moves into the area. Sterile situations are thus ephemeral developments in local areas and migrate through regions of prolonged humid climates. In the process, however, they can locally supply large quantities of quartz sands to humid streams for transportation to the sea.

Alternations of Climate — Arid to Humid

One of the common fallacies of geologic thought is the assumption that climate in a region has remained the same over a long period of time. Actually, climates are continuously changing. The Pleistocene epoch gives

---

*W. S. Chepil, Function and significance of wind in sedimentology, U.S. Dept. Agr. Misc. Pub. 970, pp. 89–94, 1965.
†J. H. Quinn, Geological implications of soil mechanics, *Arkansas Acad. Sci. Proc.*, vol. 16, 1962.

the most familiar example of climate change, but it is frequently considered to be a freak, whereas in the geologic column less drastic changes were probably commonplace. Even in historical times, marked climate changes have been seen. In portions of our midcontinent area there has been sufficient climate change since settlement in the nineteenth century to convert prairie grasslands into forest lands and to change the behavior of streams. Changes of climate are very important in the transportation of sediments, because the sedimentary material prepared by weathering and short-distance surface transportation in one climatic environment will be out of equilibrium when climate changes and will thus be subject to large-scale stripping and redeposition. It is during these climate changes that probably the largest amount of clastic material reaches the sea.

The transition from arid to humid climates is most important in supplying large quantities of coarse clastics (sands and gravels) to the sea (Garner, 1959).* In arid regions mechanical weathering produces sand and gravels plus fine materials. The wind removes the silt and clay by deflation, and sheet wash and stream action build up a blanket of the coarse materials on the land. During the transition, when the region again becomes humid, there will be no well-established drainage system, and regions of low relief will become flooded with broad sheetlike rivers which flow with high velocity along many shallow braided channels (Fig. 3-25) (Garner, 1967).† These rivers, because of the amount of rainfall, must flow through to the sea, and

*Garner, *op. cit.*, p. 1365.
†H. F. Garner, Rivers in the making, *Sci. American*, vol. 216, pp. 84–96, 1967.

Fig. 3-25. Air photo of the complex channel system of the Rio Caroni in Venezuela. *(Courtesy Dr. H. F. Garner.)*

in the process they carry vast quantities of the available loose material with them. This material will, of course, be dominantly sand and gravel. As the humid climate continues, incision of the streams establishes well-defined channels, and vegetation will take over the higher land. Humid-type weathering and erosion then take over, and the coarse materials not stripped during the sheetflood stage will be reduced by chemical weathering.

In mountainous regions a particularly large volume of coarse clastics may be produced. During arid cycles, ravines, valleys (Quinn, 1957),* and intermontane basins are filled with a deep layer of the coarse clastic materials (Garner, 1965).† When humidity returns to the region, channels can be established rather rapidly because of the steeper slopes and better-defined drainage lines of the mountain arid region. The streams begin to incise the thick clastic blanket and gradually strip out the valley fills (Fig. 3-26). Because of the great thickness of the valley fill, the humid streams continue to have an abundant supply of coarse clastics for a long time, and tremendous volumes of coarse clastics are moved to the sea before the streams incise to bedrock. This is in contrast to arid-plain regions, where the blanket of sand and gravel is relatively thin and incision to bedrock is accomplished in a relatively short time.

*Quinn, *op. cit.*, p. 158, 1957.
†Garner, *op. cit.*, p. 1350, 1959.

Fig. 3-26. Incision of a river into a deep alluvial-valley fill in a mountainous region. The Urumbamba River, Cordillera Oriental, Peru, near 9,000-ft elevation. *(Courtesy Dr. H. F. Garner.)*

Alternations of Climate–Humid to Arid

During prolonged humid climates, a deep residual weathered regolith accumulates on the land and has the well-defined zones of a typical soil profile. During the change from humidity to aridity, this regolith is stripped as vegetation dies off, but the manner of its stripping depends on the relief of the area. In areas of low relief, stripping may take the form of layer-by-layer removal of the regolith by wind and sheet runoff. The main contribution to the sea is from the B and C horizons of the soil profile. The B horizon in humid regions is the layer of accumulation of iron in the weathered materials, and the C horizon is the layer of decomposing bedrock. Removal of these materials layer by layer gives rise to red beds from the B horizon and arkoses from the C horizon if the bedrock was feldspathic. In areas of more mafic igneous rock the C horizon could give a graywacke-type sediment.

In areas of moderate relief, stripping of the residual regolith is accomplished by gullying rather than by layer-by-layer removal. The resultant sediments are a mixture of materials from all layers of the residual regolith and contain organic-rich A-horizon material, iron-rich B-horizon material, partially decomposed C-horizon material, and probably some fresh material from the bedrock. It is possible that the iron of the B horizon would be reduced by the organic material from the A horizon so that the resultant sediment would have the composition and character of a graywacke. In either case no large amount of the coarser clastic material reaches the sea, unless the sea is immediately adjacent or is encroaching upon the land mass, because of the development of intermittent and internal drainage. Illustrations of stripping of a humid residual regolith owing to destruction of plant cover can be seen in areas such as Sudbury, Ontario (Fig. 3-27), and Ducktown, Tennessee, where smelter fumes have killed all plant life for miles around. The regolith has been stripped off all hillsides by sheetwash and piled indiscriminately into the valleys, leaving the hillsides bare.

A summary of these concepts of the correlation of climate with continental and marine deposits of clastic materials is given in Table 3-4.

Glacial Transportation

Continental glacial transportation is completely unselective with regard to the size or character of materials moved. Materials moved in this manner may be either sediments from the residual regoliths developed in preexisting climatic regimens or sediments from bedrock. There should be little difference in the resistance of these two basic types of materials to abrasion or plucking, because they all would act as thoroughly consolidated materials. Residual regoliths from earlier climates would be thoroughly cemented by ice in the areas ahead of the advancing glacial movement and in many instances might be more resistant to erosion than is lithified rock. For instance,

Fig. 3-27. Stripping of the regolith due to destruction of plant cover by smelter fumes, Coniston, Ontario. The foreground shows the bare bedrock and residual very coarse clastics after all finer material has been washed away. The slope at the extreme left is the edge of the slag pile from the smelter. The black line through the alluvial flat is the remnant of a road. The Grenville metamorphic front passes through the foot of the ridge across the valley so that the low ridge in front is composed of recognizable sediments, whereas the higher ridge behind is composed of amphibolite-facies gneiss.

Table 3-4.   Correlation of Continental Climate with Continental and Marine Clastic Sediments

Climate	Character of regolith and plant cover	Material removed by running water	Effect of wind	Material contributed to sea
Humid	Heavy plant cover; deep residual chemically weathered regolith	Clay and fine silt; material in solution; organic materials	Accumulation of wind-blown dust as loess or masked in soil	Solutions; organic-rich clays and silts
		Steep slopes resulting in landsliding		Graywacke constituents
	Prolonged weathering developing sterile soil; no plant cover	Quartz sand	Deflation of dust and sand-dune development	Pure-quartz sands
Arid	Poor plant cover; dominant mechanical weathering	Sand and gravel developing: on low slopes — gravel blankets; in mountain-ous — valley-fill gravels	Deflation of silt and dust; sand	Organic-poor silts; minor sands and gravels along arid shorelines
Arid to humid	Residual blanket of arid sands and gravels; Low relief — removed by sheet floods	Sand and gravel	None	Sand and gravel
	High relief — removed by valley entrenchment	Large volumes of sand and gravel		Large volumes of sand and gravel
Humid to arid	Deep chemically weathered residual soil: Low relief — zone-by-zone stripping	Weathered and incompletely weathered B and C soil horizons	Deflation of silts and clays	Some sand, silt, and gravel, giving red beds and arkoses
	Moderate relief — gullying through all zones of soil simultaneously	Weathered, imcompletely weathered, and unweathered B and C soil horizons and bedrock	Deflation of silts and clays	Sand, silt, clay, and some gravel, giving graywacke constituents

sand cemented by ice could possibly be mechanically stronger than an identical sand cemented by calcite because of the superior crushing strength of ice over calcite. Evidence of this can be seen in some glacial tills, in that boulder-sized and -shaped masses of unconsolidated sand are present which must have been cemented by ice when deposited.

Most of the coarser-sized glacially transported materials are derived by plucking of the glacial bed. Therefore their transportation is at least in part along the sole of the glacier. They may be lifted into the glacial ice mass by thrusting of one mass of ice over another, particularly near the margins of the glacier where active ice advances by overriding stagnant ice. In the process the stagnant ice is sufficiently thickened to become active. Thus rock materials carried by the ice are subjected to extreme mechanical forces, both by grinding along the bottom and by shearing within the ice mass during overriding. Consequently, mechanically durable rock fragments are transported greater distances than are the mechanically incompetent rocks, and they are still recognizable. Shale, weakly consolidated sandstone, and schist appear to be particularly susceptible to mechanical disruption in this process. In contrast, massive igneous and metamorphic rocks and limestone are less susceptible to this type of disintegration. Pebble counts in a number of areas in the Pleistocene tills indicate that the bulk of the coarser sizes is of relatively local origin, and boulders from relatively distant recognizable source areas are less abundant and are durable rock types (Flint, 1957).* This suggests that either the bulk of the coarse clastic material transported is of local origin (a few tens to a very few hundreds of miles) or much of the coarser fragments are reduced mechanically to such dimensions by longer-distance transport that they are unrecognizable.

Melting of glaciers produces a local situation that is in many respects similar to arid regions as far as sediment transport is concerned. Vast quantities of sands and gravels are available and, along many glacial marginal regions, are probably not tied down by effective vegetative cover. Internal drainage into lakes and swamps is also common, allowing for trapping of the bulk of the coarse clastics instead of their being carried through to the sea. Thus a coarse clastic blanket is spread out beyond a glacial front, and much of the finer clastics are trapped in lakes. Except where the glacier comes to the sea, little but fine clastic material is contributed to the sea by running water.

Ice rafting of clastic materials beyond the limit of glaciation where the glacier terminates in standing water bodies is a minor contributing factor in transportation. Blocks of calved ice from the glacier margin float away, carrying with them any rock material frozen into the ice. As the ice melts, this material is dropped and added to the sediment accumulating in that area. This is important only where the dropped material is markedly different

*R. F. Flint, "Glacial and Pleistocene Geology," pp. 127–129, Wiley, 1957.

from that accumulating. Then exotic blocks are very noticeable, particularly where very coarse-sized fragments are dropped into accumulating fine-grained material. Such pebbles and boulders do not necessarily indicate glaciation, however, because ice rafting of large fragments can result simply from freezing of lakes and rivers. Freezing takes place from the surface downward, and pebbles and boulders in a shallow bed may be incorporated in the ice. In the spring melting, the boulder may be lifted off the bottom by either rise of water level or melting of ice from the top down. The incorporated rock fragment may then be carried away in the floating ice. Such exotic blocks should be fairly common in lacustrine deposits and near the mouths of rivers in the marine environment. Identical blocks could be transported by floating trees, with the rock fragments trapped in the root system.

TRANSPORTATION IN MARINE DEPOSITIONAL ENVIRONMENTS

General Statement

Transportation in the marine depositional areas is of four basic types. First, and most apparent, is the transportation along beaches and in adjacent waters by wave action. Second, is transportation below the normal wave base by bottom currents. Third, is the localized transportation by high-density turbidity currents flowing along the bottom on slopes. Fourth, is the transportation of fine materials by surface currents. Each of these has a distinct effect on the size of the materials moved and the character of the deposits formed.

Beach Transportation

Wave transportation along and adjacent to beaches is a very obvious and sometimes very violent process. The breaking of waves in shallow water converts the circular movement of water in a vertical plane to a horizontal translation of the mass of water. Erosion of bottom sediment takes place throughout the zone, from the high-water mark out to the maximum depth at which the wave motion strikes the bottom. Deposition of sediment will also take place through this same zone. The balance between erosion and deposition varies from time to time and place to place along a single beach as a result of the variations in direction of wave approach, the wavelength, and the wave height. During storm seasons, erosion is most active along the back-beach area, and deposition takes place as bar development in deeper water parallel to the shoreline. In contrast, during seasons of no storms the offshore bars are eroded and the sediments are deposited on the beach, building up a berm (Bascom, 1960).* Longshore currents move large quantities of sediment parallel to the shoreline. This is the result of waves striking the beach at an angle and moving materials up and down the beach but at the same time transporting the material along the beach.

*W. Bascom, Beaches, *Sci. American*, vol. 203, pp. 80–97, 1960.

The ultimate distribution of sediment in the beach area is probably the result of single high-energy storms, as pointed out by Gretener (1967).* Statistically, such storms are frequent enough to cause the reworking and redeposition of all sediment in the area, and undoubtedly, they account for the removal of vast quantities of sediment to deep-water environments. However, the sediment which is redistributed by such storms is largely brought to the area by other processes.

Beach erosion and transportation are very selective processes. Because of continual movement of the material and agitation of the water, the sediments are thoroughly sorted (Fig. 3-28). All fine-grained material is progressively removed in suspension in the agitated water and is carried out to sea into quieter-water zones. Rip currents are particularly effective in removing the finer beach materials. Rip currents build up when waves are abnormally high, throwing more water over the offshore bars than usual. The

*P. E. Gretener, Significance of the rare event in geology, *Am. Assoc. Petroleum Geologists Bull.,* vol. 51, pp. 2197–2206, 1967.

Fig. 3-28. Well-sorted beach sediment. The only sand-sized material present in the view is in the pockets at the water's edge; 6 ft out the sediment is all sand. Lake Superior at Great Sand Bay, Keweenaw County, Michigan.

water then moves along the beach to low points in the bars and flows out, scouring channels in the process. The outflow currents may reach velocities as high as 4 knots.

The materials transported and deposited along the beach zones originate from two sources. First, there are materials brought to the shoreline by continental transportation agents, such as river-transported sediment or windblown material. Here the character of the beach itself and the way in which the sediment is handled will depend on the balance between supply of material and transporting power of the waves along the beach. The character of the sediment under these circumstances is dependent on the agent which supplied them to the sea. The second source of sediments is from direct wave erosion of the headlands. The mechanical forces of wave impact on bedrock cause rock disintegration and the production of much coarse clastic material. Chemical weathering is minimized in such areas, because as chemical weathering begins, it loosens materials which are immediately removed by wave action. The sediment produced is dominantly by mechanical action, and its maturity or immaturity depends on the nature of the bedrock being eroded (Fig. 3-29).

Fig. 3-29. Igneous-rock pebbles derived from the Lake Shore conglomerate, which contains abundant igneous-rock pebbles, on a Lake Superior beach. The pebbles are dominantly rhyolite porphyry, and there are no outcrops of this rock type in the area, except as pebbles in the older conglomerate. Eagle River, Michigan.

Bottom-current Transportation

Sediment transportation along the bottom is carried out by currents of various origins. Bottom currents over any large area develop mainly as a response to density difference in the masses of water. Cold arctic and antarctic waters are heavier than the normal surface water of the ocean and thus sink to the bottom, moving gradually toward the equator. Thus in the South Atlantic Ocean there is a current of cold water moving northward west of the Mid-Atlantic Ridge which spills over low points along the ridge to flood the eastern basin of the Atlantic. In the Southern Hemisphere this current shows measurable velocities of over 3 cm/sec at the foot of the South American continental slope. At average velocity, water masses of the bottom require 5.5 years to flow from 48° S. Lat. to the equator (Defant, 1961).* This antarctic bottom water can be recognized as far north as 40° N. Lat. in the western Atlantic by its abnormal low temperature. In contrast, the Arctic Ocean basin is cut off from the Atlantic by a relatively shallow sill extending from North American to Greenland, Iceland, Faeroes, and Scotland. Relatively little cold bottom water flows over this sill, but what does flows down the slope to the Atlantic basins at velocities as high as 35 cm/sec (Defant, 1961).† Such a current would easily transport fine sediment settling out of the higher water to the oceanic basin floor.

Similar density currents are set up by local evaporation, thus causing increased salinity. The western mouth of the Mediterranean at Gibraltar is an example. Evaporation in the Mediterranean exceeds the influx of river water and rain. As a result a current of normal marine water of about 36 percent salinity flows into the sea at the surface, and the heavier water of more than 38.5 percent salinity sinks to the bottom. This heavy water pours back out of the sea as a bottom current through the Straits of Gibraltar and out over the floor of the eastern Atlantic to a depth of 1,000 to 1,200 m. There it spreads out as a more or less horizontal sheet of highly saline water on top of colder denser water (Defant, 1961).‡ The current through the Straits of Gibraltar can exceed 100 cm/sec (Defant, 1961).§

Strong currents develop in straits and between large islands and mainlands because of tides. An extreme example is the Strait of Messina between Sicily and Italy. The tides in the Tyrrhenian and Ionian Seas are nearly 180° out of phase. That is, high tide in one coincides in time with low tide in the other, and these phase differences extend to within 3 km of each other through the Strait. The resulting current attains velocities of 200 cm/sec and reverses twice a day (Defant, 1961).¶ Such currents primarily affect the surface water and so would be important in shoreline- or suspended-

---

*A. Defant, "Physical Oceanography," vol. I, p. 683, Macmillan, 1961.
†*Ibid.*, p. 680.
‡*Ibid.*, p. 182.
§*Ibid.*, p. 532.
¶*Ibid.*, vol. II, pp. 395–399. Macmillan, 1961.

sediment transport. They would affect only bottom sediments in very shallow areas or in river-mouth estuaries.

Recent study of the physiography of the Atlantic continental shelf of the United States (Uchupi, 1968)* shows extensive sediment transportation by currents acting on the bottom. Elongated sand waves are common in the areas of Georges Bank and Nantucket Shoals. These sand waves are up to 20 m high and 7 km long. Some in shallower water are migrating at a rate of 12 m/year, and when the tidal current flows at right angles to the sand waves, the entire crest of each wave is in violent sheet flow at least 1 m thick. In these areas the sediments undergoing transportation are being derived from underlying sediment. Further south on the coast, from New York to Cape Kennedy, there are much larger scattered sand ridges 4 km wide and some tens of kilometers long. These are thought to be the result of storm-wave action.

The effectiveness of such currents in transporting sediments probably varies very widely. Evidence from deep-sea-bottom cores indicates that fine-grained sediments are swept off the flanks and tops of suboceanic highs and deposited in the lower areas. Samples of water taken at various depths in the ocean show that the water in the first few dozen meters above the bottom contains suspended solid particles which are maintained in suspension because of eddy diffusion of bottom currents over rough bottoms (Pettersson, 1954).† Deep-sea transportation of sediments by bottom currents seems to be an established fact.

Bottom-current transportation near the shoreline and across the continental shelf does not seem to have been studied in detail. However, we may expect transportation to be much more effective in this environment than in the deep-sea environment, because currents reaching the bottom could be much stronger. The net effect of such currents should be the continual transportation of the finer-grained materials seaward. This effect can be seen by considering tide effects in a bay into which rivers flow. The inflow of the tide will have to operate against the outflow of water owing to river influx. In contrast, the outflowing tide will have the added volume of river water influx into the bay. Thus more sediment will be carried out by the outflowing tide than is brought in by the rising tide. In many areas of active wave action, fine-grained materials are being introduced or produced along the shoreline but are not accumulating there. It is therefore clear that they are being transported out of the area, probably much of them out to sea.

Turbidity-current Transportation

Submarine slumping of masses of water-saturated unconsolidated sediments down slopes of low inclination is thought to account for transportation of

*E. Uchupi, Atlantic continental shelf and slope of the United States–Physiography, *U.S. Geol. Survey Prof. Paper* 529–C, 1968.
†H. Pettersson, "The Ocean Floor," Yale University Press, 1954.

large volumes of sediments. A sedimentary pile building up by gravitative settling on a gentle slope below wave base can readily accumulate to an unstable state, particularly if the source of sediment is at the top of the slope. If such an accumulation is then disturbed by a submarine earthquake or by abnormally long-wavelength storm waves, it begins to move as a mass down the slope. As downslope motion develops, strong turbulent currents develop in the bottom water, churning up the sediment into the water and eventually developing a heavy suspension of sand and mud in the water. Such a slurry has a density greater than that of the surrounding sea water and therefore moves downslope as a current of mud called a *turbidity current*. These currents have several effects on the bottom sediments. First, the mass which initially breaks loose moves downslope. Second, the turbulence and velocity of the current produced moves far beyond the area of its initiation and put in motion the unconsolidated sediment over which it passes. And third, because of the suspended material in the water, the density of the mass is increased and may effectively buoy up larger particles, allowing them to be carried great distances. The current and its suspended load of sediments move downslope and spread out over flat bottom areas to deposit a layer of material in deep quiet-water areas. Sands and remnants of shallow-water organisms have been recovered from ocean-bottom cores at great distances from land, where no other known type of current could have carried them.

The systematic breaking of submarine cables in certain areas gives the best direct evidence of turbidity currents on the sea floor. The most familiar example is a slide triggered by an earthquake on November 15, 1929, in the Grand Banks south of Newfoundland. Cables lying within 60 miles of the epicenter broke instantaneously, but downslope to the south in the Atlantic, cables broke in sequence for a distance of 300 miles (Heezen and Ewing, 1952).* A recent restudy of the area using seismic techniques and sediment-core samples indicates that the sequence of events is complicated (Heezen and Drake, 1964).† A tremendous slump block broke loose and moved downslope as a unit or as a mass of large units, causing the initial cable breaks. Turbidity currents then developed, probably as a result of adjustment at the head of the major slump or between large units, and moved downslope through channels to coalesce as a broad front into the North American basin. Maximum velocities calculated for the turbidity currents range from 15 to 55 knots, depending on the interpretation of the data.

Surface-current Transportation

Fine-grained sediments, introduced into the ocean by rivers or by wind-blown dust, settle very slowly toward the bottom and may be carried considerable distances before the particles fall below the level of moving surface

*B. C. Heezen and M. Ewing, Turbidity currents and submarine slumps, and the 1929 Grand Banks earthquake, *Am. Jour. Sci.*, vol. 250, pp. 849-873, 1952.
†B. C. Heezen and C. L. Drake, Grand Banks slump, *Am. Assoc. Petroleum Geologists Bull.*, vol. 48, pp. 221–233, 1964.

water. Probably even more importantly, however, surface currents carry the dissolved materials of rivers to all parts of the ocean so that material dissolved during weathering in one area may be deposited by chemical or biochemical activity anywhere in the ocean basin.

When muddy river water enters the ocean, it does not immediately drop its load of sediment. Because river water is fresh, it is usually less dense than marine water, in spite of its load of sediment. Therefore it floats on the sea surface, spreading out as a thin layer. The sediments carried by the fresh water may thus be transported out from land and picked up by the marine surface currents. The rate of settling of clastic particles from water varies according to the size, shape, and density of the particle and the density and turbulence of the water. Stokes' law gives the settling velocities of spherical bodies as

$$W = \frac{2}{9} g \frac{\rho_1 - \rho_2}{u} r^2$$

where W is the settling velocity in cm/sec, g is the acceleration of gravity, ρ_1 and ρ_2 are the densities of the sphere and the liquid, respectively, u is the viscosity of the liquid, and r is the radius of the sphere. Settling velocity thus increases rapidly with increased size. Coarse particles of sand or larger size settle out of river water rapidly and are deposited near the mouth of the river. They are, therefore, not transported by surface currents. Silt- and clay-sized particles settle much more slowly, their velocities ranging from 301 m/day for coarse silt ($\frac{1}{16}$-mm diameter) to 0.001 m/day for the finest clay ($\frac{1}{8,192}$-mm diameter). The settling characteristics of clay-sized particles may be complicated by the flocculation of the particles by sea water; this causes a number of fragments to adhere to each other, producing a larger mass whose settling characteristics will be different from that of the individual fragments. Such flocculant masses settle more rapidly than the individuals would but not as rapidly as a single homogeneous mass of the same size. The extent to which clays are flocculated and the effect of flocculation on transportation and deposition are difficult to evaluate in sediments or sedimentary rocks. Studies of the sediments of the deep-sea basins show that a variety of clay-sized particles are present, some of which are clearly volcanic ash but much of which have been developed by subareal weathering and have been distributed over the sea floor by surface currents (King, 1963).*

The distribution of dissolved salts throughout the sea by surface currents is of great importance, because it divorces the character of chemical and biochemical deposits from the nature of dissolved salts derived from the adjacent land masses. It is not necessary, for instance, to call on excessive

*C. A. M. King, "An Introduction to Oceanography," p. 240, McGraw-Hill, 1963.

leaching of calcium from a land mass adjacent to actively growing calcareous reefs. In fact, the deposition and accumulation of silica seem to be at a maximum in areas of cold water (King, 1963),* whereas maximum leaching of silica on the land takes place in warm humid to tropical regions. Current distribution of dissolved salts combined with biological activity and accumulation of precipitated material operate to create uniform concentration gradients throughout the open seas where circulation is possible.

EFFECTS OF TRANSPORTATION

The effects of transportation on clastic particles are threefold: (1) the sediments are sorted according to size, (2) the particles are modified in shape and surface characteristics, and (3) the particles are modified in size. Size and shape limitations as inherited from the parent rock have been discussed previously, but these are modified by distance and the agent of transportation.

The sorting of sedimentary materials begins with the initiation of transportation so that only those sizes are moved which are available in the source area and which are capable of being moved by the first transporting agent. Thus a river capable of transporting pebbles but having only silt available will transport only silt. However, in flowing to the sea the river may pass through several source-area environments or may flow over channel fills deposited under a different climatic environment and pick up a wide size variety of materials. These will then be sorted by size according to the method of transportation, that is, suspended materials, materials in saltation, and bed load moved by traction. Clastic particles are similarly sorted by size in each transportation system which they encounter on the way to their ultimate deposition site. Wind, wave action, and marine currents each sort out those sizes they are capable of transporting and leave behind all larger sizes. Each is dependent on what is available to them. The size range in an accumulated sediment is limited by the largest size brought to the depositional site and the smallest size carried out of the depositional area by the transporting agent operating there.

The size and shape of clastic particles are modified during transportation by abrasion, impact, solution, and chemical weathering. Shape change is usually discussed in terms of two factors, roundness and sphericity. *Roundness* may be defined as a measure of the extent to which sharp corners and edges have been reduced to curves by the modification of the grain. Qualitative or descriptive terms which may be applied to this property are angular, subangular, subrounded, and rounded. *Sphericity* is the measure of the approach of a particle to the shape of a sphere. Descriptive terms which may be used to describe this property are spherical (or equidimensional), tabular, prismatic (or prolate), and bladed.

Ibid., p. 237.

Theoretically, sphericity, roundness, and size are geometrically independent functions, but practically, they are closely interrelated. This is because the effectiveness of the agents which modify the size and shape of particles depends on the transporting agent and on the size. Most rock and mineral particles are wetted by water. This means that when water is available, each particle will be surrounded by an attached film of water molecules. In large particles this film is of negligible thickness so that impact or grinding of one fragment on another causes abrasion or chipping. However, in small particles the thickness of the attached water film becomes large relative to the size of the particle so that the grains are held apart by the cushioning effect of the film. The lower limit of abrasion rounding seems to be a size of 0.1 mm; fragments smaller than this limiting size are practically unaffected by transportation but retain their inherent or inherited shape.

Sand-sized particles in the size range 0.1 to 2.0 mm abrade differently, depending on the transporting agent. Transportation by running water yields very little abrasion rounding, even with extreme distances of transport. Kuenen (1959)* investigated several methods of abrasion by running water through the use of quartz, feldspar, and limestone as crushed fragments and as sawed cubes of various sizes, and he demonstrated this lack of appreciable abrasion rounding. Thus well-rounded sand must be the result of processes other than stream abrasion. In a subsequent study, Kuenen (1960)† investigated abrasion by eolian action and found that dry quartz-sand grains are abraded at a rate a few hundred to 1,000 times that of subaqueous abrasion. Limestone cubes are abraded less rapidly than quartz cubes of the same size. It appears that the brittleness of the quartz in comparison with limestone causes the difference. A further experiment under the same conditions showed that polished well-rounded sand grains undergo practically no abrasion under eolian transportation. Apparently, shape and surface characteristics are very important.

Rounding of sand-sized grains by chemical action has been proposed as an important process. The evidence seems to indicate that this is not true but that chemical action in a river may produce a polished surface on sand grains. If chemical rounding were effective, then the smallest sizes would be most affected because of their large surface area relative to volume. Actually, sizes less than 0.05 mm seem to remain completely unrounded, disproving this hypothesis. However, sand grains immersed in a saturated solution of their own constituents become highly polished. Such polishes are seen on many sands, and the polished surfaces extend into concave sur-

---

*Ph. H. Kuenen, Experimental abrasion, Part 3, Fluviatile action on sand, *Am. Jour. Sci.*, vol. 257, pp. 172–190, 1959.
†Ph. H. Kuenen, Experimental abrasion, Part 4, Eolian action, *Jour. Geology*, vol. 68, pp. 427–449, 1960.

faces and even into cavities where abrasion could not be effective (Kuenen and Perdok, 1962).* The solubility of amorphous silica in the pH range 4 to 8.5 is approximately 125 ppm, whereas the solubility of quartz in the same pH range is of the order of 12 ppm (Fig. 3-6) (Pickering, 1962).† Amorphous silica is produced extensively in the weathering of silicates and will be present in solution in river water. Even if such water is not saturated in silica with respect to amorphous silica, it can easily be supersaturated with respect to quartz. As a result silica will be deposited on all surfaces of the quartz grains, including concavities, producing a polished surface.

Shape and size modification of fragments larger than sand size is quite noticeable even in short distances of transport. The grinding and impact of fragment on fragment quickly rounds corners and edges so that the roundness of the particle is increased without, at first, much effect on original shape and size. Breakage of fragments is not the most influential factor in size reduction at the early stages. As impact chipping and abrasion continue, all surfaces are affected, and corners are continually rounded so that sphericity is generally increased and size is gradually reduced (Kuenen, 1956).‡ Several characteristic surface textures result from stream abrasion. Most commonly the pebbles have a dull mat surface without a polish (Fig. 3-7). Polished pebbles are the result of other environments, particularly desert varnishes or the polish of gastroliths developed as gizzard stones of reptiles. A second, less common, surface feature is the development of impact cracks, which appear as conical fractures extending into the pebbles from the point of impact (Fig. 3-30).

INFLUENCE OF DEPOSITIONAL ENVIRONMENT ON SEDIMENTS

General Statement

The depositional environment is the fourth major factor to influence the character of sediments. The medium from which the particles settle out or are precipitated has a particular effect on sorting, character of bedding, nature of fossils and their state of preservation, and character of precipitated soluble materials. The factors which influence the depositional environment are both mechanical and chemical. Mechanical factors include the nature of the transporting agent, that is, air, dilute-water suspensions, muddy slurrys, or ice; turbulent versus quiet environments; velocity and consistency of direction of currents; water depth; organic activity in the

*Ph. H. Kuenen and W. G. Perdok, Experimental abrasion, Part 5, Frosting and defrosting of sand grains, *Jour. Geology*, vol. 70, pp. 648–658, 1962.
†R. J. Pickering, Some leaching experiments on three quartz-free silicate rocks and their contribution to an understanding of lateritization, *Econ. Geology*, vol. 57, pp. 1185–1206 (especially p. 1196), 1962.
‡Ph. H. Kuenen, Experimental abrasion of pebbles, Part 2, Rolling by current, *Jour. Geology*, vol. 64, pp. 336–368, 1956.

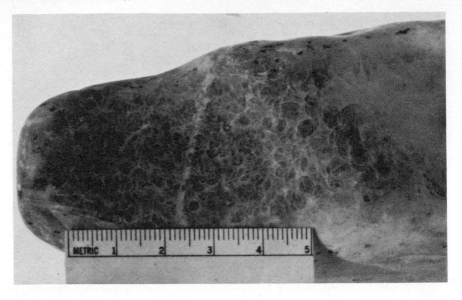

Fig. 3-30. Impact scars on a novaculite pebble. Saline River gravel, Sheridan, Arkansas.

depositional site; and time involved in sediment accumulation. Chemical factors involved include pH, oxidation-reduction potential (Eh), salinity, and temperature. Clastic sedimentary materials are particularly susceptible to the mechanical environment, although some bulk properties of the sediment are the result of the character of minor nonclastic materials which are dependent more on chemical environment. Chemical, biochemical, and evaporite sediments are most dependent on the chemical environment, but the textures of these rocks are often the result of mechanical agents. Thus in almost all sedimentary rocks, both chemical and mechanical features of the depositional environment must be considered.

DEPOSITION OF CLASTIC PARTICLES

The four fundamental agents of transportation and subsequent deposition are air currents, dilute aqueous suspensions and currents, concentrated muddy slurrys, and ice. These differ in their density and viscosity and therefore in their ability to move solid particles. As viscosity and density increase in the transporting agent, the velocity required to maintain particles in suspension decreases. At the same time the sorting of the deposited materials decreases. Wind and clear-water suspensions require relatively high velocity currents to move coarse clastic particles, but mud flows and glaciers can move very large materials at much lower velocities. In contrast, wind-deposited materials are extremely well sorted, whereas glacial deposits are essentially unsorted.

A major difference is seen in comparing turbidity-current deposits and normal shallow-water marine deposits. In a turbidity current the effective density of the water is increased as the sediment is stirred up to form a slurry. This principle is used extensively in the mineral-dressing industry in classifying crushed mineral grains by settling velocity. Two types of classifiers are used: one in which grains settle through essentially dilute suspensions — the free-settling classifiers; the other in which a dense suspension is developed — the hindered-settling classifiers. In the hindered-settling classifiers, with the solid particles dominantly quartz, the density of the suspension has been determined at about 1.5 g/cm³. The settling velocities for crushed quartz of various sizes through the two types of systems are given in Table 3-5.

In this system the suspended particles are only slightly smaller than those settling through the suspension, and markedly smaller sizes are not involved. In a turbidity current where very abundant fine-grained material is available, the density could be higher and the settling velocities lower. Thus a turbidity current with a given velocity and turbulence can maintain in suspension a maximum grain size several times larger than can a clear current of identical velocity and turbulence.

As the turbidity current moves off the slope onto the basin floor, the suspended sediment is deposited rapidly, the coarsest sizes first followed by progressively finer-and-finer-sized materials. The result is a size-graded sequence of beds deposited as current velocity and turbulence wane (Fig. 3-31). The basal unit consists of sand with admixed silt- and clay-sized materials and is in sharp contact with underlying finer-grained beds of an earlier sequence. The basal unit grades transitionally upward into an upper silt-clay- or clay-sized unit (Fig. 3-32). Characteristic bedding features are seen in each unit, as shown in Fig. 3-31 (Bouma, 1962).*

*A. H. Bouma, "Sedimentology of some Flysch Deposits," Elsevier, 1962.

Table 3-5. Settling Velocity of Crushed Quartz in Free-settling and Hindered-settling Conditions*

	Velocity, mm/sec	
Average size, mm	Free-settling	Hindered-settling
1.85	147	61.5
1.04	95	36.4
0.51	53	19.86
0.26	30	7.28
0.175	19	5.22

*Selected from R. H. Richards and C. F. Locke, "Textbook of Ore Dressing," McGraw-Hill, 1925.

If, however, a turbidity current is stopped by a barrier so that forward motion of the mass stops while the mass retains turbulence, then a second type of size grading results. All particles begin to settle at the same time, and arrival at the floor depends on the position of the particle within the mass and its settling velocity. All coarse sand throughout the mass will settle rapidly and will reach the floor with an abundant mixture of silt- and clay-sized particles which were originally near the floor. Subsequently, the bulk of the fine sand will reach the floor, again with an abundant silt and clay ad-

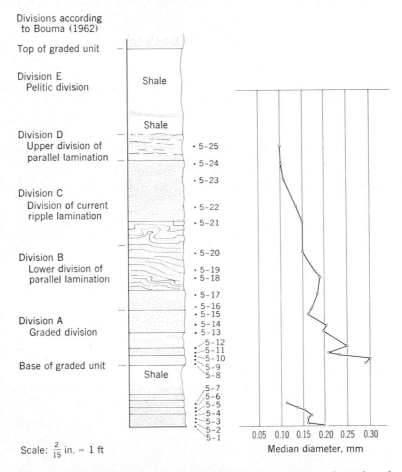

Fig. 3-31. Turbidity-current deposit. The unit is nearly 25 ft thick and can be subdivided into the five divisions of Bouma indicated at the left. Atoka formation of Pennsylvanian age on Winding Stair Mountain near Page, Oklahoma. *(Measured section and size determination by C. W. Hickcox.)*

Fig. 3-32. Turbidity-current deposits. Four successive units are visible, with their graded-sand base to the left grading up into shale to the right. Jackfork formation of Pennsylvanian age near Lake Greeson Dam, Murfreesboro, Arkansas. (*Courtesy C. W. Hickcox.*)

mixture. Next, the bulk of silt-sized materials will accumulate, again with abundant clay, and finally, the remaining clay-sized materials will accumulate. The product is a massive single unit in which the size of the largest particles present decreases upward through the unit but in which the finer sizes are present throughout the unit.

Transportation in water currents sufficiently dilute so that there is no increased buoyancy involves three mechanisms and size ranges. Fine materials are carried in uniform suspension at a distance above the floor. Such material usually represents a wide range of grain size, but normally the maximum size in such a suspension is in the range 200 to 300 microns. This size represents the maximum that can normally be carried for any distance by turbulence. The second sediment mass is carried in a graded suspension along the current floor. These grains are transported by saltation and short-term suspension due to strong eddy currents produced by irregularities of the bottom. Particles decrease in size upward in the suspension, hence the term *graded suspension*. Maximum grain size in this material is usually medium to coarse sand. The third type of transportation involves rolling or sliding along the bottom and moves grains which are too large to be picked up by turbulence.

As the current wanes, sediment begins to be deposited. The first batch consists of the rolled particles and the largest material of the graded suspension. This gives a well-sorted sand containing an admixture of scattered larger fragments (Fig. 3-33). Such sediments are often preserved as medium to coarse sandstones containing scattered granules or fine pebbles. In such a sediment there will be very little fine-grained material, because such materials are retained in uniform suspension. With further reduction of current velocity, the material carried in graded suspension is progressively deposited. This results in a sequence of well-sorted uniform sands which do not contain larger rolled grains or any appreciable fine-grained admixture. Finally, when current flow essentially ceases, the fine-grained uniform suspension settles out to form a poorly sorted mixture of fine sand, silt, and some clay. Such products may be seen in a vertical sequence, representing successive crops of sediment in one area as current fluctuates. Sandstone

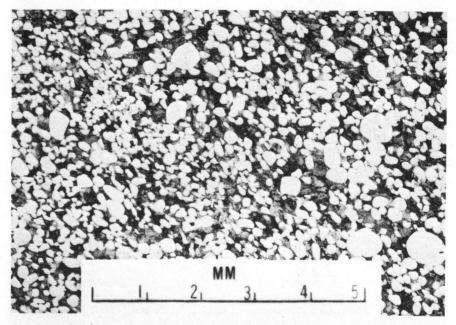

Fig. 3-33. Sandstone showing the products of a graded suspension plus larger rolled grains. The graded suspension consisted of quartz and bioclastic sand grains. Note the alinement of the larger grains diagonally across the field, probably indicating bedding. Note also the poorer rounding of the smaller quartz grains. Sandstone unit in Cotter Dolomite of Ordovician age, Beaver Dam site, Eureka Springs, Arkansas. Projection of a thin section.

beds in this type of environment are often coarse at the base, contain scattered pebbles or granules, and grade upward into progressively finer sand. The next bed will again be coarse at the base and become progressively finer upward (Fig. 3-34). The same relationship can also be seen regionally in a unit representing a short time interval. In the high-energy areas, currents will be strongest, and the sediment deposited will consist of the coarse fraction of the graded suspension plus rolled grains. Laterally, the rolled grains will be absent in quieter water, and the sediment will consist of the products of the graded suspension. Finally, in the lowest-energy zones, only the products of uniform suspension will be present.

Ice, however, can carry any size of material, an upper limit being set only by the size of the ice mass and the largest-size fragments available. Materials are deposited by melting rather than by cessation of movement; therefore the deposit is completely unsorted — all available sizes are present and are randomly distributed through the deposit.

Depth of water and turbulence influence the character of bedding and the organic content of the sediment. In shallow agitated water, particularly where affected by waves, sediments will rarely be thin-bedded or laminated. The continual vibration and reworking destroy such features, and even in normally quiet shore areas, the infrequent major storms crossing the area

Fig. 3-34. Graded bedding in thin sand units. The 6-in. scale rests on the coarse base of one unit parallel to the bedding. Lorrain quartzite of Precambrian age near Desbarats, Ontario.

will cause extensive erosion and consequent rapid redeposition of sediments and destroy such features (Hayes, 1967).* Thin lamellar bedding can develop only in quiet waters where it cannot be destroyed. Therefore the only fossils preserved in agitated zones are the thick-shelled varieties, and even they are disarticulated, broken, and abraded if the wave action is too strong. In quiet deeper waters, delicate organisms can live and can be preserved. They are frequently found in quiet-water deposits as articulated shells, often in their living positions. Burrowing and bottom-scavenging animals have an additional influence, in that even in quiet water they may turn over the fine-grained sediments, homogenizing them and destroying fine laminations developed during deposition (Ginsburg, 1957).†

Current action and persistence of current directions influence clastic sediments in the development of characteristic structures on the surfaces of the sediments. These may be preserved by subsequent burial if there is sufficient difference in the character of the sediment which buries the surface features. Ripple marks of both current and oscillation types, cross bedding, and scour marks are familiar examples. Scour marks and ripple marks are undoubtedly developed in many sediments where they are not preserved. Cross bedding is much more commonly preserved, because it is usually a larger-sized feature and is therefore much more difficult to destroy. Such features are useful in determining the current directions in developing paleogeographic interpretations of sedimentary basins.

The greatest influence of time on clastic sediments is in the rate of accumulation or sedimentation. Commonly, there is no evidence of rate of accumulation preserved in a sediment, but such evidence is occasionally seen. Fossil leaf prints which cut across bedding (Fig. 3-35) and standing tree stumps buried in sediments are occasionally seen in continental sediments and shoreline lagoonal deposits. Such features must indicate very rapid accumulation of sediment, because they would have to be buried before the plant material could decay to be preserved. The time factor in sedimentation can also be seen in the varve type of bedding, where paired layers of sediment of contrasting composition or texture are supposedly developed for each yearly climate cycle. The best examples are found in glacial-lake deposits, where the summer cycle is represented by silt to fine sand and the winter cycle by clays.

Unconsolidated sediments frequently undergo a downhill slumping where they accumulate underwater on gentle slopes. Such slump structures give a variety of appearances, depending on the proportions of coarse mater-

*M. O. Hayes, Hurricanes as geological agents, South Texas Coast, *Am. Assoc. Petroleum Geologists Bull.*, vol. 51, pp. 937–956, 1967.
†R. N. Ginsburg, Early diagenesis and lithification of shallow water carbonate sediments in South Florida, in: R. J. LeBlanc and J. G. Breeding (eds.), *Soc. Econ. Paleontologists and Mineralogists Spec. Pub. 5*, pp. 80–99, 1957.

Fig. 3-35. Fossil leaf curved to cut across bedding in fine sandstone.

ials to clays and on the thickness of the individual beds. Where the sediment consists dominantly of thick beds of coarse-grained sediment resting on fine-grained materials, the sand units will thicken downhill, cutting across underlying beds to develop a thick wedge-shaped mass and giving the appearance of channel fill on the uphill side. Fine bedding within the slumped mass is intricately contorted, and the fine sediments at the base of the slumped unit are contorted or slickensided. A second common form of slump develops where the beds are a few inches to a very few feet thick and alternate between sand and shale. Here slumping causes the development of close folds in the sand beds and flowage of the mud in between (Fig. 3-36). A final form is the "pull-apart" structure, which develops when thin units of sand are interbedded in clay. When slumping takes place, the sand beds are pulled apart into plates, which frequently become curled up on the edge to make dish-shaped masses. The clay flows between these units and develops a highly contorted flow structure. Such slump structures are clearly indicative of deposition on a sloping surface, and the direction of movement, and therefore the slope, can often be interpreted.

DEPOSITION OF CHEMICAL SEDIMENTS

The materials dissolved during chemical weathering are transported through to the sea and are distributed by marine currents to all parts of the ocean.

Fig. 3-36. Depositional slump structure. Soft-sediment downslope sliding of calcareous sand on clay. Note the pinching of the clay (now highly fissile shale) up into the axis of the fold. Also note the erosional truncation of the folds by the overlying cross-bedded calcareous sandstone. Bloyd Formation of Pennsylvanian age, Brentwood, Arkansas.

There they are deposited as a result of chemical precipitation, biochemical precipitation, or evaporation. Chemically and biochemically precipitated materials settle to the bottom but are subject to resolution on the way to the bottom. Thus the materials which accumulate are those materials precipitated and not redissolved. Resolution is particularly important in the case of biochemically precipitated materials, because while the organisms are alive, their hard parts are protected by organic tissue or are rebuilt as material is redissolved. On death of the organism, this protective covering is removed and the hard parts are subject to solution. Thus silica-precipitating organisms may live in great numbers in an environment in which no silica can accumulate.

Quantitatively, the two most important biochemically precipitated materials are silica and calcium carbonate. Silica is precipitated by some sponges, by radiolaria, and by diatoms, each of which produces minute opaline silica tests or spicules. Calcium carbonate is precipitated by most of the larger invertebrates, by foraminifera, and by some algae. Calcium carbonate is precipitated organically as either calcite or aragonite and contains variable amounts of magnesium carbonate. The aragonite structure can accept essentially no magnesium, whereas biogenic calcite can contain up to 30 percent magnesium carbonate (Revelle and Fairbridge, 1957).* Table 3-6 tabulates

*R. Revelle and R. Fairbridge, Carbonates and carbon dioxide, in: Treatise on ecology and paleoecology, *Geol. Soc. America Mem.* 67, vol. 1, pp. 239–296 (especially pp. 270–271), 1957.

Table 3-6.   Mineralogy and Chemistry of Biochemical Precipitates*

Type of organism	CaCO₃ type	% MgCO₃	Other mineral matter
Marine algae			
Diatoms			Opal
Green algae	Aragonite	< 1	
Brown algae	Aragonite		
Red algae			
Coralline	Calcite	7–30	
Other	Aragonite	< 1	
Golden brown algae			
Coccoliths	Calcite	< 1	
Protozoa			
Radiolaria			Opal
Foraminifera	Calcite	0.3–16	
Porifera			
Siliceous			Opal
Calcareous	Calcite	5–14	
Coelenterata			
Hydrozoa	Aragonite		
Alconaria			
Heliopora	Aragonite		
All others	Calcite	7–15	
Zoantharia	Aragonite	< 1	
Bryozoa	Calcite and/or aragonite	< 4	
Brachiopoda	Calcite	< 4	Chitinous
Echinodermata	Calcite	5–17	
Mollusca			
Pelecypoda			
Pectin and oysters	Calcite	1.3–2.8	
All others†	15–100% aragonite	0.09–1.6	
Gastropoda†	Aragonite	0.08–2.4	
Cephalopoda	Aragonite	< 0.5	
Arthropoda	Calcite	1–11	Chitinous and calcium phosphate

*Data from R. Revelle and R. Fairbridge, Carbonates and carbon dioxide, in: *Treatise on marine ecology and paleoecology, Geol. Soc. America Mem.* 67, vol. 1, chap. 10, 1957.
†Within single species the aragonite to calcite ratio varies; in general, the higher the water temperature, the higher the aragonite

the mineralogy and magnesium carbonate contents of organic contributions to sediments. In shallow-water environments, carbonates or silica of biogenic origin will accumulate if life is sufficiently abundant so that the precipitated material is buried more rapidly than it can redissolve. Water temperature may be a factor in the solution of carbonates, in that temperature controls the solubility of carbon dioxide gas. Cold waters contain more dissolved gas and therefore increase the solution of carbonates as bicarbonates.

On the continental shelf, chemical sediments accumulate in most environments but retain their identity only where they are not masked by larger volumes of land-derived detrital material. Extensive carbonate sediments are most commonly developed adjacent to low-lying humid land areas where vegetation prevents the erosion and removal of anything except fine clastics. In areas of higher relief on adjacent land masses, such as volcanic islands, carbonates build up as reefs fringing the island, and clastic sediments are trapped between the land and the reef.

Carbonate deposits of organic origin take several basic forms. The most obvious is organic reefs. Modern reefs are composed of a rigid calcareous framework of interlocking or organically cemented corals and coralline algae. This framework controls the accumulation of clastic carbonate sediments in and around itself to build up a topographically high area on the sea floor. The framework skeletons generally make up 10 percent or less of the total mass. Modern reefs are largely restricted to tropical seas in warm shallow water. However, Teichert (1958)* notes the existence of coral reefs in the North Atlantic as far north as 70° N. Lat. on the coast of Norway. In the geologic past a greater variety of organisms acted as framework builders than do so at present (Fig. 3-37). These framework organisms build a reef core which is characterized by being massive, poorly bedded, and commonly very porous. As the core builds up into the wave zone, it is subjected to fragmentation and wave erosion, and the margins of the reef are buried in a complex of inclined reef-flank beds. These beds are composed of fragments of reef core plus abundant fragments of other organisms which lived in the reef but were not framework builders. Algae commonly cement such material into resistant masses which can be differentiated from the core only by the attitude of the beds.

A second type of organically derived deposit is blanket deposits, originally composed of disarticulated, broken, or abraded shell material. Such units are of great lateral extent compared with their thickness. Fragmented material is usually well sorted according to size and often shows current-depositional features. Shell fragments (Fig. 3-38), crinoid columnal segments, bryozoan chips, etc., are commonly oriented in horizontal

*C. Teichert, Cold- and deep-water coral banks, *Am. Assoc. Petroleum Geologists Bull.*, vol. 42, pp. 1064–1082, 1958.

Fig. 3-37. Reef-core limestone — biolithite. Coral, encrusting bryozoa (fine cellular structure), and dense algal limestone (dark uniform gray) forming the framework of a reef mound. Pitkin formation of Mississippian age, Wesley, Arkansas.

Fig. 3-38. Brachiopod biosparite with all valves and fragments parallel to the bedding. Rochester, New York. *(Ward's Universal Rock Collection.)*

positions of maximum stability. Fine-grained materials are often completely removed so that the cement is coarsely crystalline calcite (Fig. 3-39). Cross bedding and large-scale ripple marks are often present where deposition was under strong current conditions, and graded bedding is often present in single units where currents have changed with time. Quartz sand is a very common minor constituent in such deposits. The requirements for such a development are shallow water strongly agitated by wave action, abundant life, and a paucity of land-derived clastic materials (Fig. 3-40).

The origin of the fine-grained calcareous muds, silts, and fine sands which indicate no evident organic source is still a problem. Both chemical and biochemical origins for these materials have been proposed, and probably, both are correct for special conditions. Chemical precipitation of fine needles of aragonite occurs on a portion of the Bahama Banks in the lee of Andros Island. Here the water is very shallow, is warmed by the sun, and is subject to excessive evaporation. Loss of carbon dioxide by warming and supersaturation by evaporation result in the precipitation of the aragonite needles 1 to 10 microns long (Illing, 1954).* Similar-appearing lime mud can be produced by disintegration of organically precipitated carbonates. Algae, particularly *Halimeda*, precipitate aragonite needles within their tissues and as an external coating precipitated as a result of the removal of carbon

*V. L. Illing, Bahaman calcareous sands, *Am. Assoc. Petroleum Geologists Bull.*, vol. 38, pp. 1–95, 1954.

dioxide from the water (Fig. 3-41). This material forms a spongy deposit of weakly coherent material which disintegrates into a mud on death of the plant (Lowenstam and Epstein, 1957).* Regardless of its organic or inorganic origin, this material is of great importance in the formation of many lime-stones. Illing (1954) describes the accretion of such material into silt- and fine-sand-sized aggregates, which have been termed *bahamite* by Beales (1958).† On the mildly wave-agitated shallow bank, the aragonite needles coalesce into loosely aggregated clumps. Sometimes a sticky organic sub-stance acts as an initial binder, but with time the aggregates cement them-selves into coherent grains. Such fine-grained lime muds and sands are essentially devoid of macrofossils and are very pure calcium carbonate. Beales has described several limestones of Paleozoic age which, apparently, were originally bahamite sediments.

Chemically precipitated aragonite muds are also a source of the abun-dant ooliths of the Bahama Banks. Ooliths are sand-sized grains, usually containing a nucleus of a fragment of shell or mineral matter about which two or more concentric layers of fine-grained carbonates have been built up by accretion. In most modern ooliths the layered portion of the body is composed of aragonite needles in a random tangential arrangement. In ancient ooliths this aragonite has recrystallized to radially oriented calcite fibers. Recrystallization of calcite may either preserve or destroy the con-centric structure so that ancient ooliths show concentric or radial structure or both about the nucleus (Fig. 3-42). The nucleus may be a shell fragment, a sand grain (very commonly quartz), or a microcrystalline carbonate aggre-gate of the bahamite type. In some the original nucleus was so small as to be undetectable after recrystallization. Carozzi (1960)‡ visualizes two factors involved in oolith formation. First, a current which is capable of moving nucleus grains into a region of aragonite precipitation, and second, local agitation which keeps the transported grains moving while they are built up by accretion into an oolith. The current action supplying nuclei may be greater or less than the agitation. If the current is stronger, then the larger grains transported into the ooliting environment will not become concentri-cally coated, whereas the smaller grains will all grow to an essentially uni-form size. Freeman (1962)§ has described irregular-shaped ooliths on the Texas coast which have been built in a quiet-water nonagitated environ-ment. Hence the essentially spherical form of most ooliths would indicate agitation during accretion.

*H. A. Lowenstam and G. Epstein, On the origin of sedimentary aragonite needles of the Great Bahama Bank, *Jour. Geology*, vol. 65, pp. 364–75, 1957.
†F. W. Beales, Ancient sediments of Bahaman type, *Am. Assoc. Petroleum Geologists Bull.*, vol. 42, pp. 1815–1880, 1958.
‡A. V. Carozzi, "Microscopic Sedimentary Petrography," pp. 241 ff, Wiley, 1960.
§T. J. Freeman, Quiet water oolites from Laguna Madre, Texas, *Jour. Sed. Petrology*, vol. 32, pp. 475–483, 1962.

Fig. 3-39. Coquina—biospar limestone. A sediment of broken, abraded, and well-sorted shell fragments from which all fines have been removed. The sediment is incompletely cemented by coarse crystalline calcite. St. Augustine, Florida. *(Ward's Universal Rock Collection.)*

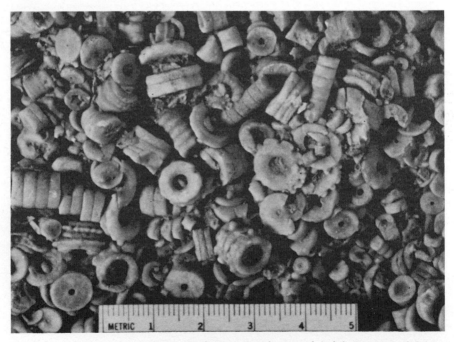

Fig. 3-40. Well-sorted and largely disarticulated crinoidal debris—crinoidal biosparrudite. The specimen gives evidence of abundant marine life with very little influx of land-derived clastic material. Boone Formation of Mississippian age, Springdale, Arkansas.

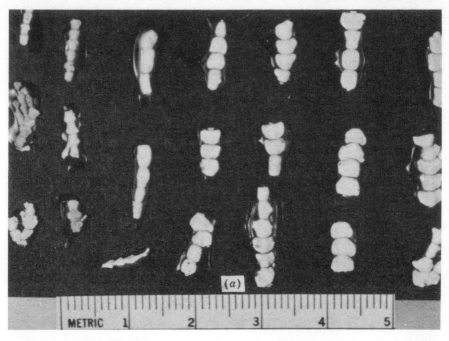

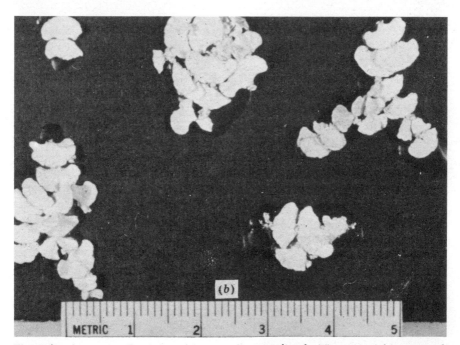

Fig. 3-41. Aragonite precipitated by the algae *Halimeda*. The material is very soft and spongy. The sticklike colonies (top) are from Moro Bay, California; the leaflike colonies (bottom) are from Barbados, B. W. I.

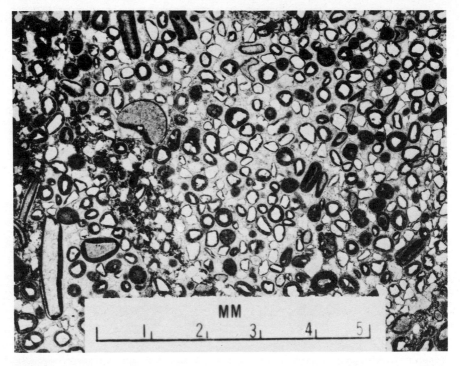

Fig. 3-42. Ooliths developed on quartz-sand grains—oomicrite. Some ooliths show concentric structure, whereas some show radial structure. The matrix is microcrystalline calcite. Brentwood limestone of Pennsylvanian age, Fayetteville, Arkansas. Projection of thin section.

The problem of nonclastic silica in sediments is a difficult one. Silica is brought to the sea in true solution as silicic acid, H_4SiO_4. Average sea water contains 1 to 2 ppm silica, whereas most large rivers contain 3 to 50 ppm. What becomes of this excess is a major problem. It cannot be flocculated as a colloid by the electrolytes of the sea, because it is in true solution and well below the saturation limit of around 125 ppm. Two alternatives are open to explain the low amount in the sea. First, living organisms can account for the removal of much silica, but their preservation after death in such strongly undersaturated sea water is a problem. Second, silica can be removed from the sea by the authigenic regeneration of clay minerals (Mackenzie et al., 1967).* At present, silica is accumulating in the sea as the dominant constituent of sediments in only two environments: one as belts of radiolarian ooze in very deep cold arctic and antarctic waters, where carbonates are re-

*F. T. Mackenzie, R. M. Garrels, O. P. Bricker, and F. Bickley, Silica in sea water—Control by silica minerals, *Science*, vol. 155, pp. 1404–1405, 1967.

dissolved, and the other in the equatorial belt of the Pacific Ocean, where cold nutrient-rich bottom water wells upward to the surface and supports an abnormally high organic population. The occurrence in the geologic column of thick-bedded chert units and abundant cherty limestones presents difficulties.

The occurrence of cherty limestones can probably be accounted for by the accumulation of biogenic silica in the sediments and its subsequent reorganization during diagenesis. Radiolaria and diatom tests and siliceous sponge spicules are all minute and would be rapidly buried in accumulating coarser bioclastic carbonates, thus preventing their immediate resolution. In contrast, the thick-bedded cherts and novaculites are associated with graywacke and shale and appear to represent deep cold-basin deposits and probably very slow accumulation of silica. Some of these deposits indicate an original biogenic source for the silica, whereas others contain no evidence of biogenic origin. Several sources for the abnormal silica concentration have been offered. Many cherts are associated with contemporaneous volcanics, and it is thought that the silica was derived by direct reaction between sea water and hot lavas or associated submarine hot springs. Such silica either could be precipitated directly or could result in a greatly increased population of diatoms or radiolaria. A second source is through the submarine weathering of slowly accumulating volcanic ash; all other constituents are presumably being removed in solution (Goldstein and Hendricks, 1953).[*] Related to this is the possible source of silica from the alteration of volcanic ash to bentonite, but this seems to be a quantitatively insufficient source for cherts associated with bentonite. A final possible source is through prolonged chemical weathering of an essentially peneplaned area. Silica would be dissolved in this environment because of the breakdown of silicate minerals to form lateritic soils, and very little clastic material would be carried to the sea. A geochemical differentiation of the dissolved materials contributed to the sea would have taken place in the depositional environment.

The deposition of the quantitatively minor chemical sediments is controlled by the Eh of the depositional environment. This is particularly important in the accumulation of iron and carbon in sediments, but it is also important in the formation of sulfur and the oxidation state of manganese. An ion is oxidized when it loses electrons and is reduced when it gains electrons. The Eh is the affinity of a substance for electrons in comparison with a standard, hydrogen. A negative Eh means that a reducing condition exists, that there is a deficiency of electrons absorbing ions. In the oxidizing environment, iron is deposited as the insoluble ferric oxide or hydroxides hematite and limonite, losing three electrons to oxygen. As electron-absorbing ions

[*]A. Goldstein and T. A. Hendricks, Siliceous sediments of the Ouachita facies in Oklahoma, *Geol. Soc. America Bull.*, vol. 64, pp. 421–441, 1953.

become less abundant, a reducing condition develops, and iron is reduced to the ferrous state, having lost only two electrons. Ferrous iron may be deposited as the carbonate siderite or as sedimentary iron silicates, such as glauconite or chamosite. In still more intense reducing conditions, sulfur will also be reduced from the six-valent state, and ferrous iron will be deposited as the sulfides pyrite, marcasite, or the amorphous form melnikovite.

Carbon from organic sources is also strongly influenced by the Eh of the depositional environment. There are two direct effects: first, in an oxidizing environment the organic matter is broken down to give carbon dioxide and other gases. Second, where there is a paucity of oxygen the normal bottom-scavenging animals will not be able to live because of toxic conditions resulting from the accumulation of organic wastes. Thus if organic material accumulates faster than it can be consumed by chemical and biological breakdown, a reducing condition develops, the carbonaceous materials accumulate, and the detrital sediments settle into the area. Black carbonaceous sediments indicate that organic material is being introduced into the depositional site faster than it is being destroyed. Commonly, such an environment requires stagnant-water conditions in the basin of deposition, but this is not required.

The environment within marine sediments is always reducing below some certain levels, because available oxygen dissolved in the trapped water will ultimately be used up by chemical or biological (bacterial) activities. However, the iron of these sediments will not necessarily be reduced because of a lack of reducing agent. Carbonaceous materials are the strongest reducing agent in the sediments; but under normal oxidizing environments, where ferric iron is deposited, the organic materials are also oxidized or removed by scavengers and therefore are not present to reduce the iron.

Deposition of Evaporites

Deposition of soluble salts by evaporation is generally considered to be a separate process from the deposition of chemical and biochemical sediments, because the environment required is different. As marine water is evaporated, the dissolved salts are progressively concentrated to the point of saturation and are deposited. In normal marine waters the first material to be deposited is calcite, followed by dolomite. The amount of these materials is very small, and evaporation of 1,000 m of sea water gives only a few centimeters of limestone. Gypsum or anhydrite is the next salt to reach saturation. The alternative deposition of gypsum or anhydrite depends on water temperature and salinity. Gypsum appears when the salinity has reached 3.35 times normal at 30°C, and anhydrite occurs at higher salinities (Deer, Howie, and Zussman, 1962).* When the water has been concentrated to

*W. A. Deer, R. A. Howie, and J. Zussman, "Rock Forming Minerals," vol. 5, p. 208, Longmans, 1962.

one-tenth of its original bulk, halite reaches the saturation point and begins to precipitate. Normally, evaporation does not reach the point where the still more soluble salts of potassium and magnesium are deposited; evaporation to 1.54 percent of the original volume is required for this deposition. The salts precipitated at this ultimate evaporation include sylvite (KCl), polyhalite [$K_2Ca_2Mg(SO_4)_4 \cdot 2H_2O$], epsomite ($MgSO_4 \cdot 7H_2O$), and others.

The concentration of sea water by evaporation and the production of thick deposits of salts present a problem. Ocean water contains, on the average, only 3.5 percent dissolved salts. Therefore complete evaporation of shallow arms of the sea can develop only thin layers of salt. The development of thick salt beds has been explained by evaporation of marine water in shallow basins connected to the sea by restricted channelways. The Gulf of Kara-Bogaz-Gol on the Caspian Sea is the type example. Here the Gulf is cut off from the Caspian Sea by a bar through which there is a restricted channelway. Evaporation in the Gulf maintains water level below that of the sea so that a current flows continuously into the Gulf and is there evaporated. In this manner great volumes of water can be evaporated without requiring great depths of water at any time.

An additional factor which is probably of considerable importance has been recently recognized. That is, large volumes of salts are transported inland for great distances by the wind. As bubbles of wind-developed froth on storm waves at sea break, they throw minute droplets of sea water into the air. These are picked up by the wind and are evaporated to form minute salt crystals which may be carried great distances before they settle out into vegetation or are washed out of the air by rain (Gorham, 1961).* In humid regions these salts are flushed out by rain water and carried away by the streams. In arid regions these salts are carried down by runoff and concentrated in adjacent basins. The process may add salts to isolated marine arms beyond that added by the influx of normal marine waters. It is interesting to speculate on the possibility that with a critical amount of rainfall in subarid regions, the lime introduced into the soil, which appears as caliche, may not be wind-transported marine salts, the lime having been concentrated by the selective removal by rain water of the more soluble salts.

Quantitatively Minor Primary Sediments

Primary iron-rich sediments are locally important and economically very important, since they have acted as the protore of the large iron ore deposits of the Precambrian Lake Superior region. The sediment, as deposited, may have been banded chert and siderite, chert and iron oxides, or iron silicates. In tropical humid regions, iron oxides and silica are important constituents in river waters, much more so than in temperate humid regions, and are con-

---

*E. Gorham, Factors influencing supply of major ions to inland waters, with special reference to the atmosphere, *Geol. Soc. America Bull.*, vol. 72, pp. 795–840, 1961.

tributed to the sea in sufficient quantities to produce the iron-rich sediments. No special source, such as volcanics, need be called on.

The precipitation of iron in modern sediments is controlled by Eh, pH, activity of sulfur, and abundance of organic matter. Curtis and Spears (1968)* conclude from laboratory and field studies that only ferric oxides can precipitate from normal depositional waters. This takes the form of primary hematite or limonite. Ferrous compounds other than pyrite can be developed only from sediment pore water, below the sediment-water boundary where reducing conditions develop. The minerals formed here include pyrite, pyrrhotite, magnetite, siderite, and iron silicates such as chamosite. Iron for the development of these minerals is derived from primary ferric precipitates. Pyrite and siderite are the more common products, pyrite developing where organic decay supplies sulfur and siderite developing where there is a paucity of sulfur but available carbon dioxide from organic decay. The requirements for the formation of ferrous silicates would be low carbon dioxide and sulfur activities and availability of silica in the pore solutions. In stagnant-water bottom environments with much decaying organic matter, the primary precipitate is pyrite rather than the ferric oxides.

Siderite and pyrite occur commonly as concretionary bodies in sediments, and pyrite commonly replaces fossils. This occurrence accords with their being diagenetic minerals. Limonite, siderite, and chamosite all occur in ooliths in some sediments. Curtis and Spears cite the evidence and conclude that both siderite and chamosite are either secondary minerals in the ooliths or, in the case of chamosite, the ooliths are diagenetic and not the product of gentle current action as in the case of calcite ooliths. In contrast, limonite ooliths are definitely primary sedimentary features. The thermodynamic relationships show that the introduction of ferrous solutions into a calcite-water system results in the solution of calcite and the precipitation of ferric hydroxides. Thus the oolitic or fossiliferous hematite ores of the Clinton formation of Silurian age can be in part the primary precipitation of oolitic limonite or the diagenetic replacement of calcite by limonite. The apparently primary nature of the chert-siderite or ferrous silicate associations of the Precambrian remain a puzzle in the light of Curtis and Spears' work. One possible answer may be that the atmosphere in middle Precambrian time was quite different from that of today so that slightly reducing conditions were readily attained in the depositional environment. A second possibility is that the banded iron formations are continental in origin, developing in lakes subject to thermal stratification and annual turnover (Govett, 1966).†

*C. D. Curtis and D. A. Spears, The formation of sedimentary iron minerals, *Econ. Geology,* vol. 63, pp. 257–270, 1968.
†G. J. S. Govett, Origin of banded iron formations, *Geol. Soc. America Bull.,* vol. 77, pp. 1191–1211, 1966.

Glauconite is an iron-rich clay of the illite group and an important sedimentary product of some environments. It generally seems to form from preexisting silicates, such as detrital biotite or pelleted clay minerals, by reaction with sea water or by direct colloidal precipitation. Reducing conditions are required, and very slow sedimentation in the depositional site is a necessity. If sedimentation is at a normal rate, the primary detrital silicate minerals are buried before they can be converted to glauconite. The ratio of ferric to ferrous iron is rather constant and magnesia is required, which would indicate critical Eh and salinity for its formation (Deer, Howie, and Zussman, 1962).*

Primary-bedded phosphatic sediments may have several origins. In areas of nondeposition of clastic materials, phosphatic bones and teeth may accumulate to form a bone bed. Oolitic and pisolitic structures of many phosphorites indicate that their deposition was chemical rather than biochemical. The common association of phosphorites with black shale and chert, and the common occurrence of iron silicates or carbonate, suggest an environment of deposition which is not a normal marine one. Neutral to slightly acid conditions, a reducing environment, and very slow sediment accumulation seem to be requisite. Although the sea contains only a small percentage of dissolved phosphate, over a long period of time and under stable conditions it is sufficient to account for such bedded phosphorites.

INFLUENCE OF POSTDEPOSITIONAL MODIFICATION ON SEDIMENTS

General Statement

Once a sediment has been deposited, it is still subject to internal reorganization, both chemically and mechanically. Cements may be introduced from outside the sediment or may develop by solution and redeposition of essential primary constituents of the sediment. The minerals of the primary sediment may react with each other or with solutions, either those trapped as the particles accumulated or those introduced at a later date, to form new minerals. The grains may be reorganized mechanically as a result of the accumulating weight of sediment on top of them, or they may be reorganized and the constituents separated by chemical action. Each of these processes tends to work toward the conversion of unconsolidated particles into lithified sedimentary rock.

For purposes of discussion, these processes are divided into five main classes. (1) Compaction, which involves mechanical rearrangement of grains to decrease volume by elimination of pore space. (2) Cementation, which involves induration of the sediments by cementing grains together either with material introduced from outside the sediment or by solution and re-

*W. A. Deer, R. A. Howie, and J. Zussman, "Rock Forming Minerals," vol. 3, p. 40, Longmans, 1962.

deposition of primary materials of the sediment. (3) Recrystallization, which involves the structural reorganization of the sedimentary minerals; frequently, new minerals result from growth in place in the sediment. (4) Chemical unmixing of complex sediments, which involves the segregation of minor constituents into nodules, concretions, and similar structures. (5) Metasomatism, which is the wholesale modification of the chemical composition and mineralogy of a sediment as a result of the introduction and removal of materials by migrating solution from outside the sediment. In most rocks several types of processes are operative at the same time and are intimately interrelated. The processes will be discussed individually.

Compaction

Reduction of volume by the physical crushing and closer packing of grains is particularly important in the lithification of shale. As clay flakes settle out of water, they do not tend to come into close packing because of their very slow settling velocities, their large surface area, which allows adhesion of one grain to another, and their relatively large adsorbed-water films. Freshly deposited marine clays may have as much as 80 percent pore space filled with water. As the mud is buried by accumulation of more material, this water is squeezed out and porosity is reduced to as little as 5 percent. Reduction of pore space brings the clay flakes into closer contact; as a result they adhere to one another, and this process by itself will cause lithification. Generally, other processes are involved as well, including recrystallization or modification of the clay minerals and cementation as a result of migrating fluids. Compaction is not important in the coarser clastic sediments, because their original porosity is not nearly as great as that of clay and because the nature of the grains does not result in adhesion of grain to grain.

Cementation

Induration is accomplished in cementation through the filling of pore spaces by mineral material which may be either introduced from outside the sediment at any time in its history or developed by solution and redeposition of materials from the sediment. Cementation is the most important process of lithification of the coarser clastics, particularly clean sands and sand-pebble accumulations. The most important mineral cements are calcium carbonate (usually as calcite), silica (most commonly as quartz but also as opal, chalcedony, or chert), and iron oxides. Less common cements are pyrite, gypsum, barite, anhydrite, and asphaltic materials. In many dirty sands, such as produce graywacke, clay is an interstitial material and lithification is effected by the recrystallization of the clay. In the purer sandstones, cementation frequently involves two cements: each introduced independently at different times or one replacing another. In such cases the character of the sandstone will vary from point to point in the rock.

Carbonate cementation may be accomplished either by introduction of carbonate in solutions migrating through the rock or by recrystallization of carbonates deposited with the sand. The origin of the cement can sometimes be determined by the texture of the rock. Where the carbonate was deposited with sands as clastic grains of limestone or shell fragments, recrystallization of these fragments to form a crystalline cement leaves the nonsoluble sand grains floating in the cement rather than in contact (Fig. 3-43). The insoluble sand grains, which were originally supported by the detrital carbonate, are no longer in contact with other sand grains. Such texture is most commonly seen where carbonate cement is minor and patchy. Floating sand grains are not normally developed where the cement either is introduced or was deposited originally as an interstitial carbonate mud just filling the pore spaces. Introduced and primary interstitial cements may be either fine-grained material or large crystals formed of calcite and including many sand grains. The sand-calcite crystals of the South Dakota Badlands are an extreme example of coarsely crystalline cement (Fig. 3-44). Desert roses, or barite rosettes, are similar, except that the cement is coarsely crystalline barite rather than calcite.

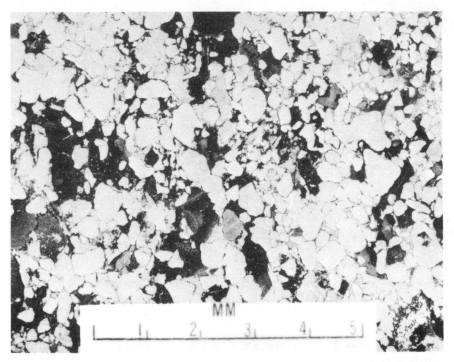

Fig. 3-43. Sandstone with remnant bioclastic fragments supplying calcite for the cement. Note the floating quartz grains, left center, and the penetration of the fossil fragments by the sand. Winslow Formation of Pennsylvanian age, Mountainburg, Arkansas. Projection of thin section.

Fig. 3-44. Sand-calcite crystals. Large euhedral calcite crystals developed during the partial cementation of a sand. Badlands, South Dakota.

Silica cements may likewise be either introduced from outside sources or developed within the sand by solution and redeposition. Megascopically recognizable silica cements are usually in the form of secondary enlargements of clastic quartz grains. If silica is to be deposited from solution onto quartz grains, it will commonly be deposited in structural continuity with the nucleus grain. If this secondary enlargement is not sufficient to fill the pore spaces, then euhedral crystal faces may be developed on the sand grains. Under ideal conditions, such as sometimes develop where quartz grains are secondarily enlarged in limestones, a doubly terminated euhedral quartz crystal will develop. Normally, there is too much mutual interference of the growing grains to develop perfect euhedralism in sandstones. If there is an abundance of silica available so that the voids are completely filled, then no crystal faces will be evident due to complete mutual interference.

The origin of the silica for cementation is usually not megascopically determinable. It may be introduced from external sources, such as connate waters forced out of associated shales during compaction. Many shales contain a considerably larger proportion of silica than the clay minerals can accommodate. This silica may be present either as quartz grains or as colloids which can migrate with the connate water during compaction. Also, diagenetic changes in the clays as a result of reaction with dissolved salts in the connate water liberate colloidal silica for cementation of adjacent sands. The second major possible source of silica is from the pressure solution of quartz-sand grains at points of contact under the load of accumulating sediments. At points of contact between grains, silica may dissolve because of high local pressure and be precipitated in the interstitial spaces because of lower pressure. The effects of pressure solution are visible microscopically in some sandstones as microstylolitic grain boundaries. Such structures are not visible megascopically, but the process can supply the silica needed for the complete filling of pore spaces. A third source of silica cement is from siliceous organisms, such as sponge spicules, deposited with the sand. Sponge spicules are opaline silica and therefore are highly susceptible to solution in the proper pH range.

Ferruginous cements are not as common as it would appear from the iron-stained character of many sandstones. Only a very small percentage of iron oxide in the form of hematite or limonite is necessary for a strong color effect. Primary iron cements are introduced in subaqueous environment as carbonate or sulfides. These, in turn, may be oxidized subareally to give limonite or hematite. Primary iron oxide cements are common in continental deposits, where iron oxide colloids are developed by weathering and are transported downward by groundwater.

Replacement of one cement by another introduces many complexities into the cementation history. Such relationships only rarely can be identified megascopically; they require detailed thin-section studies of contact relationships. Replacements of silica and calcite interchangeably have been reported. In general, calcite cementation involves corrosion and etching of quartz-sand grains, whereas silica cementation does not. Thus frosting and etching of quartz grains cemented by secondary silica may indicate an earlier carbonate cementation. Carbonate cements seem to be susceptible to replacement by a variety of minerals, including hematite and barite. Barite roses appear to be a replacement product.

Recrystallization and Authigenesis

Recrystallization of primary minerals is particularly important in the lithification of limestone and shale. Carbonate sediments are frequently out of equilibrium with their new environment as they accumulate and, as a result, will quickly begin to recrystallize. Much organic carbonate is in the metastable form aragonite. As long as it was renewed or protected by the living organism it was retained as aragonite, but on death of the animal, aragonite will begin to recrystallize into the stable form calcite. Similarly, on burial even to very slight depths, calcite encounters different Eh and pH environments as well as solutions of different compositions from those of the immediate depositional environment. It therefore reacts chemically, recrystallizing in such a manner as to reestablish equilibrium with the new environment. During these processes, crystals will regrow to develop inter-locking boundaries and to reduce pore space so that the material becomes indurated. The microscopic characteristics of the recrystallization products are described by Folk (1965).* The presence of intraformational conglomer-ates in limestone and of obviously transported but essentially contem-poraneous limestone fragments in limestone matrixes indicates that such induration can develop while burial was still shallow enough so that wave action could erode and move these materials. The recrystallization of the limestone may continue for a long time after deposition, or it may not even be initiated until long after deposition. Chalk is an example of an essentially unindurated limestone. It has been suggested that its lack of induration is due to an absence of aragonite in the initial sediment to initiate recrystalliza-tion. Carozzi (1960)† reports that chalks are composed dominantly of calcite coccoliths plus secondary foraminifera and minor contributions from other organisms so that the sediment is originally very poor in aragonite.

Recrystallization of clay materials in sediments is also due to a lack of equilibrium between the deposited mineral grains and their new chemical environment. Most clay minerals developed initially in a subareal environ-ment in the presence of very dilute solution. In the marine environment the composition and concentration of the sea water are quite different, initiating changes in the mineralogy. Once deposited, there is again a change of chemical milieu with respect to Eh and pH, and as compaction sets in, there are changes in composition and concentration of the connate waters. As a result of all these changes the clay minerals are continuously subjected to recrystallization and modification of composition, and the very fine-grained nature of the materials increases their chemical activity. In the crystalliza-tion process, interlocking of grains results in lithification. Under high magnifications this is quite apparent, especially in argillaceous sandstones, such as graywackes, where clay flakes can commonly be seen penetrating sand grains.

Authigenesis is the process of development of mineral grains in a sediment by crystal growth in place. The resulting crystals are frequently quite euhedral and show no evidence of abrasion. The materials necessary for the authigenic growth are present in the sediment or in the associated solutions, and reactions between these materials to form equilibrium assem-blages produce the new materials. Many authors consider the quartz added to sand grains during secondary enlargement to be authigenic. The silica necessary for growth was originally present as colloidal material, as bio-genically precipitated opaline material, or possibly in solution. Growth

---

*R. L. Folk, Some aspects of recrystallization in ancient limestones, in: L. C. Pray and R. C. Murray (eds.), Dolomitization and limestone diagenesis, a symposium, *Soc. Econ. Paleontologists and Mineralogists Spec. Pub. 13*, pp. 14–48, 1965.
†A. V. Carozzi, "Microscopic Sedimentary Petrography," p. 261ff, Wiley, 1960.

takes place on the quartz-sand grains in place and is limited only by mutual interference with adjacent grains. Doubly terminated euhedra of quartz formed in this manner are common in some limestone. A similar process can develop authigenic overgrowths of calcite on single-crystal clastic fragments such as echinoid or crinoid plates. New minerals can also develop authigenically without a detrital nucleus to initiate growth. Perfectly euhedral delicate little crystals of the titanium oxides, particularly brookite, develop in this manner in some sandstone, and single crystals and crystal groups of pyrite and marcasite in many sediments originally rich in organic material are probably authigenic.

One important authigenic mineral encountered in sandstones and coarser clastic sandstones is glauconite. This mineral occurs as bright-green rounded or lobate granules of sand size, frequently having much the same shape as some foraminifera. Glauconite is quite rich in iron and potassium. The high iron content makes the sediments become readily iron stained, and the high potassium content has led to the use of glauconite concentrates in dating sediments by the potassium-argon method. Much has been written about the origin of glauconite, and the present consensus of opinion is that it is the result of authigenic reorganization of biotite or clay minerals, particularly illite, on the sea floor. The environment for glauconitization seems to be one of very slow accumulation of sand-sized sediments under reducing or very mildly oxidizing marine conditions. High organic activity seems to be essential, particularly an abundance of mud-ingesting animals. Glauconite is so easily recognized and frequently reaches such high concentration in sands that it is the most important accessory mineral in many stratigraphic-correlation problems.

Chemical Unmixing

Subsequent to deposition, sediments composed of several constituents of differing solubilities may be partially or completely unmixed by the action of solutions migrating through them. Concretionary structures in the rock are the common result of this unmixing. Three types of concretionary structures of this origin are most common: chert nodules or lenses in limestone, segregation of carbonate cement in clean sandstone into spherical balls or euhedral crystals (Fig. 3-44), and irregular or nodular calcareous concretions in shale (Fig. 3-45). In each of these the segregated materials were probably disseminated throughout the sediment and were then concentrated by the action of solutions which were initially trapped in the sediment or which subsequently migrated through the sediment. Sujkowski (1958)* presents a mechanism for the segregation of these materials which seems to fit the occurrence of concretionary bodies. According to Sujkowski, dispersed

*Z. L. Sujkowski, Diagenesis, *Am. Assoc. Petroleum Geologists Bull.*, vol. 42, pp. 2692–2717, 1958.

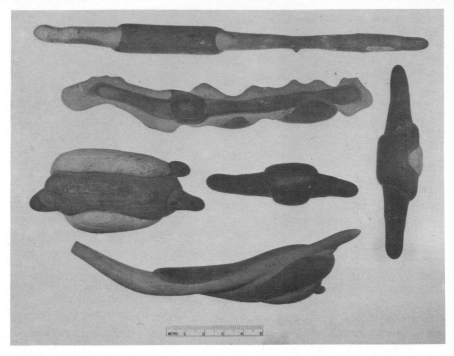

Fig. 3-45. Bizarre calcareous concretions developed in glacial lake-bed clays. West shore Waterbury Reservoir, Camels Hump quadrangle, Vt. *(Courtesy Dr. R. H. Konig.)*

materials are dissolved in the connate waters of the sediment, aided by pressure increase due to loading and to gas evolution from decaying organic material. Disseminated fine-grained carbonate and opaline organic debris, such as sponge spicules and radiolaria, are particularly susceptible to solution. At the same time, iron compounds and sulfates are reduced. During this stage the solution is not able to migrate and remains in place; however, the sediment attains some degree of coherence. At this point joints develop, allowing pressure reduction due to gas escape and allowing directional migration of solutions to low-pressure areas. As a result of pressure reduction the solubilities of materials change and probably the pH of solutions also changes so that the dissolved constituents are redeposited as segregated bodies. In the case of dissolved silica, the silica passes into a hydrous sol stage. In the case of carbonates in solution, the loss of carbon dioxide from solution causes direct precipitation. Subsequent uplift of the sediment or retreat of the sea allows the flushing out of connate solutions by fresh water and causes a marked change in pH so that the colloidal silica coagulates to a gel stage as distinct segregations. Dehydration and recrystallization produce chert nodules.

A similar segregation is frequently shown in ferruginous sediments as a result of subareal weathering. Here tubelike or hollow-box structures are developed by the oxidation of iron to limonite along joints or at the margins of clay-ironstone concretions. As leaching and oxidation take place, the iron migrates in solution to the surface of the body, or to fractures, and is precipitated as limonite, forming a hard cement for the surrounding sand. The original iron-cemented sand or concretion may be completely leached, leaving unconsolidated sand or clay inside the limonite-cemented structure (Fig. 3-46). Case-hardening of some ferruginous-cemented-sandstone outcrops may develop in the same manner.

Metasomatic Modification

Many sediments show evidence of having had their entire composition modified by the introduction and simultaneous removal of major constituents of the rock by migrating solutions. Four clear examples of this process are the dolomitization of many limestones, the silicification of limestones as exemplified by siliceous oolites, the development of oolitic and fossiliferous hematites, such as the Clinton iron ores, and the development

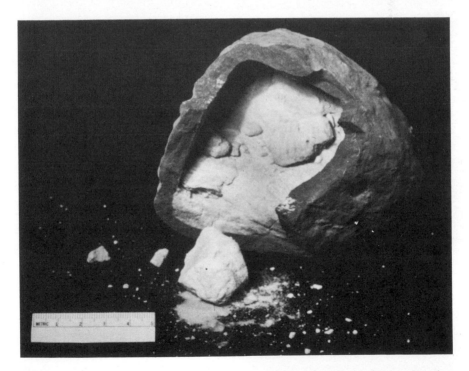

Fig. 3-46. Case-hardening by ferruginous cementation during weathering. Iron has migrated to the surface of the mass, leaving the interior sand uncemented. Ordovician, Beaver Dam Reservoir, Arkansas.

of some phosphorites from limestone. In each case the material of the original rock has been partially or completely removed and new minerals have developed, frequently by a volume for volume replacement as evidenced by the perfect preservation of the original structures and textures. The source of the replacing materials is not always clear. It may be a result of reaction of freshly deposited sediments with marine water adjacent to and penetrating the sediment. Metasomatism may also be caused later by percolating connate water from adjacent sediments or from groundwater. In some examples, metasomatism has been the result of ascending hydrothermal and juvenile water.

Modern development of dolomite was not found until 1961 and 1962. Prior to this, dolomite was not thought to be developing in the marine environment. The search for modern dolomite was prompted by the development of a sound theory of dolomitization by Adams and Rhodes (1960),* based on a concept of reflux action proposed by King (1947).† Since Adams and Rhodes proposed their mechanism, three areas have been discovered, in the Netherlands Antilles, the Persian Gulf, and the Bahamas (1965),‡ where dolomite is currently developing. In each of these areas dolomite is being formed from porous calcite and aragonite sediments a few inches above mean high-tide level in hot dry climates. Sea water is thrown up on the sediment banks by storms or abnormal high tides. There it evaporates to a great extent in the next few days to develop warm magnesium-rich brines. Since these brines are heavier than the solutions normally saturating the sediments, they move downward in the sediments, displacing the normal connate water and reacting to develop dolomite from the calcium carbonates. Replacement is very selective, following porous zones and often cutting across bedding. In none of these examples is the dolomite development very thick or very extensive really, but in each, carbon 14 dates of the sediments give ages of 3,300 years or less so that they are modern developments.

King's (1947) concept of a reflux action was developed to explain the low ratio of halite to anhydrite in the Permian basin of Texas and New Mexico. He envisioned cool normal sea water entering a basin over a barrier reef and being progressively evaporated as it flowed over the surface. Ultimately, the water became sufficiently heavy by increased salinity so that

---

*J. E. Adams and M. L. Rhodes, Dolomitization by seepage refluxion, *Am. Assoc. Petroleum Geologists Bull.*, vol. 44, pp. 1912–1920, 1960.
†R. H. King, Sedimentation in Permian Castile Sea, *Am. Assoc. Petroleum Geologists Bull.*, vol. 31, pp. 470–477, 1947.
‡See three papers in: Dolomitization and limestone diagenesis, a symposium, *Soc. Econ. Paleontologists and Mineralogists Spec. Pub. 13.* K. S. Deffeyes, F. J. Lucia, and P. K. Weyl, Dolomitization of Recent and Plio-Pleistocene sediments by marine evaporite waters on Bonaire, N. A.; L. V. Illing, A. J. Wells, and J. C. M. Taylor, Penecontemporary dolomite in the Persian Gulf; E. A. Shinn, R. M. Ginsburg, and R. M. Lloyd, Recent supratidal dolomite from Andros Island, Bahamas.

it sank to the bottom of the basin and flowed back seaward below wave base and flowed out either along channels through the barrier or through the barrier itself if it was sufficiently permeable. King called this return flow of heavy saline water the *reflux*, and it is similar to the deep current flowing out of the Mediterranean Sea, except that in the Permian it was far more dense, being sufficiently saturated to precipitate anhydrite and halite.

Adams and Rhodes (1960) added to this concept the infiltration of the heavy reflux brine into porous reef and bioclastic limestones and resulting dolomitization. In a normal open-shelf environment without a barrier, the reflux current would flow seaward unimpeded. No high salinity would build up, and a simple bottom current would develop down the shelf margin into the deep sea. As a result, regardless of evaporation, the bottom water would not become dense enough to replace the connate water of the sediments and no dolomitization would result. However, with a barrier present, the reflux current would be impeded to pond on the basin floor and build up salinity and density. Thus it would develop a high density to the point where it would displace connate water in porous bioclastic lime-stone and reef rocks and flow through permeable zones to return to the open sea as a very slow seepage. The very warm dense brine, high in magnesia and low in oxygen and carbon dioxide, would be a highly chemi-cally active solution in contact with calcium carbonates and would react to produce dolomite.

Whether this process would be effective in producing the extensive dolomite of Proterozoic and early Paleozoic ages, where there are no recognizable barriers, is debatable. It would probably be as applicable to the reef dolomite of Silurian time associated with the evaporites of Michigan and Ohio as it is with the Permian of west Texas. However, the author is inclined to believe that the dolomites of Proterozoic age, at least, are related to the extensive metasomatic granitization of mafic volcanics evidenced in the Precambrian shield. Magnesia released in this process was probably carried to the surface as hot springs and caused dolomite, rather than calcite, to be the stable primary carbonate sediment.

A final possible origin of dolomite is the selective leaching of calcite from biogenic high-magnesium calcite deposits. As shown in Table 3-6, many invertebrates and algae contain appreciable proportions of magnesium carbonate in their hard parts. In porous and permeable deposits, diagenesis could remove a portion of the more soluble calcium carbonate, enriching the sediment in magnesium carbonate. Simultaneous or subsequent recrystallization could produce dolomite. This selective solution process is probably not affected by the overlying sea water, since many old reefs in the present sea show no appreciable dolomitization.

Elsewhere, metasomatic dolomite has been developed as a result of hydrothermal modification. The dolomites associated with the Tri-State zinc district are an example. Here cherty limestone has been mineralized at low temperatures (115 to 135°C), and magnesia, silica, and sulfides have been introduced. The main zone of dolomitization is peripheral to the inner zones of silicification and sulfide deposition.

Metasomatic silicification of limestone and shale is also common. This silicification may, as in the case of the Tri-State district, be the result of hydrothermal activity, but it is also a common product of low-temperature sedimentary processes. The development of siliceous oolites is the clearest example. In these rocks the oolitic texture is clearly preserved by chert of varying textures and by zones of inclusion in the chert. Preservation of original textures is frequently so detailed that concentric and even radial structures of the original carbonate oolite are preserved (Fig. 3-47). All stages of replacement can frequently be seen in a series of selected samples. The origin of the silica for this metasomatic replacement is an open question. Probably most of it was derived from adjacent sediments and was introduced by connate-water or groundwater solution.

Iron oxides and calcium phosphates may be developed by metasomatic replacement of carbonate materials in some unusual marine environments. The preservation of carbonate invertebrate remains such as foraminifera bryozoa and brachiopods by hematite or limonite and the preservation of crinoidal debris by collophane clearly indicate replacement. Field relation-

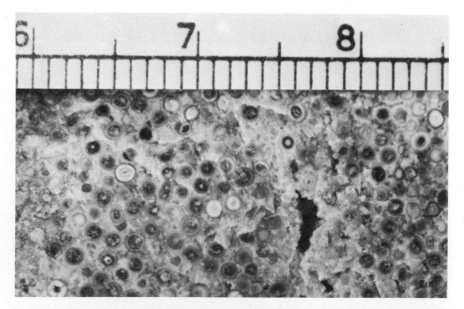

Fig. 3-47. Chert replacement of oolite. The chert has perfectly preserved the concentric structure of the original ooliths. State College, Pennsylvania. *(Ward's Universal Rock Collection.)*

ships of these materials indicate that replacement took place on the sea floor immediately after deposition and before burial. Evidently, when sea water is saturated with iron or phosphorus and deposition of normal sediments is slow enough, there may be a volume for volume replacement involving solution of carbonates, particularly when they are very fine-grained or aragonitic or highly porous from organic tubes as in the crinoidal material, and simultaneous deposition of iron oxides or calcium phosphates.

Metasomatic replacement may either completely destroy finer textures of the parent rock or preserve them in sharp detail. Dolomitization and silicification most commonly destroy original textures. Thus dolomites are usually fine-grained to crystalline rocks characterized by a marked paucity of fossils and other primary textures. Similarly, complete silicification usually destroys microtextures, resulting in structures and mottled rocks containing vestiges of organic structures. Metasomatic replacement is most clearly demonstrated in those quantitatively minor examples where there is perfect preservation.

4 METAMORPHIC PROCESSES

General Statement

Metamorphic rocks are those which have been modified in a solid state in response to change in the physical or chemical environment below the zone of weathering and cementation. The change may be mineralogical, textural, or both and is accomplished by the development of new minerals, the disappearance of preexisting minerals, or the recrystallization of pre-existing minerals. All processes, however, take place in a solid rock without the intervention of a silicate-melt phase. Solutions are probably often involved, but at no time do they make up more than a small fraction of the volume of the reacting mass.

During metamorphism the bulk composition of the rock may remain essentially unchanged, or it may be drastically changed. This allows a subdivision of metamorphism into two fundamental types: (1) isochemical, or closed-system, metamorphism which involves no change or little change in bulk composition; (2) allochemical, or open-system, metamorphism which involves marked changes in bulk composition and is frequently termed metasomatism. Strict isochemical metamorphism is probably a rarity, because in most rocks, some compositional change is inevitable as one mineral assemblage forms from another. For example, a change in volatile content is involved in developing anhydrous minerals from hydrous minerals or in the reaction of carbonates with other minerals to form silicates. Similarly, there is a common introduction of minor volatiles, such as fluorine or boron, where metamorphism is intimately associated with igneous activity. Thus the term *isochemical metamorphism* must be used somewhat loosely to allow for compositional changes involving loss of volatiles or minor gain of volatiles. Allochemical, or metasomatic, metamorphism will be limited here to those processes where there has been appreciable introduction or removal of the main rock-forming oxides or wholesale introduction of volatiles.

The major problem of the study of metamorphic rocks is twofold. (1) The character of the rock prior to metamorphism must be determined. To understand the historical and stratigraphic development of a metamorphosed area, the original character of the rock is of prime importance and can be deciphered only by a thorough understanding of the possible mineral assemblages for each typical bulk composition. In this deciphering, the mineralogy of the rock is the main indicator of parent-rock character. (2) The character and environment of metamorphism must be determined. This is particularly important in reconstructing the tectonic and intrusive history of a region. The mineralogy of the rock is important here, in that different mineral assemblages are stable in different pressure-temperature environments. However, the texture of the resulting metamorphic rock, including intergranular relationships and grain orientation, is of prime importance. In the following discussion the environment of metamorphism and its causes are considered. Following this, the processes of crystal growth in the solid environment of resultant textural relationship in the metamorphosed product is examined. Finally, the changing mineral assemblages representative of different important parent-rock groups are outlined.

METAMORPHIC ENVIRONMENT

Solid Environment

The most important environmental factor of metamorphism is that the changes take place in a solid. This means that all new crystals must make room for themselves as they grow by replacing preexisting minerals or by physically shoving aside the adjacent minerals. Texture and intergranular relationships are thus strongly controlled by the ability of different minerals to make room for themselves. The problems of nucleation, growth, and euhedralism of crystals are all influenced by the solid environment. In addition, the materials needed for growth of crystals must be transferred through solid rock or along intergranular boundaries to the site of the growing crystal. The solid character of the mass limits the radius through which these needed materials may be transported to the crystal, and thus it indirectly controls the preservation or destruction of primary textural and structural features of the rock. The driving energy for metamorphic processes comes dominantly from temperature and pressure changes.

Temperature Changes

Change in rock temperature can be brought about in a number of ways. The most obvious source of heat is magmatic intrusion adjacent to the area undergoing metamorphism. Crystallization of the magma body must involve loss of heat to adjacent rock masses. In the case of large intrusions, a great amount of thermal energy must be involved. However, the apparent effects of such heating of wall rock are remarkably limited in many examples. The zone of obvious thermal modification of wall rocks is frequently very narrow compared with the size of the magma body and the amount of thermal energy which must have been lost. This is due to two causes: (1) rocks are very poor conductors of heat (Jaeger, 1957),* and (2) much heat is lost to the surface through the escape of volatiles up fractures in the roof rock.

A second source of heat is the geothermal gradient. As a result of deep burial in geosynclines, sediments are brought into zones of elevated temperature, possibly several hundred degrees above surface temperatures. In the discussion of temperatures within the earth (Chap. 1), it was pointed out that the most likely temperature at a depth of 100 km is in the range 1100 to 1200°C. Extrapolating this value upward to the surface gives a reasonable value on the order of 500°C for the base of the continental crust. Temperatures in the base of geosynclinal piles of sediments may reach this value, or possibly higher ones, because of the heat-blanketing effects of the sediments themselves. Shales have considerably lower thermal conductivity across the bedding than do massive igneous rocks. Therefore, as heat is conducted toward the surface through the crust, it escapes upward through the sediments at a slower rate than it is being supplied from below through the crust. In addition, some types of sediments selectively remove radioactive elements from sea water, and abnormal radiogenic heating is added to the nonconducted heat accumulation to develop abnormally high temperatures at the base of the geosynclinal pile. This is possibly the most important heat source in regional metamorphism.

A minor source of heat is the frictional development during deformation of rock masses. Evaluation of such a source is difficult, although the available evidence indicates that it is small. Mylonitized zones along the soles of great thrust faults should be expected to show effects of frictional heating. In most examples, however, mylonites show little thermal effect developed at the time of mylonitization. Heating as a result of subsequent metamorphism or hydrothermal activity is common, but the lack of thermal effects associated with thrust faults does not mitigate against its importance in other environments. Thrust faults are commonly shallow-depth phenomena having direct access to the surface. Therefore heat may be lost as fast as it is generated by thermal conductivity of the rocks or by the circulation of water through the fractured materials. Greater amounts of heat may be generated by friction during the more plastic flow of material at the depth of normal metamorphism.

A final source of heat is from depths within the mantle through the unknown causes of orogeny. Convective overturn of the mantle would supply heat for metamorphism as well as for magma generation. The streaming of hot gases from depth, as visualized by Eardley (1957)† (Fig. 2-4), would also supply heat to the crustal rocks and geosynclinal sediments

*J. C. Jaeger, The temperature in the neighborhood of a cooling intrusive sheet, *Am. Jour. Sci.,* vol. 255, pp. 306–318, 1957.
†A. J. Eardley, The cause of mountain building–An enigma, *Am. Scientist*, vol. 95, pp. 189–217, 1957.

for metamorphism. As yet, there is no dominantly recognizable source of thermal energy for metamorphism wholly adequate to account for the great regionally metamorphosed terranes.

Hydrostatic Pressures

Hydrostatic pressure is that pressure exerted uniformly on a point from all directions because of the load of the overlying material. In dealing with rocks in the crust of the earth, two distinct types of hydrostatic pressure must be recognized: lithostatic pressure is load pressure developed as a result of the weight of the column of rock to the surface; fluid pressure is the pressure in the rock of the fluid present in connected channelways to the surface or to shallower depths within the crust. The pressure of the fluid is dependent on the height of the column of fluid and not on the column of rock. Fluid pressure in the rock may be quite different from lithostatic pressure at the same depth: at a depth of 10,000 ft, lithostatic pressure in rock of average density 2.6 would amount to roughly 9,600 psi, whereas fluid pressure in rock of average density 1.0 would be roughly 4,300 psi. Fluid pressure in rock is much more variable than this simple example would indicate, inasmuch as it depends on the nature of connected channelways in the rock. As a consequence fluid pressure may approach rock pressure in some circumstances (Hubbert and Rubey, 1959; Bredehoeft and Hanshaw, 1968).*

In many metamorphic reactions the pressure of water vapor or carbon dioxide is most important in controlling the direction of reaction. The partial pressure of either of these volatiles, or of any other gas phase involved in a reaction, will be equal to the fluid pressure only if there is one type of volatile present. This is not normally the case in metamorphism so that partial pressure of a volatile involved in a reaction will be the total fluid pressure times the mole fraction of the volatile involved in the reaction, and it will normally be less than the lithostatic pressure. Lithostatic pressure increases at a rate of about 250 to 300 bars/km so that the range in metamorphism is 1 to 10,000 bars.

Directional Pressure, or Shear

Because of the nonfluid nature of rocks, directed pressures may be developed in rocks during deformation. Such pressures are extremely important in controlling the fabric and mineral orientation of many metamorphic rocks. It is the directional pressure which is particularly effective in the orientation of mineral grains with marked growth anisotropy, as discussed on the next two pages.

*M. K. Hubbert and W. W. Rubey, Role of fluid pressures in mechanics of overthrust faulting, *Geol. Soc. America Bull.*, vol. 70, pp. 115–166, 1959; J. D. Bredehoeft and B. B. Hanshaw, On maintenance of anomalous fluid pressures: I. Thick sedimentary sequences, *Geol. Soc. America Bull.*, vol. 79, pp. 1097–1106, 1968.

The origin of directed pressures is obscure, but it is inevitable in the uplift of geosynclinal prisms. Kay (1951)* shows the idealized predeformational cross section of the Appalachian geosyncline from Maine to eastern New York; a miogeosynclinal basin east of Vermontia is approximately 100 miles wide and contains 25,000 ft of sediment. Using this as an example, and considering the curvature of the earth, the compressional effects of uplift can be seen. The equatorial circumference of the earth gives 69.1766... miles to the degree. Therefore $1\frac{1}{2}°$ would give an arc distance of 103.765 miles. Referring to Fig. 4-1, in which the $1\frac{1}{2}°$ angle has been greatly exagger-

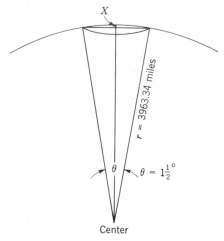

Fig. 4-1.

ated for easier visualization, it is seen that by piling on sediments, the circumference will be reduced to a chord. The depth of sediments necessary for this is

$$X = r - r\cos\frac{\theta}{2}$$
$$= 3963.34 - 3963.34 \times 0.999914$$
$$= 0.34 \text{ miles, or } 1,795 \text{ ft}$$

Therefore subsidence below this depth of sediment will require a lengthening of the crust under the sediments. Upon uplift of the geosyncline, this process will be reversed, and the basal sediments must be compressed to some degree. The crustal shortening involved can be calculated and in this model amounts to $\frac{1}{2}$ mile, involving 0.8 percent reduction in cross-sectional area during the uplift. Although a shortening of this magnitude could readily be accommodated by compaction, it would still give a directional component of pressure.

*M. Kay, North American geosynclines, *Geol. Soc. America Mem.* 48, p. 26, 1951.

The major cause of compression and directed pressure in the rocks, however, must be related to the deeper-seated causes of orogeny. Convective overturn of a portion of the mantle, thrusting in the mantle due to radial contraction of the earth, or asymmetric convective movement of fluids as visualized by Eardley (1957)* would all produce directed pressures in the crust.

The magnitude of directed pressures involved in metamorphism is difficult to estimate and undoubtedly varies at different stages of metamorphism. During lower grades of metamorphism, at low temperatures and pressures, the relaxation of stress by recrystallization flow and chemical mobility in the rock is slow and shear stresses may be high, on the order of 2,000 to 3,000 bars. At high grades of metamorphism, under high temperatures, recrystallization flow and chemical mobility in the rock may be rapid enough so that shear stresses are accommodated as fast as they develop, and the resulting directional pressure may be insignificant compared with hydrostatic pressure.

Metasomatic Agent

Some change of bulk composition of rocks undergoing metamorphism is normal, and there are probably very few instances in which it has not occurred. In a strict sense metamorphic rocks can be divided into two main groups on the basis of change or no change of composition: isochemical, or closed-system, metamorphism, and allochemical, or open-system, metamorphism. Strict closed-system metamorphism, in most cases, would allow little formation of new minerals, since in many mineral reactions there must be some gain or loss of material. Thus the progressive changes

$$\text{Chlorite} + \text{calcite} \rightarrow \text{hornblende} \rightarrow \text{pyroxene}$$

or

$$\text{Clay mineral} \rightarrow \text{muscovite} \rightarrow \text{potash feldspar} + Al_2SiO_5 \text{ mineral}$$

involve progressive dehydration and, in the first example, loss of CO_2. Similarly, the reaction

$$CaCO_2 + SiO_2 \rightleftharpoons CaSiO_3 + CO_2$$

involves loss of CO_2. Most such reactions are reversible, and if the evolved gas cannot escape, equilibrium will be reached quickly and the reaction will not proceed. Thus the concept of a closed system must be broadened to allow for bulk-compositional changes involving loss of volatiles evolved in metamorphic reactions.

Open-system metamorphism involves the introduction or removal of constituents or both from the rock so that the bulk composition is changed

*Eardley, *op. cit.*

appreciably with respect to one or more components. Such a process is commonly termed *metasomatism*. The agents of metasomatism most commonly appear to have been fluids or gases migrating through fractures or along intergranular boundaries or intragranular flaws and imperfections. Some metasomatic processes have been explained by migration of ions over long distances through rocks. Ion migration is certainly important in the adjustment of individual crystals or, perhaps, even small crystal groups to changing metamorphic environments, but it is probably not effective over large areas because of the very slow rate of such ionic diffusion. The nature of the fluids involved in the transfer of material in metamorphism can only be guessed at. Undoubtedly, the main components of the fluid phase are water and carbon dioxide, but because of high temperatures and pressures and because of the nature of the channelways they occupy, the water and carbon dioxide behave very differently from dilute solutions as found in surface water or weathering environments. Pressure and temperature through most of the range of metamorphism are above the critical point of water so that water must exist as a dense-gas phase. In addition, since channelways involved are mostly very small, the residual surface charge on crystals affects the fluid state. Because most rock-forming minerals are dominated volumetrically by oxygen, crystal surfaces have a net negative residual charge which adsorbs dipolar water molecules in a film of mono-molecular or polymolecular thickness. In the metamorphic environment, such a film would be in a constant state of formation and destruction so that its behavior would be very different from that of a macrobody of water. Diffusion of activated ions through this film could be very rapid and must involve simultaneous diffusion of different ions in opposite directions. In a metamorphic reaction such as

$$ABX + CY \rightarrow ACX + BY$$

ACX will form only if C diffuses toward the growing crystal and, simultaneously, B diffuses away. For example, in the reaction

Muscovite + Al-poor chlorite → biotite + Al-rich chlorite + quartz
+ water

there must be a simultaneous migration of aluminum from muscovite to chlorite and of iron and magnesium from chlorite to mica.

Several sources of the fluids involved can be recognized. In meta-sedimentary rocks, connate solutions deposited with the sediment or migrating from adjacent sediments undergoing metamorphism will fill all available channelways in the rock. Such solutions not only will supply the fluid phase but also will supply those ions in the original connate waters for reaction with the original minerals to cause change in the composition of solid phases. Albitization of mafic and intermediate volcanics deposited with geosynclinal sediments is often a result of just this type of process. Saline connate waters escaping from adjacent marine sediments contribute sodium to the volcanics, albitizing the original plagioclase (Waters, 1955).* Such a process would also require introduction of silica and simultaneous removal of lime, alumina, and, perhaps, some ferromagnesian materials. Hietanen (1967)† describes the development of scapolite in metamorphosed calcareous shales near the Idaho batholith and explains its high content of chlorine as being due to salines in the sediments.

A second source of fluids is progressive dehydration of the rock through the formation of less hydrous or anhydrous minerals from hydrous minerals. Evolution of water through such reactions is a progressive process throughout all grades of metamorphism. As long as any hydrous minerals remain, a rock undergoing metamorphism can never be completely dry. If the rock loses water, new reactions will proceed, releasing water until all hydrous minerals are destroyed Probably very few such reactions involve the loss of only water; other ions will also be liberated which may migrate into another body of rock and modify its composition. Dehydration may continue to the highest temperatures of metamorphism, and water so released may contribute to the partial fusion of the rock (Lundgren, 1966).‡

Dehydration of intrusive magmas not only is a source of fluids for metamorphism of adjacent rock but also is a source of many ions for alteration of bulk composition. Much of the water lost by magmas undoubtedly escapes through fractures and vents of various types, as evidenced by associated hydrothermal veins, volcanism, and explosion pipes, but there is abundant evidence of wall-rock metasomatism around many magma bodies. The effect of volatile escape clearly involves metasomatic introduction and removal of materials; this is evidenced by boron causing tourmalinization and fluorine developing muscovite and topaz in place of potash feldspar or resulting in the development of fluorite, fluorapatite, or members of the chondrodite-humite group of minerals in calcareous rocks. Fluids escaping from a parental granite magma have, in the opinion of some authors, caused extensive metasomatic granitization of roof rocks. This subject will be considered later in this chapter.

A final, and perhaps most important, source of metasomatic solutions is the depths of the lower crust or upper mantle. The same processes which apparently cause orogeny, large-scale plutonic intrusion, and widespread

---

*A. C. Waters, Volcanic rocks and the tectonic cycle, *Geol. Soc. America Spec. Paper 62*, pp. 703–722, 1955.

†A. Hietanen, Scapolite in the Belt series in the St. Joe-Clearwater region, Idaho, *Geol. Soc. America Spec. Paper 86*, 1967.

‡W. L. Lundgren, Muscovite reactions and partial melting in south-eastern Connecticut, *Jour. Petrology*, vol. 7, pp. 421–453, 1966.

metamorphism would allow the degasing of the lower crust and upper mantle. If these very hot gases were not involved in magmatic activity, they would have to escape upward by permeation through the rocks. In the process they would react with those rocks, developing mineral phases which were stable in the particular physical and chemical environment. The gases would probably be siliceous and alkaline so that the phases developed would be the minerals of the granite family. This process may be the important one in the origin of metasomatic granites.

Types of Metamorphism

Metamorphism is classified on the basis of the relative importance of the four metamorphosing agents just discussed. Metamorphic change in rock primarily as the result of temperature elevation is called *thermal*, or *contact*, metamorphism. This typically takes place immediately adjacent to intrusive igneous masses, particularly those emplaced at relatively shallow depths under a low hydrostatic pressure. Heat is derived from the adjacent magma body and frequently is accompanied by an accession of volatiles and trace elements from the magma. Hydrostatic pressure is normally low, seldom appearing to exceed 3,000 bars. Shearing pressure is normally unimportant, except as a local response to differential thermal expansion in rocks having a primary directed fabric or in local shear zones resulting from the upward magma movement. The contact thermal aureole is normally limited in width from a few feet to a few thousand feet and is often selective in the type of rock which is affected. Metamorphism and metasomatic alteration are often more extensive in calcareous rocks than in other rock types in the same contact zone.

Kinetic, or *dynamic*, metamorphism primarily involves shear stresses without appreciable elevation of temperature. Hydrostatic pressure is frequently low, since this type of metamorphism is most frequently a shallow-depth phenomenon. Shear stresses applied to rocks at greater depth usually involve elevated temperature, at least as a result of the geothermal gradient. The response of rocks to purely kinetic metamorphism depends on their rigidity or plasticity, and rocks can be classified as competent or incompetent, depending on their behavior in this environment.

The most important type of metamorphism involves elevated temperatures and high hydrostatic pressures, with or without shearing stresses. Because both temperature and pressure are important, this is frequently called *dynamothermal* metamorphism, and it is the common type of metamorphism of the great axial areas of deformed geosynclines; therefore it is more commonly called *regional* metamorphism. When shearing stresses are unimportant and pressure appears to have been entirely hydrostatic as a result of the overlying column of rock, the process is often called *load* metamorphism. Regionally metamorphosed terranes usually consist of a series of more or less distinct zones representing progressive mineralogical and textural changes as a response to increasing intensity of environment. Distinctive minerals or mineral assemblages which are apparently pressure or temperature sensitive or both have been recognized in certain rock types which make definitions of different intensity zones possible.

Rocks metamorphosed at high temperatures and pressures appear at the surface at low temperatures and pressures. It is obvious, then, that they were subject to a period of declining metamorphic intensity and that mineral or textural adjustment was able to take place during the interval. Such effects are called *retrograde* metamorphic effects. It is obvious that retrograde metamorphism is not as effective as progressive thermal or regional metamorphism; otherwise, high-grade metamorphic rocks would never persist to surface environments. The reason for this is at least twofold. First, the rate of chemical reactions is temperature-dependent, and during declining temperatures, the chemical-reaction rate is much lower than it is during rising temperatures. Second, most metamorphic reactions require the presence of an intergranular fluid film for the transfer of material, as discussed previously. During progressive metamorphism most fluids have been driven out of the rock so that water is not available for ion migration or for the formation of low-temperature hydrous minerals. Minor retrograde effects are common in many rocks but are frequently difficult to distinguish from chemical-weathering effects. Sericitization of feldspars, chloritization of ferromagnesian minerals, uralitization of pyroxene, and serpentinization of olivine and pyroxene are common incipient retrograde products.

The term *autometamorphism* is applied to the modification of igneous rocks in the solid state during the initial cooling phase. Serpentinization and steatitization (development of talc) are common autometamorphic effects in ultramafic igneous rocks. In these processes there is extensive change in bulk composition, involving introduction of much water and removal, or at least migration, of the divalent metals calcium, magnesium, and iron. Extensive veining by the carbonates is common. The autometamorphic processes grade into the normal deuteric alteration of the more important magmatic rocks. Thus incipient sericitization or chloritization could be considered either a normal deuteric alteration or a metamorphic process. Normally, if such processes do not involve any extensive change in mineralogy or bulk composition, they would be considered igneous processes rather than metamorphic processes.

The final major type of metamorphism involves marked compositional change; it is therefore called *metasomatism*. This may occur in any temperature and pressure range and with or without accompanying shear. Therefore a wide variety of environments can be involved. Metasomatic changes associated with contact metamorphism involve relatively low pressures and temperatures that decrease away from the contact, so a wide range of temperature may be present. This often results in a distinct zonation of the rock with respect to the character and amount of introduced material. Probably the most important metasomatic environment is that associated with regional metamorphism and involves development of rocks having a mineralogy and texture approaching that of the granite family. The environment is a deep-seated one, under high hydrostatic pressure and at elevated temperature. Shearing stresses are important in some cases. One of the major discussions in modern petrology is whether the associated rocks of true granitic texture are the end product of this type of metasomatism or whether they are the original cause and source of the granitizing material. Most petrologists agree that the transitional types between a true granitic texture and a true metamorphic texture are the result of metasomatic alteration.

METAMORPHIC MODIFICATIONS

GROWTH OF SINGLE CRYSTALS

Crystallization in the Solid Environment

Textural relationships between the minerals of metamorphic rocks are dependent on the processes of crystal growth in the solid environment. Several factors influence the place, the rate, and the orientation of growing crystals and their resultant size and euhedralism. In igneous rocks crystallizing from a magma, these features are interpreted as indicating the order in which the crystals formed from the magma. In metamorphic rocks no such interpretation is possible, since all minerals are, at least theoretically, crystallizing or recrystallizing simultaneously. The character of the grains depends on the way and the ease with which the different minerals make room for themselves in the solid, the effects of textural or structural features inherited from the parent rock, and the effects of the various pressures involved in metamorphism.

Three distinct processes can be recognized as enabling growing crystals make room for themselves. These have been defined by Ramberg (1952)* as secretionary growth, concretionary growth, and replacement. Most minerals probably grow by a combination of these processes rather than by any one individually, but each may play a dominant role in certain minerals or environments.

Secretionary Growth

Secretionary growth takes place in a void in a rock. Voids may be of primary origin, such as intergranular space in coarse clastic rocks, or may be developed by nonhydrostatic pressure in the rock. Brace, Paulding, and

*H. Ramberg, "The Origin of Metamorphic and Metasomatic Rocks," University of Chicago Press, 1952.

Scholz (1966)* find that during compression of granite under high confining pressures, open fractures develop as a result of directional compression which is considerably less than the crushing strength of the rock. A resulting volume increase takes place which may be as much as twice the purely elastic volume change. If the rock is fractured, an increase of volume inevitably results, and pressure differentials are set up. Fluids present in the rock migrate to such low-pressure areas and carry dissolved materials with them. If the solution was saturated with any constituent while dispersed through the rock, the reduction of pressure will cause supersaturation, and crystals will begin to form. They will continue to grow by simultaneous solution of their constituents in the rock and by ionic migration through the fluid to the growing crystal in the cavity (Gresens, 1966).† The process can be visualized (Fig. 4-2) by assuming that a quartz crystal

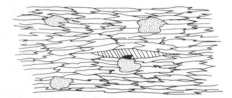

Fig. 4-2. A tension fracture has opened (cross hatched) adjacent to a quartz grain. Nearby quartz grains (stippled) are under lithostatic pressure. Quartz will migrate through fluids from the grains in the rock to the grain in the cavity to cause it to grow (solid).

under lithostatic pressure is in the rock and is in direct contact through a fluid film with a second quartz crystal growing in the cavity. The crystals in the cavity will be exposed to the pressure of the fluid, which is usually less than the lithostatic pressure. The stability of the two crystals depends on their relative free energy. Free energy increases with increased pressure, because pressure does work on a crystal which tends to break down the crystal structure. Therefore the crystal under lithostatic pressure is at a higher energy state than the one under only fluid pressure. Reactions or migration of materials always take place in such a direction as to minimize the free energy of the system, and thus the crystal in the cavity will grow at the expense of the crystal in the rock. Once the cavity is filled, pressure differences will no longer exist and the process will stop.

*W. F. Brace, B. W. Paulding, and C. Scholz, Dilatency in the fracture of crystalline rocks, *Jour. Geophys. Research,* vol. 71, pp. 3939–3953, 1966.
†R. L. Gresens, The effect of structurally produced pressure gradients on diffusion in rocks, *Jour. Geology,* vol. 74, pp. 307–321, 1966.

Secretionary growth in a rock often results in segregation of different minerals into different zones through a process called *metamorphic differentiation*. If a rock is homogeneous throughout and is subjected to shearing stresses such that it fractures, the rock may be progressively segregated into two contrasting mineral assemblages. Each time a fracture develops, certain constituents will migrate through the fluids to grow in the fracture, at least in part, by secretionary growth. The fluid phase must contain all the constituents of the rock but not all in the same proportions. Some materials will be relatively more soluble than others and will therefore tend to migrate preferentially to the fractures. The remaining material in the rock will thus be depleted in the most mobile constituents. After repeated fracturing, the rock may develop a banded structure, segregating differing mineral compositions into more-mobile and less-mobile phases (Fig. 4-3). This effect is seen even at relatively low grades of metamorphism, for instance, in the segregation of quartz lenses and bands in mica schists derived from originally uniform argillaceous rocks.

Concretionary Growth

Concretionary growth involves crystals making room for themselves by literally forcing aside adjacent crystals. In the discussion of the mechanical-weathering effects of growing crystals, it was pointed out that a crystal can exert a force up to its crushing strength on the adjacent rock. This process is effective in metamorphism, except that then there is no release of stress by rupture of the surrounding solid medium. Nucleation of a new crystal does not occur immediately at the temperature and pressure at which the new mineral becomes thermodynamically stable. Some degree of supersaturation of the milieu in the constituents of the newly stabilized minerals is required before nucleation. Supersaturation results in an excess of free energy in the system. As the crystal now begins to grow, its free energy will be less than that of its surroundings. Therefore the crystal will grow against the adjacent crystals and tend to shove them aside. This will result in an increased pressure on the crystal, thereby increasing its energy state. Crystal growth will stop when the free energy of the crystal is equal to the free energy of the surrounding fluids, which are supersaturated in the constituents of the growing crystal. Therefore the extent to which a crystal will grow by concretionary processes depends on the degree of supersaturation of the surroundings in the constituents required for nucleation of the growing crystal.

The process of concretionary growth is usually assisted by other factors. The materials for the growing crystal will, in most metamorphic environments, be derived from nearby grains of other, now unstable, minerals. Therefore the outward shove of the crystal is accommodated, at least in part, by the simultaneous destruction of the adjacent grains. Further, in

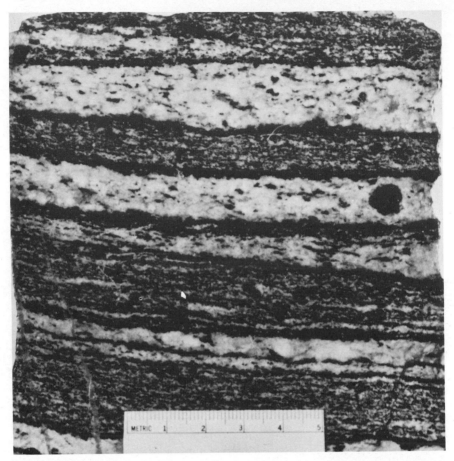

Fig. 4-3. Banded gneiss showing selective movement of the light-colored constituents. The concentration of the dark minerals along the margin of the dark bands is the result of depletion of those zones in quartz. The dark round grain is a garnet. Grenville, Georgian Bay area, Burwash, Ontario.

many metamorphic reactions the higher-temperature mineral assemblage, which will be the assemblage growing in progressive metamorphism, is denser than the lower-temperature assemblage, so there will be a volume reduction of the mass which will aid in making space for the growing crystals.

One important development of concretionary growth is augen structure (Fig. 4-4). Consider a crystal growing against a load pressure; if the surrounding rock is not too plastic and has some continuity perpendicular to the load pressure, the growing crystal will wedge the surrounding material apart, leaving low-pressure areas between the layers adjacent to the crystal. The growing crystal can now grow into these zones, or other minerals

Fig. 4-4. Augen structure. A large porphyroblast of potash feldspar has developed in a gneiss, wedging the rock open. Coarse crystalline quartz has occupied by secretionary growth the void so created. Grenville, Georgian Bay area, Parry Sound, Ontario.

may grow in these zones by secretionary growth. The result is the development of a lens-shaped body having its major dimensions perpendicular to the load pressure. Such bodies are called *augen*, in reference to their eye-shaped cross section. Augen structures are commonly developed around large garnets, where they have grown in a mica schist. The garnets may be either spherical or flattened ovoids, and the low-pressure areas in the corners of the eye are often filled with quartz. Apparently, quartz can grow more rapidly by secretion, because of its high mobility, than the garnet can grow in the same environment. Another common example of augen structure occurs in the metasomatic development of alkali feldspars in granitized transition zones and gneisses. Large blocky feldspar crystals form in augen gneisses, again with the corners of the eyes most commonly filled with quartz.

Replacement

The third process of crystal growth is replacement, in the sense that the volume occupied by the old mineral is now occupied by a new mineral. This is not a simple process even in the simplest types of reactions. For example, with increasing temperature at constant pressure, kyanite is replaced by sillimanite, since the two minerals have the same composition. However, they do not have the same density and unit volume. Kyanite has a unit-cell volume of 306.08 Å^3, whereas sillimanite has a unit-cell volume of 324.70 Å^3. Therefore, for a given volume of kyanite to be replaced by an equal volume of sillimanite, there must be a simultaneous diffusion of the constituents of the crystal away from the growing sillimanite.

In more complicated reactions there must be even more complicated relationships, involving migration of material both toward and away from the growing crystals. Since migration in both directions is required, this must be accomplished by diffusion of ions and not by flow of fluids. The driving force is free-energy gradients which operate in such a direction as to reduce free energy in the total system to a minimum. Probably a strict volume for volume replacement is relatively uncommon in metamorphism involving rising temperature and pressure, because too many other factors are involved in the environment. In retrograde metamorphism there is frequently evidence of such a replacement, as seen, for instance, in pseudo-morphs of chlorite after garnet, amphibole after pyroxene, muscovite after andalusite, etc.

Crystal growth, particularly where replacement is important, often results in inclusions of unincorporated minerals in the growing crystal. Where inclusions are numerous, this results in a sieve texture of the crystal called *poikiloblastic* texture (Fig. 4-5). If a large crystal is to grow in a fine-grained solid rock, the composition of the original rock volume which will be occupied by the crystal will, in almost every case, be different from the crystal. There will be excesses of certain constituents and deficiencies of others. The deficiencies will be supplied by ionic migration. The excess material may be removed by the same migration process, or it may be physically shoved aside by concretionary action of the growing crystal, or it may be included in the growing crystal as discrete mineral grains. It is this third possibility which gives rise to the poikiloblastic texture. The ability of the growing mineral to shove aside or incorporate inclusions varies with the nature of the growing crystal, the nature of the possible included grains, and the stage of development of the growing crystal. Andalusite in the early stage of growth is apparently incapable of shoving aside inclusions and becomes dark colored from minute included grains.

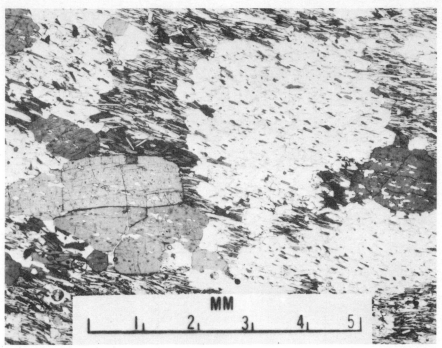

Fig. 4-5. Poikiloblastic porphyroblasts in schist. Tourmaline (large dark crystals) and microcline (large white crystals) both include abundant grains of quartz and micas from the groundmass. Note that the included crystals are smaller than the same minerals in the groundmass but have the same orientation. Note also the well-developed crystal shape of the tourmaline in contrast to the irregular microcline. Custer, South Dakota. Projection of a thin section. (*Ward's American Rock Collection.*)

As growth of the andalusite continues and the prism faces become well defined, it becomes capable of clearing aside possible included materials, and the characteristic chiastolite cross develops. Cordierite, however, contains inclusions throughout the crystal, and does not appear to be capable of clearing aside foreign mineral grains which cannot be accommodated by its composition. Other commonly poikiloblastic minerals are garnet, chloritoid, biotite, staurolite, and albite.

Crystalloblastic Series

The ability of a mineral to make room for itself varies quite widely. An empirical series of minerals of different strengths has been developed by observation of intergranular relationships. Such a series is called a *crystalloblastic series* (Becke, 1904).* It implies that, most commonly, a mineral high in the series can make room for itself at the expense of minerals lower

*F. Becke, Über Mineralbestand und Structur der kristallinischen Schiefer, *Internat. Geol. Cong. Rept.*, vol. 4, 1904.

in the series. Four crystalloblastic series have been set up, differing in character of metamorphic environment and bulk composition of the rock, as follows.

Thermal metamorphism	
Argillaceous and quartzo-feldspathic rocks	*Calcareous and mafic igneous rocks*
Garnet, tourmaline, sillimanite	Wollastonite, grossularite, apatite
Magnetite, andalusite	Magnetite, epidote
Micas, chlorites	Olivine, pyroxene, dolomite
Cordierite, quartz, feldspars	Scapolite, plagioclase, micas
	Tremolite, vesuvianite, calcite
	Quartz, potash feldspars

Regional metamorphism	
Argillaceous and quartzo-feldspathic rocks	*Calcareous and mafic igneous rocks*
Iron ores	Garnet, pyrite, tourmaline
Garnet, tourmaline	Epidote, olivine, pyroxene
Staurolite, kyanite, sillimanite	Hornblende
Micas, chlorites	Dolomite, albite
Quartz, feldspars	Micas, tremolite, chlorites, talc
	Calcite, quartz, potash feldspars

The crystalloblastic strength of a mineral relative to the adjacent grains is not necessarily related to either size or euhedralism but merely to its ability to make room for itself by concretionary growth. Quite commonly the minerals with high crystalloblastic strength are also relatively large and euhedral. Euhedralism is related to another little-known property of crystals, *form energy*. Size is dependent on many factors, including ease of nucleation in the particular environment, rate of nucleation and crystal growth, amount of the mineral allowed to form as a result of bulk composition (Jones and Galwey, 1964),* size and distribution of the minerals from which the new crystal will form, the surface tension between adjacent grains, and probably, other unidentified factors (Rast, 1968).†

DEVELOPMENT OF METAMORPHIC TEXTURES

Inherited Textures

Since metamorphic rocks are developed from preexisting rocks, many textural features are inherited by the metamorphic rock. However, such

*K. A. Jones and A. K. Galwey, A study of possible factors concerning garnet formation in rocks from Ardara, Co. Donegal, Ireland, *Geol Mag.*, vol. 101, pp. 73–93, 1964.
‡N. Rast, Nucleation and growth of metamorphic minerals, in: W. S. Pitcher and G. W. Flynn (eds.), "Controls of Metamorphism," pp. 73–102, Wiley, 1965.

textures are more or less rapidly lost because of the migration of materials through the rock and, particularly, because of shearing displacement during metamorphism. As new crystals begin to develop, they draw their constituents from a zone of parent materials around the point of growth. This zone may be conceived of ideally as a sphere of affect around the crystal. Its actual shape, however, will depend on the shape of the crystal and on any directional fabric in the rock or directional factor in the metamorphic environment. As the intensity of metamorphism increases, the sphere of affect of each growing crystal will become larger because of the greater mobility of materials, and inherited textures within the sphere of affect will be gradually obliterated.

This effect can best be seen by considering a thinly bedded sediment with incompatible compositions in adjacent beds. A good example is seen in the metamorphism of the bedded chert and siderite iron formations of the Precambrian. Starting with $\frac{1}{2}$-in. beds of alternating chert and siderite, metamorphic reactions require the formation of an iron silicate (Fig. 4-6a). At first the reaction will take place only at the bedding-plane contacts so that the rock will be composed of beds of chert and siderite separated by a thin zone of iron silicate mineral (Fig. 4-6b). For further reaction to take place, iron must migrate into the chert and silica must migrate into the siderite. Carbon dioxide must, of course, migrate out of the system for the reaction to go to completion. As the reaction proceeds, both chert and siderite are converted into iron silicate, and the thickness of the silicate zone depends on how far ions can migrate (Fig. 4-6c). When the sphere of migration has a radius of one-half the thickness of the original beds, there will be complete homogenization to form a single silicate mass, and both chert and siderite, and therefore the bedding, will be destroyed, assuming appropriate proportions of the two to begin with (Fig. 4-6d). Thus, inherited features will be destroyed when the sphere of effective migration has a radius of one-half the minimum dimension of the original texture.

Shearing aids greatly in the destruction of primary textures because of the physical movement of masses relative to each other and because of

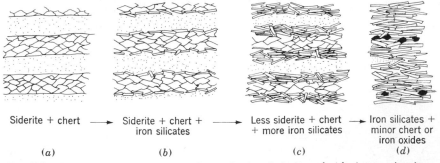

Siderite + chert ⟶ Siderite + chert + ⟶ Less siderite + chert ⟶ Iron silicates +
                    iron silicates   + more iron silicates   minor chert or
                                                                    iron oxides

(a)   (b)   (c)   (d)

Fig. 4-6. Idealized metamorphism of thin beds of chert and siderite to give iron silicate and ultimately destroy all evidence of bedding. Chert–stippled; siderite–crude rhombus pattern; iron silicate–elongated needle pattern.

greater chemical mobility of materials under shear. This is particularly effective when the direction of shearing rupture is at an angle to any directional fabric of the parent rock. This means that in regional metamorphism, where shear is an important part of the environment, inherited features will be destroyed more rapidly than in thermal metamorphism, where shear stresses are unimportant. Shear has two effects. First, it results in physical displacement of primary features which will cause granulation and distortion. The effect on bedding in a metasediment is often like that of sliding a deck of cards so that each card is offset from its neighbors. This results in an apparent change in bedding direction and an apparent thinning of beds, as in shear or slip folding. In the case of thinly bedded materials, it will quickly destroy much of the evidence of bedding. Shear has the second effect of causing fractures which result in pressure differentials and, consequently, secretionary growth of the more mobile constituents results. The migration of materials will further mask primary structures.

Sedimentary rocks contain many textural features which may be inherited by metamorphic rocks. Bedding is the most obvious characteristic and can often be recognized throughout the entire range of metamorphic intensity, particularly where beds are of contrasting composition (Fig. 4-7). The thickness of the bedding is the controlling factor, along with compositional differences between beds. Complete metamorphic recrystallization

Fig. 4-7. Preservation of bedding with metamorphism. The original sediment consisted of graded silt and clay units. The silt units have been metamorphosed to a fine-grained micaceous quartzite and the clay units to a staurolite-mica schist. The contact between two units shows clearly about 2 cm above the scale. Cobble from the north shore of Lake Huron.

can convert uniform shales to an essentially homogeneous product, but intercalated calcareous or siliceous layers will still be recognizable as units with textural or compositional differences. One major problem is the distinction between bedding and those compositional variations resulting from metamorphic processes. Secretionary growth along shearing directions results in compositional banding not parallel to bedding which may be confused with bedding.

Diagenetic concretionary structures in sediments are very commonly preserved in low to middle grades of metamorphism, but they are progressively destroyed so that only large structures will persist to high grades. Concretions are always of contrasting composition from their host rock so that there will be marked mineralogic differences after metamorphism. A concentric structure usually results which has a reaction product between the concretion and its host. Thus chert concretions in limestone, after metamorphism, consist of a central core of recrystallized quartz surrounded by a lime silicate zone embedded in marble (Fig. 4-8). If the concretion is small enough, it will be entirely converted to lime silicate and will appear as a more or less diffuse patch in marble. Other features of bedding, such as ripple marks, small-scale cross bedding, and flow scour and grooves, may be preserved. However, very similar appearing features may be developed by

Fig. 4-8. Diopside and tremolite developed between a chert concretion and the surrounding dolomite. The mineral sequence across the two specimens, from left to right, is; dolomite marble with scattered tremolite, a 1-cm band of tremolite, coarsely bladed diopside, and recrystallized massive chert. Randville dolomite near Randville, Michigan.

plastic deformation during metamorphism; therefore, caution must be used in interpreting such features as primary.

Fossils are rarely preserved in metamorphic rocks beyond the lowest grades, and when they are, they are deformed. In noncalcareous rocks the fossils originally present usually have a marked chemical contrast with the rock so that reactions take place, forming new minerals and diffusing the fossil material throughout a larger zone. All organic textures are usually rapidly destroyed. Fossils may be recognized in some metamorphosed calcareous rocks in spite of extensive recrystallization. This is particularly true of echinoidal and crinoidal material, where each plate is a single calcite crystal. The sutured contacts between crinoid stem plates are frequently preserved, and the radially distributed inclusions representing pore spaces in echinoid spines can occasionally be seen. The best environment for preservation of fossils in metamorphic rocks is in relatively thin limestone beds, lenses, or pods in shale. Here the plasticity of the shale during metamorphism accommodates the stresses and the limestone is relatively undeformed, although it will commonly be broken up into podlike bodies (Konig and Dennis, 1964).* In addition, the low-grade phyllite and schist formed from the shale appear to create a closed system in the limestone pods, preventing the escape of carbon dioxide and therefore preventing reactions which destroy fossil textures. Deformed fossils often are useful in working out stress relationships in metamorphic rocks.

Pebbles and conglomerates are commonly preserved during metamorphism, particularly where the pebbles are of durable rock types such as sandstone or granite. The major materials of such pebbles are stable throughout the range of metamorphic reactions (Fig. 4-9). Limestone or shale pebbles, in contrast, are relatively unstable and commonly have a matrix of similar composition so that they are rendered unrecognizable. Pebbles undergo plastic deformation, modifying their shape or orientation, and may be completely recrystallized, but their boundaries frequently remain sharp (Walton, Hills, and Hansen, 1964).† The shape and orientation of the pebbles can often be used to determine direction and magnitude of stresses by applying the principles of the strain ellipsoid. Caution must be exercised, however, because of the original lenticular shape of some pebbles and the parallel imbricate structure of some conglomerates.

Metaigneous rocks frequently preserve details of the original igneous texture. Amygdules and phenocrysts are often preserved as patches of minerals of contrasting composition from the groundmass. Amygdules

---

*R. H. Konig and J. G. Dennis, The geology of the Hardwick Area, Vermont, *Vermont Geol. Survey Bull.*, No. 24, 1964.
†M. Walton, A. Hills, and E. Hansen, Compositionally zoned granitic pebbles in three metamorphosed conglomerates, *Am. Jour. Sci.*, vol. 262, pp. 1–25, 1964.

Fig. 4-9. Metamorphosed conglomerate. The granite and vein-quartz pebbles are readily recognizable; however, the chlorite-schist pebbles are nearly unrecognizable in the groundmass. Huronian basal conglomerate, Negaunee, Michigan.

often appear in low-grade metamorphics as highly lenticular bodies of chlorite or carbonate and the shape can often be used to solve structural deformation problems. When thorough recrystallization has taken place or there has been extensive shearing, such small-scale igneous textures are lost. Igneous bodies such as dikes, sills, and flows are usually recognizable.

Metamorphism develops characteristic textures and structures which will be described subsequently. Polymetamorphism, or modification by two different metamorphic processes or epochs, will, as a consequence, show some evidence of both phases. These can often be recognized in the resulting rock as two distinct metamorphic fabrics superimposed on each other, particularly where the second phase of metamorphism is of a lower intensity or does not involve as thorough a recrystallization as the earlier phase. Mechanically crumpled schistose rocks may indicate two epochs of deformation. Similarly, regionally metamorphosed rocks may be subsequently affected by thermal metamorphism. Frequently, however, where both epochs have involved thorough recrystallization, they can be identified only by detailed microscopic study of grain orientation as involved in petrofabric analyses.

New Metamorphic Textures

Several factors control the texture of the metamorphic rocks as new minerals develop. The two most important are rock composition, including the resulting new minerals which must grow, and character of the pressure environment, which influences grain orientation. On the basis of composition, rocks may be classified into two main categories: monomineralic and polymineralic. That is, rocks whose total composition can be accommodated essentially by one mineral, and rocks whose composition is such that several minerals are required to accommodate all the rock constituents. Mono-mineralic rocks, such as pure quartzite or calcite marble, must have a relatively simple textural character, inasmuch as all grains have equal crystalloblastic strength and, therefore, an equal ability to make room for themselves at the expense of a neighbor. Under such a situation, solution of grains takes place at points of contact between the grains according to Riecke's principle, and material is redeposited in the low-pressure intergranular interstices until all voids are filled and all grains are in contact throughout their borders. This process usually results in the development of a uniform grain size throughout the rock, and all grains have interlocking or sutured contacts. Smaller grains are dissolved and added to larger grains, because the small grains have a greater surface area per unit of volume than do larger grains. The resulting texture is called a *mosaic* texture (Fig. 4-10).

Fig. 4-10. Mosaic texture in marble. Uniform grain size with highly irregular grains. The smaller grains are extensions of larger grains out of the section. Projection of a thin section.

In polymineralic rocks, however, the different minerals have differing crystalloblastic strength — differing ability to make room for themselves at the expense of others in the solid environment. Therefore the different minerals in the rock will be present in contrasting grain sizes or contrasting degrees of euhedralism, developing a *crystalloblastic* texture (Figs. 4-11

Fig. 4-11. Crystalloblastic texture. Hexagonal prismatic porphyroblasts of corundum in a syenite gneiss. Note the range of size and euhedralism of all minerals. Dillon, Montana. (*Ward's Universal Rock Collection.*)

and 4-12). If all the minerals have about the same crystalloblastic strength, then the texture may be similar to the mosaic of a monomineralic rock, particularly if all the minerals approach an equidimensional habit. However, if the minerals show a wide range of strength, or have tabular or prismatic habits, then the texture of the metamorphite will be strongly controlled by the nature of the pressure involved in the system. This is in large part a result of a factor known as *growth anisotropy* in the crystal (Ramberg, 1952).*

Growth anisotropy is the preferential growth of crystals in certain directions. If every crystal grew equally rapidly in all directions, then every crystal would be a sphere. Spherical shape is approached in certain crystals, such as the combined dodecahedron and trapezohedron of the garnets or the rounded hexoctahedral form of diamonds, but acicular, bladed, and tabular crystals depart widely from a spherical shape. This latter group of crystals, therefore, obviously does not grow with equal ease or equal rates in all directions but, instead, shows a marked preferred growth direction, or growth anisotropy. In tabular or flaky minerals, such as the micas, it is easier to add new ions to the crystal structure by the lateral extension of

*Ramberg, op. cit., p. 115.

Fig. 4-12. Crystalloblastic texture. Large staurolite crystals in a biotite-quartz groundmass.

already existing sheet structures than it is to initiate new sheet structures. Similarly, in minerals with chain structures, such as the amphiboles, it is easier for the crystal to grow by addition of ions to already existing chains than it is to initiate new chains. Therefore amphiboles grow most rapidly along the c crystallographic direction and usually appear as acicular or bladed crystals with well-developed large prism zones. We can thus say that a crystal is dominantly bounded by its slowest-growing faces, and the greater the tendency a mineral has to diverge from a spherical shape, the greater is its growth anisotropy.

Growth anisotropy has an important corollary in solution anisotropy. That is, if a crystal is put into an environment in which it will dissolve, it does not dissolve at the same rate in all directions, but the direction which shows the most rapid growth also shows the most rapid solution. This can be shown nicely by dissolving a quartz sphere in hydrofluoric acid. The sphere does not simply become a progressively smaller and smaller sphere; solution is most rapid in the c crystallographic direction, somewhat less rapid at the negative ends of the a crystallographic axes, and almost negligible at the positive ends of the a axes. [Penfield performed this experiment, and the resultant form is shown in Figs. 529 and 530 by Dana (1932).*] This agrees with the typical prismatic growth habit of quartz. Growth and

*E. S. Dana, "A Textbook of Mineralogy," 4th ed., revised by W. E. Ford, Wiley, 1932.

solution rates are sensitive to pressure, because pressure controls free-energy states of both the crystal and the surrounding fluid.

In a polymineralic rock under simple hydrostatic pressure, ideally all crystals of each mineral species are equally free to grow, regardless of their orientation. This results in a completely random orientation of all crystals throughout the rock, and the resulting texture is called *decussate* (Fig. 4-13).

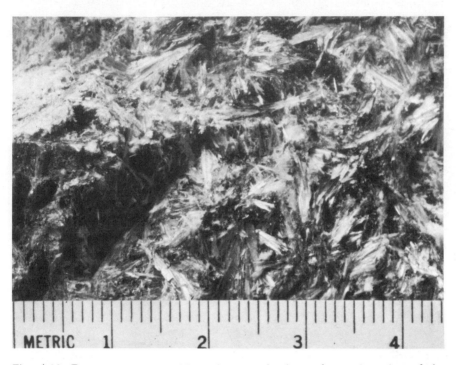

Fig. 4-13. Decussate texture. Note the completely random orientation of the fibrous crystals. Anthophyllite schist, Cashiers, South Carolina. (*Ward's Universal Rock Collection.*)

For example, if tremolite is to grow in a marble under purely hydrostatic pressure environment, the individual tremolite crystals can assume any orientation with the rock. Decussate textures are therefore characteristic of many thermally metamorphosed rocks.

In contrast, however, where a directional pressure is involved, such as in regional or load metamorphism, an entirely different situation exists. Because the pressure is directed, faces of crystals which are perpendicular to the stress will be under higher pressure than other faces which are parallel to the stress direction. Therefore, because increased pressure increases the free energy of the crystal face, there will be a free-energy difference between the stressed and unstressed faces. If the surrounding fluids are under hydrostatic pressure, then conditions may be such that the stressed crystal faces dissolve, whereas the nonstressed crystal faces grow. In this environment consider two crystals of one mineral having marked growth and solution anisotropy; one crystal A has its maximum growth or solution rate perpendicular to the applied stress, whereas the other crystal B has its minimum growth or solution rate perpendicular to the applied stress, as in Fig. 4-14. Both crystals will dissolve on the stressed face $A1$ and $B2$, but face $A1$ will dissolve less rapidly than $B2$ because of solution

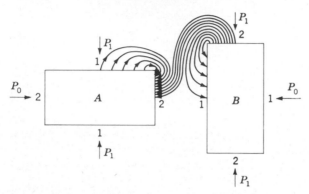

Fig. 4-14. Growth anisotropy. Two crystals A and B exposed to a directed pressure. The arrows show that net transfer of matter extends from B to A. (*After H. Ramberg, The Origin of Metamorphic and Metasomatic Rocks, The University of Chicago Press, 1952.*)

anisotropy. Similarly, both crystals will grow on the nonstressed faces $A2$ and $B1$, but $B1$ will grow less rapidly than $A2$ because of growth anisotropy. Therefore crystal B will dissolve more rapidly than it grows and will disappear at the expense of crystal A, which is growing more rapidly than it is dissolving. In such a situation, preferred orientation of crystals having strong growth anisotropy must result such that their direction of maximum growth rate is oriented perpendicularly to the applied stress.

The majority of directed fabrics in metamorphites have been developed as a result of directed pressure; however, there are other, less important, origins for parallelism of minerals. A directional fabric may be inherited from the parent rock and thus be independent of pressure environment. Many argillaceous rocks have preferred orientation through parallelism of clay or micaceous flakes or through thin lamellar bedding. In such rocks new micaceous minerals may grow in parallel orientation as a replacement of or pseudomorph after the clay orientation (Fig. 4-15). Many thermally metamorphosed argillaceous rocks show schistosity from an environment which should give a decussate texture. Similarly, a preferred crystal orientation may develop as a result of reaction between two incompatible units during metamorphism.

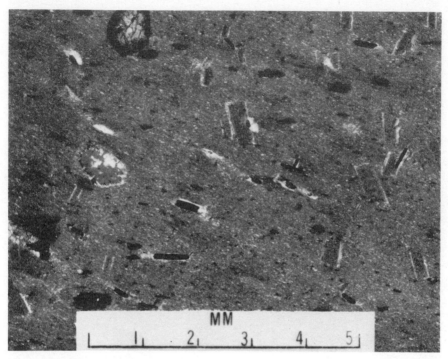

Fig. 4-15. Schistosity inherited from the parent sediment. The phyllitic groundmass shows a strong planar element across which the porphyroblasts cut. Note the narrow bleached zone around each biotite porphyroblast. Waits River formation, Mackville, Plainfield quadrangle, Vt. Projection of a thin section. (*Courtesy Dr. R. H. Konig.*)

Kinetic-metamorphic Textures

Kinetic metamorphism is the response of rocks to shear stress without appreciable elevation of temperature and, consequently, with little recrystallization. Rocks can be classified into two groups according to their behavior in this environment. Some rocks behave as competent masses; that is, they resist shear up to stresses sufficient to cause rupture. Other rocks behave as incompetent masses and undergo flow and realinement of material. The behavior of larger units depends on the proportion of competent and incompetent material.

Competent rocks fail by rupture, resulting in fragmentation. In the kinetic-metamorphic environment, two distinct phases of rupturing can be recognized which are gradational into each other. They are brecciation and mylonitization. The first stage is the development of widely spaced fractures with little or no movement of the fragments relative to each other. As deformation proceeds, the fragments are further reduced in size and are moved relative to each other. The movement results in grinding, and the rock becomes a breccia—a mixture of rock fragments embedded in a matrix of crushed-rock material. Most often the uncrushed blocks are highly angular but are of such shapes that they could not be fitted back together. As movement and grinding continue, the fragments may become rounded and develop into a pseudoconglomerate. As crushing continues with movement, the fragments may be completely reduced to fine-grained material and ultimately, almost to a powder. This product is called a *mylonite*. Such zones of rock powder are often encountered in fault zones, which are areas of dominant kinetic processes. Mylonites themselves are soft and pulverulent, but because of their high porosity and permeability, they are frequently cemented or recrystallized into hard flinty rocks by the action of circulating water.

Incompetent rocks are those which respond to stress by flowing plastically (Fig. 4-16) or by developing an oriented fracture pattern. Slaty

Fig. 4-16. Plastic flowage in incompetent rocks. Disharmonic folding in a limestone-siltstone sequence of beds. Bruce Limestone, Bruce Mines, Ontario.

cleavage in the argillaceous rocks is the most familiar product. However, the origin and nature of slaty cleavage is still highly uncertain. It appears as more or less closely spaced fractures oriented nearly perpendicular to the applied stress. It may develop initially as shear fractures at an angle to the applied stress and, subsequently, undergo rotation into the perpendicular orientation by slippage on the fractures, as in a twisted deck of cards. Conversely, it may develop in its final position perpendicular to stress, in which case its origin is even more obscure. Rotation of flaky minerals, deformation of equidimensional grains into lenticular shapes, and recrystallization of flaky minerals into new orientations are all possible processes involved in the development of slaty cleaveage. Each of these is normally on a microscopic to submicroscopic scale and not identifiable megascopically.

Mixture of competent and incompetent rocks under shear results in the development of boudinage and similar structures. The competent beds first fracture, and then the incompetent material is squeezed into the fracture (Fig. 4-17). As a result the resistant beds are pulled apart into a series of disconnected blocks completely surrounded by the plastic material. As deformation continues, the strain is taken up by flowage, and the blocks are moved apart without further appreciable rupture. The behavior of any one rock type varies widely, depending on the nature of the forces involved and the types of associated rocks. Generally, sandstone and limestone behave as competent material when associated with argillaceous rocks. Massive igneous rocks, such as dikes, behave as competent materials when associated with shale or limestone.

METAMORPHIC PRODUCTS

Zone and Facies Concepts

The bulk composition of the parent rocks and the environment and metamorphism are the two controlling factors in the mineralogy of metamorphosed rocks. Two important approaches have been used extensively to study these relationships. The earlier of these in modern metamorphic petrology is the zone concept, and the more recent is the facies concept. Each has its own place in the study of metamorphic rocks. The zone concept attempts to subdivide metamorphic products on the basis of the appearance of certain critical minerals in the rocks as a result of increasing intensity of metamorphic environment. For this approach it is necessary to study through the changing mineralogy of each fundamental compositional rock type as function of the evident increase of temperature or pressure or both. In following a rock type, such as the argillaceous rocks, from the margin into the center of a regionally metamorphosed geosynclinal prism, certain apparently temperature-pressure-sensitive minerals appear in the rocks in order as the intensity of metamorphism increases. In the argillaceous

Fig. 4-17. Development of boudinage structure. A 1-ft-thick bed of quartzite in phyllite develops a boudin by plastic flow of the phyllite into fractures in the quartzite. Devils Lake State Park, Baraboo, Wisconsin.

rocks these minerals are chlorite, biotite, almandite, staurolite, kyanite, and sillimanite. By assuming, then, that all metamorphites derived from argillaceous rocks which contain almandite but not staurolite, kyanite, or sillimanite have been metamorphosed to the same general range of temperatures and pressures, the field geologist can map a zone representing one range of metamorphic intensity. Similarly, rocks containing biotite but not almandite may be mapped as another, lower-intensity, zone of metamorphism. A map can thus be developed which shows lines of equal intensity of metamorphism known as *isograds*.

The difficulties of the zone concept are numerous (Atherton, 1965).[*] First, there is no assurance that the appearance of the critical mineral is indicative of a specific pressure-temperature range. Garnet, for instance, may become stabilized at a specific pressure-temperature environment at one point, while elsewhere at a higher temperature it could be stabilized at a lower pressure. Thus the garnet isograd merely indicates the line along which garnet first appears and should not imply the same temperature and pressure along that line. A second difficulty encountered is the effects of composition of the minerals. Extensive substitution of one ion for another is common in most metamorphic minerals, and this substitution affects the stability of the minerals at different pressure and temperature. For instance, the manganese garnet spessartite becomes stable at a lower temperature than does the iron garnet almandite. Therefore in manganiferous sediments, garnet may appear before biotite and at a lower intensity of metamorphism than almandite in normal argillaceous sediments. A third difficulty is that rocks of differing bulk compositions respond to metamorphism with markedly different mineralogies and reaction rates. Under an open system with respect to volatile escape, impure limestone reacts more rapidly than the argillaceous rocks so that quite often limestones are completely recrystallized to lime silicate-bearing marbles, whereas associated argillaceous rocks still show the effects of only low-grade reactions. As a result mapping of isograds must be based on the response of one rock type to metamorphism throughout the area. Inasmuch as argillaceous rocks are the most abundant sediments in the geosynclinal areas and do respond to metamorphism with several readily recognized index minerals, isograd mapping is usually based on the argillaceous rocks.

The facies-concept attempts to identify all the equilibrium-mineral assemblages possible in rocks of widely different bulk composition in each pressure-temperature environment. There are, of course, an infinite number of pressure and temperature combinations possible in the metamorphic environment, but these are grouped into ranges over which certain critical

[*]M. P. Atherton, The Chemical significance of isograds, in: Pitcher and Flynn, *op. cit.*, pp. 169–202.

mineral assemblages are stable. All the possible stable-mineral assemblages which can exist through a given range of environmental conditions belong to one facies. The facies concept is much more specific in the identification of environmental factors than the zone attempts to be. The specific values of pressure and temperature have been only partially identified as yet, but ranges of values are fairly well known (Hietanen, 1967).* The method of facies study involves the assumption that in a small area, all the rocks have been subjected to nearly the same pressure and temperature conditions. All the mineral assemblages in rocks of all bulk compositions in the area thus represent local equilibrium assemblages of one facies.

The facies concept is of much more detailed applicability than the zone concept, inasmuch as it may be applied to all rocks regardless of bulk composition. Naturally, certain bulk compositions are more useful in establishing the facies of metamorphism of a region than others because of the greater variation in mineralogy of their metamorphic equivalents. The mafic igneous rocks and the argillaceous sediments are the most diagnostic for the common range of metamorphism. A further advantage of the facies method of study is that small variations of composition of the parent rock can be recognized, especially with respect to the potash content of shale. Two difficulties are inherent in this approach as far as megascopic studies are concerned: first, it is frequently necessary to determine the composition of certain minerals to place the rocks in the proper facies, and second, the usual method for presentation of the data of facies obscures the progressive mineralogical changes in any one bulk composition of parent rock. The composition of plagioclase present in the rock is often the main clue to the facies of metamorphism. Inasmuch as plagioclase in metamorphic rocks is frequently untwinned, it is difficult to recognize megascopically and, of course, difficult to estimate the composition of megascopically. In addition, other critical minerals often occur as very small grains which can be easily masked in the rock by the more-abundant and larger grains. Epidote is a very important mineral in facies classification but is difficult to recognize megascopically because of small grain size and masking by associated micaceous minerals. Facies data are usually given with the aid of one or more schematic three-component diagrams for each facies and subfacies. The most-used diagrams are the ACF and AKF triangles in which the important minerals are plotted at their appropriate position. A represents alumina and ferric iron in excess of that required for alkali feldspars, C represents lime in excess of that required by the carbon dioxide content for calcite, F represents ferrous iron, magnesia, and manganese oxides. Separate ACF diagrams thus must be developed for rocks having excesses or deficiencies

*A. Hietanen, On the facies series in various types of metamorphism, *Jour. Geology*, vol. 75, pp. 187–214, 1967.

of silica, potash, or both. AKF diagrams are generally used for rocks having a deficiency of lime to show the mineralogical effects of variable potash contents in the clay minerals or in alkali feldspar. Fyfe, Turner, and Verhoogen (1958)* recognized 10 facies, three of which are subdivided into subfacies (Fig. 4-18), and Hietanen (1967)† recognized 13 facies, five of which are subdivided into subfacies (Fig. 4-19). This makes it difficult to trace through the changing mineralogy for any one rock type. Therefore, in the following discussion, diagrams have been developed from the facies diagrams and from the classical work of Harker (1932)‡ which attempt to trace the changing mineralogy of each of five major rock types. The five major compositional types to be studied are:

*W. S. Fyfe, F. J. Turner, and J. Verhoogen, Metamorphic reactions and metamorphic facies, *Geol. Soc. America Mem. 73*, 1958.
†Hietanen, *op. cit.*
‡A. Harker, "Metamorphism," 2d ed., Methuen and Co., 1932.

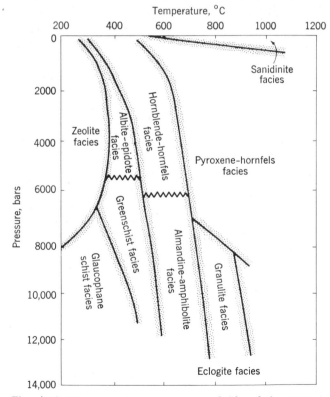

Fig. 4-18. The pressure-temperature fields of the metamorphic facies according to W. S. Fyfe, F. J. Turner, and J. Verhoogen. (*Geol. Soc. America Mem. 73, 1958.*)

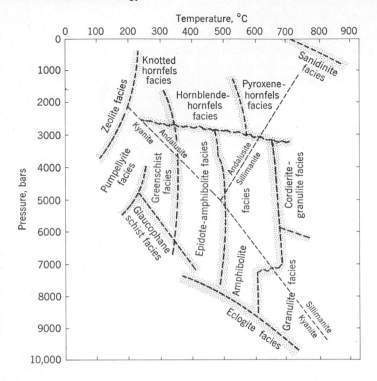

Fig. 4-19. The pressure-temperature fields of the metamorphic facies according to A. Hietanen. (*Jour. Geology, vol. 75, 1967.*)

1. Argillaceous rocks
   a. Potash-rich clays
   b. Potash-poor clays
2. Calcareous rocks
   a. Limestones
   b. Dolomites
3. Mafic igneous rocks
4. Quartzo-feldspathic rocks, including arkosic sands and granite-family rocks
5. Highly magnesian rocks — dominantly the ultramafic igneous rocks

Quartz sandstone and graywacke are not treated as a separate unit, because their mineralogical assemblage depends on the character of the interstitial material and cement. If there is appreciable interstitial clay, as in graywacke, metamorphism is like that of argillaceous rocks having excess silica. If the cement is calcareous, sandstones behave as calcareous rocks having excess silica. The less-common rock types are not treated here.

METAMORPHISM OF ARGILLACEOUS ROCKS

Regional Metamorphism of Argillaceous Rocks

For the present purposes the argillaceous rocks may be subdivided into two major groups: (1) the more normal shales, in which clay minerals are dominantly potash-bearing of the illite group of clays, and (2) the potash-poor shales, in which there is an appreciable lateritic contribution of kaolinitic clays, bauxite, and iron oxides. In addition to one or the other of these constituents, the mineralogy of shales may be generalized by variable amounts of chloritic (ferromagnesian) clays; detrital and colloidal silica, minor calcite as cement, fossils, etc.; and detrital accessory minerals, such as rutile needles, tourmaline, zircon, etc. Soda is usually present in shales as absorbed ions, detrital albite, or connate solutions, but inasmuch as the only important metamorphic soda-bearing mineral in these rocks is albite, soda is generalized in the charts as detrital albite, regardless of its actual state in the sediment. The mineralogy of shales is thus summarized in the left column of Table 4-1. In working progressively to the right across the chart, higher-grade-metamorphic assemblages of these rocks are indicated. Once a mineral appears in the rocks, it is carried across through higher facies without repeated naming. Thus only the new minerals, as they appear, are entered in the chart.

The first noticeable change in the normal potash-bearing shales is the recrystallization of illitic and chloritic clays to give recognizable flakes of muscovite and chlorite. At the same time calcite reacts with the clay minerals to form epidote or zoisite, depending on the availability of iron. Quartz, albite, and accessory minerals recrystallize at the same time, and all evidence of clastic texture is rapidly lost. The next step is the reaction of muscovite and chlorite to form biotite. The reaction here is a progressive one so that all three minerals may coexist. This is because of the variability of composition of both chlorite and biotite with respect to the ferrous iron to magnesia ratio and the alumina content. Probably, the ferrous iron to magnesia ratio in the chlorite decreases as an iron-rich biotite forms. Chlorite, persisting through the last state, becomes unstable next, reacting to form almandite garnet. Again, this is not an instantaneous type of reaction but a progressive one and involves most of the minerals of the rock. The first garnets to form have a higher ferrous iron to magnesia ratio than the chlorite from which they form. Therefore magnesia will be liberated, requiring a change in the ferrous iron to magnesia ratio of biotite. Any change in amount of biotite involves change in potash content, which, in turn, affects muscovite. In addition silica is required to convert chlorite into garnet. Thus with each change in mineral assemblages, all the mineral phases present are involved to some degree.

At higher grades of metamorphism, new reactions take place involving the feldspars. First epidote becomes unstable and reacts, picking up alumina and silica to form the anorthite molecule. Alumina in these

Table 4-1. Regional Metamorphism of the Potash-bearing Argillaceous Rocks

Zone-concept terms	Chlorite zone	Biotite zone	Garnet zone	Staurolite-kyanite zones		Sillimanite zone	
		Greenschist facies		Almandite-amphibolite facies			
Facies-concept terms (Fyfe, Turner, and Verhoogen)	Muscovite-chlorite subfacies	Muscovite-biotite subfacies	Almandite-epidote subfacies	Staurolite-almandite subfacies	Kyanite-almandite subfacies	Sillimanite-almandite-muscovite subfacies	Sillimanite-almandite-orthoclase subfacies

Mineral stability ranges:

- Quartz →
- Potash-bearing clays → Muscovite → Muscovite → ... → Quartz
- (Detrital K feldspars) → (Microcline) → ... → Sillimanite
- Biotite → ... → Orthoclase
- Chloritic clays → Chlorite → Chlorite → ... → (Biotite)
- Albite → ... → Almandite → Almandite
- Calcite → Epidote → ... → Plagioclase → Plagioclase

Facies-concept terms (Hietanen)	Muscovite-chlorite subfacies	Muscovite-biotite subfacies	Biotite-almandite subfacies	Staurolite-almandite subfacies	Staurolite-kyanite subfacies	Kyanite-almandite subfacies	Sillimanite-muscovite subfacies	Sillimanite-feldspar subfacies
	Greenschist facies			Epidote-amphibolite facies		Amphibolite facies		

rocks probably comes from muscovite so that a small amount of a potash feldspar is probably also developed. These molecules, of course, go into solid solution in any previously existent albite to form plagioclase. Again, this reaction is progressive, because the plagioclase composition is temperature-dependent exactly as it is in igneous rocks. Thus, if there is more lime present than can be accommodated by the plagioclase stable at a given temperature, the lime-bearing mineral epidote will coexist with plagioclase. As temperature rises, plagioclase will become more calcic and epidote will disappear. The final reaction in rocks of this bulk composition takes place at high temperature. Muscovite becomes unstable and breaks down to form orthoclase and sillimanite. The appearance of the aluminum silicate sillimanite is not fundamentally due to an excess of alumina but to a deficiency of alkali, because the alumina content of these rocks is very similar to that of igneous rocks. Alkali was lost by leaching during chemical weathering and stored in the sea as dissolved salts. A complementary reaction takes place at higher temperatures in the slow reaction between biotite and sillimanite to form additional orthoclase and almandite. Again, slow changes in the stable ferrous iron to magnesia ratios of biotite and garnet control the rate and extent of the reaction.

The potash-deficient argillaceous rocks undergo a somewhat more complicated series of reaction (Table 4-2). A deficiency of potash in a clay indicates thorough leaching during the weathering cycle. Thorough leaching during weathering involves marked enrichment in the soil in alumina, or ferric iron, or both and weathering toward a laterite. Therefore, in these potash-deficient shales an important factor is the ratio of ferric iron to alumina. Inasmuch as the metamorphic environment is generally a reducing one, the iron is reduced dominantly to the ferrous state and the substitution of Fe^{3+} for Al^{3+} is not important. At the first stages of metamorphism, three new minerals may appear in important quantities, depending on the iron to alumina ratio. The most important and diagnostic of these is chloritoid, a high-iron high-alumina mineral. Because it can accommodate very little of either magnesia or potash, it is usually accompanied by chlorite and as much muscovite as the low potash content of the rock will allow. In abnormally alumina-rich rocks, pyrophyllite may occur with muscovite and chloritoid, but the pyrophyllite would probably be unidentifiable megascopically. In abnormally iron-rich sediments, hematite or magnetite become quantitatively important. As metamorphism progresses, biotite is inhibited from forming at the stage in which it appears in the more normal potash-bearing shales. This is because chloritoid ties up the great bulk of available iron so that chlorite is relatively iron-poor. In the very alumina-rich sediments, kyanite appears in place of pyrophyllite, and in the more iron-rich sediments, almandite may appear in place of chlorite.

At the middle grade of metamorphism, chloritoid becomes unstable

Table 4-2.   Regional Metamorphism of the Potash-deficient Argillaceous Rocks

Zone-concept terms	Chlorite zone	Biotite zone	Garnet zone	Staurolite zone	Kyanite zone	Sillimanite zone	
		Greenschist facies		*Almandite-amphibolite facies*			
Facies-concept terms (Fyfe, Turner, and Verhoogen)	Muscovite-chlorite subfacies	Muscovite-biotite subfacies	Almandite-epidote subfacies	Staurolite-almandite subfacies	Kyanite-almandite subfacies	Sillimanite-almandite-muscovite subfacies	Sillimanite-almandite-orthoclase subfacies

Quartz
(Pyrophyllite)
Muscovite
Kaolinitic clays
Chloritoid
Chlorite
Hematite → Magnetite
Albite
Epidote
Calcite
Chloritic clays
Iron oxides

(Kyanite) → Kyanite → Sillimanite

Biotite
Staurolite
Almandite
Plagioclase

Quartz
Sillimanite
Orthoclase
(Biotite)
Almandite
Magnetite
Plagioclase

Facies-concept terms (Hietanen)	Muscovite-chlorite subfacies	Muscovite-biotite subfacies	Biotite-almandite subfacies	Staurolite-almandite subfacies	Staurolite-kyanite subfacies	Kyanite-almandite subfacies	Sillimanite-muscovite subfacies	Sillimanite-K feldspar subfacies
	Greenschist facies		*Epidote-amphibolite facies*			*Amphibolite facies*		

and is replaced by staurolite. This reaction involves the liberation of considerable iron, taken up by muscovite in the production of biotite, which appears for the first time. At this same stage epidote undergoes the reaction described earlier, and plagioclase appears as epidote disappears. Next staurolite becomes unstable, breaking down to form kyanite and almandite. Most commonly this is the first appearance of kyanite in the rocks. At higher temperature or lower pressure, kyanite inverts to sillimanite, and finally, what little muscovite has survived yields orthoclase and sillimanite, and biotite progressively yields almandite and sillimanite. The mineralogy of the highest-grade product is identical to that of the normal potash-bearing shales. However, the paucity of potash is expressed in markedly different proportions of the minerals. Almandite, sillimanite, and quartz dominate with very little orthoclase and biotite where the potash content was low, whereas in normal shales the ultimate product is a quartzo-feldspathic rock with accessory garnet and sillimanite.

Regional metamorphism may take place at lower than normal pressures and results in a somewhat modified sequence of minerals over what is described above. The environment of such a system is one in which high temperatures are attained at shallower depths than is the general case. Therefore the pressure range is 3 to 5 kbars rather than the more general 4 to 7 kbars (Hietanen, 1967).* The first area in which this low-pressure environment was recognized is in the Abukuma plateau of Japan (Miyashiro, 1958; Shido, 1958),† and such sequences are now referred to as the *Abukuma-type facies series*, in contradistinction to the *Barrovian-type facies series* named for the classical work of Barrow (1893, 1912)‡ in Scotland. Mineralogically, the main characteristics of the low-pressure series are the appearance of andalusite in lieu of kyanite and the appearance of cordierite. Chloritoid does not appear, staurolite is less common, and almandite-rich garnet is delayed in its appearance until high grades of metamorphism of argillaceous rocks. Andalusite appears at low temperatures and persists to high temperatures, where it is replaced by sillimanite. Cordierite likewise can appear at low temperatures, but persists to the highest temperatures.

Thermal Metamorphism of Argillaceous Rocks

The thermal metamorphism of argillaceous rocks can be divided into four distinct facies. The zone concept has not been applied specifically to thermal

*Hietanen, op. cit., fig. 1.
†A. Miyashiro, Regional metamorphism of the Gosaisyo-Takanuki district in the central Abukuma plateau, *Tokyo Univ. Fac. Sci. Jour.*, sec. 2, vol. 11, pt. 2, pp. 219–272, 1958; F. Shido, Plutonic and metamorphic rock of the Nakoso and Iritono districts of the central Abukuma plateau, *ibid.*, pp. 131–217.
‡G. Barrow, On an intrusion of muscovite-biotite gneiss in the southeastern Highlands of Scotland, and its accompanying metamorphism, *Geol. Soc. London Quart. Jour.*, vol. 49, pp. 330–359, 1893; Excursion to the east of Scotland. I. On the geology of Lower Dee-side and the southern Highland border, *Geologists' Assoc. Proc.*, vol. 23, pp. 274–290, 1912.

metamorphism. Again, the proportion of potash in the clay minerals is important in controlling the mineral assemblage at each grade, but the situation is sufficiently distinct to treat all the rocks in one table (Table 4-3). Andalusite or cordierite or both are the first new minerals to appear in megascopically identifiable crystals as recrystallization begins. They are both characteristic of rocks which were initially deficient in potash and represent the excess iron, magnesia, and alumina which cannot be accommodated by recrystallization of clays to sericite and chlorite. They appear as irregular ovoid spots crowded with inclusion, whereas recrystallization is so incomplete that primary detrital characteristics are still recognizable and the main mass of the rock is still microcrystalline. Muscovite and biotite, with or without excess chlorite, become recognizable at a slightly later stage, and with their appearance, the rock is completely recrystallized. At this point muscovite, andalusite, cordierite, and biotite are the dominant minerals, but only three of them can occur in any one rock. Thus the associations muscovite-biotite-andalusite and muscovite-biotite-cordierite represent relatively high-potash rocks and differ in the alumina to iron plus magnesia ratio. Similarly, the associations muscovite-andalusite-cordierite and andalusite-cordierite-biotite represent low-potash rocks, differing only in the alumina to iron plus magnesia ratio. Associated with these minerals are quartz, albite developed from any soda-bearing primary minerals, and epidote developed from calcareous material initially present. By this stage andalusite crystals have generally developed well-defined prisms, often showing a chiastolite cross.

The next stage of advancing metamorphism is difficult to distinguish megascopically in these rocks. The main criterion is the instability of epidote and the consequent formation of plagioclase of widely variable anorthite content. Epidote in the lower grades usually occurs as small granules which are not usually identifiable megascopically. The plagioclase is characteristically untwinned or very poorly twinned so that it is also difficult to recognize megascopically. Two minor changes may occur at this grade, depending on the bulk composition. If the parent rock is abnormally rich in potash (or if some potash is introduced metasomatically from the adjacent intrusion causing the metamorphism), then muscovite can become partially unstable and give rise to a potash feldspar, usually microcline, and additional andalusite. At the same time a new ferromagnesian mineral rich in iron may appear. Almandite may appear if the metamorphism is at depth under high pressure, or anthophyllite may appear if metamorphism is at shallower depths. The reason for the appearance of these minerals is that cordierite can accommodate only a limited substitution of ferrous iron for magnesia in this temperature range so that an iron-rich magnesia-poor mineral may be required to balance out the iron to magnesia ratio of the parent rock. Hornblende, the characteristic mineral of this facies, rarely occurs in these rocks because of their low lime content.

Table 4-3. Thermal Metamorphism of Argillaceous Rocks

Facies	Albite-epidote-hornfels facies	Hornblende-hornfels facies	Pyroxene-hornfels facies	Sanadinite facies
Quartz	→Quartz ——————	↑	→Quartz	→Tridymite
Potash-bearing clays	Sericite→Muscovite	Microcline; Muscovite	→Orthoclase	→Sanadine
Kaolinitic clays	Andalusite	Andalusite	Andalusite→Sillimanite	→Mullite
	Cordierite	↑	→Cordierite	→Cordierite
	Biotite		(Biotite)	
Chloritic clays	Chlorite	Chlorite →(Anthophyllite or almandite)	→Hypersthene	→Hypersthene
Iron oxides	Hematite→Magnetite	↑		→Magnetite
Albite	Albite	→Plagioclase	→Plagioclase	→Plagioclase
Calcite	→Epidote			

Under most normal contact-metamorphic environments, the highest grade is represented by the appearance of pyroxenes. In argillaceous rocks, because of the low lime content, the pyroxene is usually hypersthene formed by the breakdown of biotite. A potash feldspar is formed simultaneously from the potassium and aluminum ions of the biotite and is usually orthoclase. If muscovite and biotite have both persisted to this grade, then they react to give cordierite and orthoclase. Muscovite in excess of this reaction breaks down to give andalusite and orthoclase, and subsequently, all andalusite inverts at higher temperature to sillimanite. Thus a quartzofeldspathic rock with cordierite, hypersthene, and sillimanite is the usual ultimate product.

Xenoliths of argillaceous rocks incorporated in high-temperature mafic magmas may show still another facies which is often accompanied by partial fusion. Within the glass, minute crystals of tridymite derived from quartz, sanidine derived from alkali feldspars, mullite derived from sillimanite, and cordierite, hypersthene, and plagioclase may occur. Such materials are found only in lava flows and very shallow depth intrusions where rapid cooling can preserve the identity of the xenoliths without simply incorporating the material in the magma by assimilation. As a result equilibrium is frequently not attained, and many of the lower-temperature minerals may be present as well. Commonly, the newly formed minerals are microscopic in size so that this facies is of little importance in the megascopic study of metamorphites.

One important variant of the thermally metamorphosed argillaceous rock is that group abnormally deficient in silica as well as alkalies. Here corundum and spinel may appear as important minerals along with the more-normal minerals, and quartz will be absent. Under the extreme conditions of a metamorphosed laterite, the product will be corundum and magnetite in the mixture known as *emery*.

METAMORPHISM OF CALCAREOUS ROCKS

In the metamorphism of calcareous rocks, two problems are immediately important: (1) the percentage and nature of impurities in the rock, and (2) the ability of carbon dioxide gas evolved in the reactions to escape. Ordinarily, limestones and dolomites are thought of as relatively pure calcite and dolomite or as intermediate mixtures between the two. In such rocks metamorphism presents a relatively simple problem. The rocks are monomineralic or, at most, consist of two minerals between which there should be no reaction. Therefore metamorphism takes place as a simple recrystallization, with loss of sedimentary features and a gradual coarsening of texture to give uniform-grained calcite or dolomite marbles. Even in this purest form the metamorphism of dolomite is complicated at high temperature by the breakdown of dolomite to give calcite plus periclase (MgO)

with the evolution of carbon dioxide gas. In the presence of water vapor, periclase hydrates to become the magnesium hydroxide brucite. Therefore even pure dolomite may become a calcite-periclase or calcite-brucite marble. Most limestones and dolomites, however, are not pure rocks but contain variable proportions of admixed detrital quartz, clay, or chert so that metamorphism may give rise to some lime or lime-magnesia silicates, with or without alumina and iron. Depending on the nature and proportion of the admixed impurities, marbles can develop very complex mineral assemblages, and minerals characteristic of other parent rock types may appear.

The development of the lime and lime-magnesia silicates by reaction with carbonates is pressure sensitive, in that a volatile gas phase must be able to escape for the reactions to go to completion. If the evolved carbon dioxide cannot escape, calcite or dolomite will recrystallize along with the impurities without reaction; micaceous minerals recrystallized from clays, quartz, and carbonates are associated without any intermediate reaction products. The control of reaction versus nonreaction is frequently quite delicate and abrupt so that reaction may have taken place along bedding planes, giving lime silicates, whereas within the same bed there may have been no reaction. Similarly, in tracing along a single bed, an abrupt change can be seen between reaction and no reaction, where no visible textural or structural difference can be recognized between the two parts of the bed. In the following discussion, only those situations in which reactions have taken place will be considered.

Thermal and regional metamorphic products of limestone are very similar mineralogically due to the plastic behavior of carbonates under high pressures. This results in a reduction of shear stresses within the rock undergoing regional metamorphism so that the difference in the two environments is primarily a difference of hydrostatic pressure. Charts for thermal and regional metamorphism are presented in Tables 4-4 and 4-5.

Tremolite and epidote are the diagnostic minerals of the lowest grade of metamorphism. Tremolite develops from any initial magnesium-bearing mineral, which in a limestone, may be minor dolomite or chloritic or montmorillonitic clays. Where magnesia is derived from dolomite, silica also must be available. Epidote accommodates the alumina from the clays. Muscovite will also be present if the clay admixture in the parent sediment is potassic. In thermally metamorphosed rocks, talc may appear before tremolite because of the low pressure and ready availability of water from an adjacent magma.

The middle grades of metamorphism are characterized by diopside, plagioclase, and lime garnets. Tremolite becomes unstable in the presence of calcite and quartz and reacts to form diopside. Epidote, in the presence of albite, begins to break down to form anorthite molecules which, in turn, go

Table 4-4. Regional Metamorphism of Impure Limestones

Facies terms (Fyfe, Turner, and Verhoogen)	*Greenschist facies*	*Almandite-amphibolite facies*		*Granulite facies*
		Staurolite-almandite subfacies	*Sillimanite-almandite subfacies*	

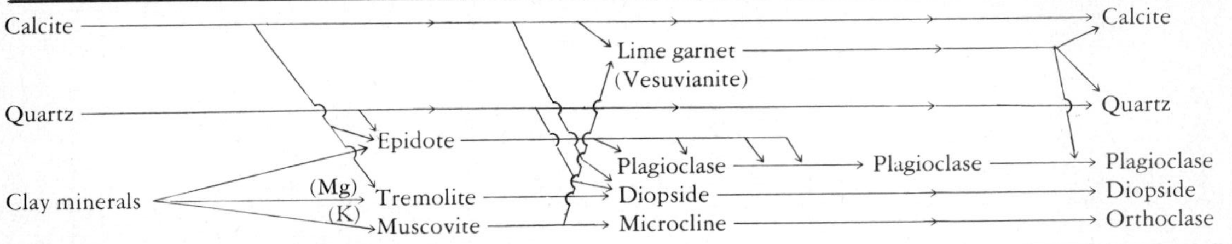
Calcite
Quartz
Clay minerals
Epidote
(Mg)
Tremolite
(K)
Muscovite
Lime garnet
(Vesuvianite)
Plagioclase
Diopside
Microcline
Plagioclase
Calcite
Quartz
Plagioclase
Diopside
Orthoclase

Facies terms (Hietanen)	*Greenschist facies*	*Epidote-amphibolite facies*	*Amphibolite facies*	*Granulite facies*

Table 4-5. Thermal Metamorphism of Impure Limestones

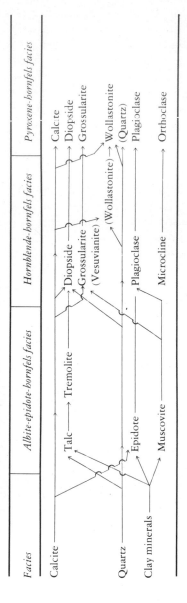

into solution in albite to give plagioclase. In regional metamorphism this is a progressive reaction, in that the plagioclase stable is in the andesine range of compositions. Therefore the rate of breakdown of the epidote is controlled by the amount of albite originally present so that epidote and plagioclase can coexist. At a slightly higher grade, any composition of plagioclase becomes stable so that all epidote disappears. Inasmuch as most limestones have very little albite initially, the plagioclase at this stage is commonly anorthite or is very close to it. In thermal metamorphism the instability of epidote is accentuated by low pressure so that it disappears more abruptly, and anorthite becomes stable at a somewhat lower temperature. The lime garnet present is usually andradite or grossularite, depending on the ferric iron to alumina ratio. The two garnets form a solid-solution series and will also accommodate any manganese present in the spessartite molecule. In both thermal and regional metamorphism, vesuvianite may also be present or may proxy for the garnet if hydroxide and fluorine ions are available. Alumina for the development of garnets may in part be derived from the breakdown of muscovite to produce a potash feldspar, usually microcline.

At the highest grades of metamorphism the difference of pressure in the thermal and regional metamorphic environments is most apparent. The reaction calcite plus quartz to yield wollastonite under the low pressure of the thermal environment and the breakdown of lime garnet to give calcic plagioclase under the high pressure of the regional environment are the distinctive features. In regional metamorphism calcite and any quartz remaining from earlier reactions coexist without reaction. Associated with them are plagioclase, usually near anorthite in composition, diopside, and minor orthoclase. Scapolite may proxy for plagioclase if a volatile phase containing carbon dioxide and chlorine is present. In thermally metamorphosed rock the characteristic assemblage is wollastonite, diopside, grossularite with minor feldspar, and calcite. This is the familiar skarn of igneous contacts, where a silica deficiency of the limestone can be made up by desilication of the magma.

A series of progressive step reactions between dolomite and its main impurity quartz was worked out by Bowen in 1940* on a theoretical basis. Subsequent studies of these rocks have proved Bowen's ideas to be essentially correct. The first six of Bowen's 13 steps, along with the percentage of quartz required in the rock necessary for each step indicated, are presented in Table 4-6. The last seven steps have been omitted here, because they occur at only very high temperatures and involve unfamiliar minerals.

The first step in the anhydrous environment involves reaction of dolomite and quartz to form tremolite and calcite. If the original rock contained 37.5 percent SiO_2, dolomite and quartz should be exhausted simultaneously,

Table 4-6. First Six Steps in the Reaction between Dolomite and Quartz

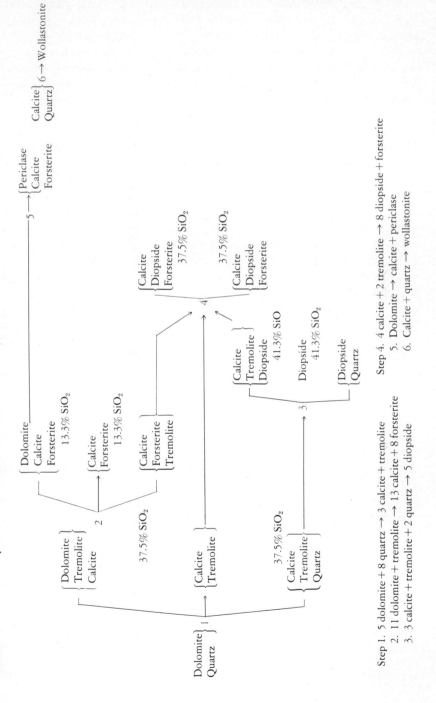

Step 1. 5 dolomite + 8 quartz → 3 calcite + tremolite

2. 11 dolomite + tremolite → 13 calcite + 8 forsterite

3. 3 calcite + tremolite + 2 quartz → 5 diopside

Step 4. 4 calcite + 2 tremolite → 8 diopside + forsterite

5. Dolomite → calcite + periclase

6. Calcite + quartz → wollastonite

and the rock will consist of three parts calcite and one part tremolite. Normally, either dolomite or quartz will be left over from the reaction. In the presence of water vapor, tremolite will be preceded by the appearance of talc and calcite. The next reaction takes place in those rocks which were initially silica-deficient so that dolomite persisted through the formation of tremolite. Here dolomite and tremolite react to form forsterite and more calcite. If the initial rock contained 13.3 percent SiO_2, dolomite and tremolite will be exhausted simultaneously, and the rock will be a calcite-forsterite marble. In less-siliceous rocks dolomite will persist, and the assemblage will be dolomite-calcite-forsterite. In rocks with more than 13.3 but less than 37.5 percent SiO_2, tremolite will be left over from the reaction, and the assemblage will be calcite-tremolite-forsterite.

The next reaction takes place in the highly siliceous rocks, in which quartz persisted, when calcite, tremolite, and quartz react to form diopside. Calcite and tremolite are required for this reaction in the same proportions in which they were produced from dolomite and quartz. Therefore, either quartz will be exhausted, leaving calcite and tremolite plus diopside, or calcite and tremolite will be simultaneously exhausted, leaving quartz plus diopside. If the rock contained 41.3 percent SiO_2, all three would disappear to leave a pure diopside rock. Three assemblages are thus possible: quartz-diopside in rocks containing more than 41.3 percent SiO_2, diopside in rocks containing 41.3 percent SiO_2, and calcite-tremolite-diopside in rocks containing between 37.5 and 41.3 percent SiO_2. The two forsterite-bearing assemblages are still stable, as is the very special pure calcite-tremolite rock.

The last reaction of prime importance involves the disappearance of tremolite. Calcite and tremolite react to form diopside and forsterite. This reaction takes place in all rocks having more than 13.3 percent and less than 41.3 percent SiO_2. A precisely balanced reaction is difficult to write, but the ratio of calcite to tremolite used in the reaction is approximately 2:1. The ratio of calcite to tremolite produced from lower-temperature reactions is 3:1 or greater. Therefore tremolite will be used up, whereas calcite remains in excess, and the product of the reaction will be calcite-diopside-forsterite. The variable silica content of the rock will be accommodated by varying proportions of diopside and forsterite.

Two additional reactions may take place at still higher temperatures. Dolomite becomes unstable, producing calcite and periclase or brucite, as mentioned earlier. This, of course, will occur only in very silica-deficient rocks, where dolomite has been in excess of that required for earlier reactions. A final reaction of importance is that between calcite and quartz to form wollastonite. Under equilibrium conditions in the first four reactions, calcite and quartz cannot coexist at the temperature required for this reaction. However, in actual metamorphic conditions, silica could be introduced into calcite marbles at this stage and allow the formation of wollastonite. Thus wollastonite-diopside-forsterite skarns may be developed.

The addition of clay and other impurities to the dolomite adds some complexity to the system. At low temperatures, epidote or muscovite or both are the characteristic aluminous minerals. By the time diopside becomes stable, epidote is unstable and again yields plagioclase. With it, phlogopite becomes the stable potash mineral because of the abundance of available magnesia. Finally, with the disappearance of tremolite, potash feldspars become stable and are formed with diopside from phlogopite. Phlogopite, however, may persist to high grades of metamorphism if the rock is deficient in silica.

METAMORPHISM OF MAFIC IGNEOUS ROCKS

Mafic igneous rocks are abundant in the geosynclinal environment and are therefore important in metamorphic rocks. In fact, two of the facies of regional metamorphism are named for the characteristic mafic igneous equivalents: the greenschist and the almandite-amphibolite facies. Two other facies representing special environments are also named for mafic igneous equivalents: the glaucophane schist and the eclogite facies. Therefore a thorough understanding of this group of rocks is essential. The progressive mineralogical changes for regional metamorphism are outlined in Table 4-7.

Mafic igneous rocks consist essentially of labradorite, diopsidic augite, and titaniferous magnetite. Other minerals may be present, but they are sufficiently unimportant quantitatively that they may be ignored for a first approximation. The first step of metamorphism is the breakdown of these high-temperature primary minerals into low-temperature minerals. Labradorite breaks down to give albite and epidote. Because of the low silica content of epidote, some silica may be set free to form quartz. Simultaneously, pyroxene gives chlorite with or without actinolite. Both these minerals are silica-deficient with respect to pyroxene so that quartz accompanies them. Considerable variability develops in the rocks at this stage, particularly in relation to the availability of volatiles. In the presence of water vapor, magnesia may be tied up as talc. Similarly, if carbon dioxide is available, lime may be tied up as calcite or other carbonates may form. The minor potash in the feldspars of the original rock may cause the appearance of muscovite, but it is usually masked by the far more abundant chlorite. Biotite becomes stable at a slightly higher temperature and is the first readily recognizable potassium-bearing mineral.

The second major stage of metamorphism is the appearance of hornblende and almandite. The transition from actinolite to hornblende involves a greatly increased complexity of the composition of the mineral, with particularly important increases in iron and aluminum content. This

Table 4-7.   Regional Metamorphism of Mafic Igneous Rocks

Facies terms (Fyfe, Turner, and Verhoogen)	Greenschist facies		Almandite-amphibolite facies			Granulite facies
	Muscovite-chlorite subfacies	Muscovite-biotite subfacies	Epidote-almandite subfacies	Staurolite-quartz subfacies	Sillimanite-almandite subfacies	Granulite facies

Mineral transitions (left to right):

- Albite ————→ Plagioclase (Andesine) ————→ Plagioclase (Labradorite) ————→ Plagioclase
- Labradorite ←→ Epidote
- Quartz ————————————————————————→ (Quartz)
- Augite → Actinolite → Hornblende ————————→ Hornblende (Diopside) → Hypersthene, Diopside
- Chlorite → Biotite → Almandite ————————————→ Almandite-pyrope
- Biotite ————————————————————→ Biotite → Orthoclase
- Titaniferous magnetite ⟨ Magnetite, Sphene ————————————→ Magnetite

Facies terms (Hietanen)	Muscovite-chlorite subfacies	Muscovite-biotite subfacies	Biotite-almandite subfacies	Staurolite subfacies	Muscovite-sillimanite subfacies	Sillimanite-orthoclase subfacies	Hypersthene-pyral garnet-sillimanite subfacies
	Greenschist facies		Epidote-amphibolite facies	Amphibolite facies			Granulite facies

material is derived from chlorite and epidote, and at the same time, chlorite yields the iron-alumina garnet almandite. There is therefore a marked increase in amphibole. The rock thus becomes hornblende-albite-epidote-quartz, with minor almandite, biotite, sphene, and magnetite. The next stage is the breakdown of epidote to form plagioclase with albite. As before, this is a progressive change with a gradual decrease of epidote and increase in the anorthite content of the plagioclase. The rock is now an amphibolite dominated by hornblende, plagioclase, and quartz.

The final stage of metamorphism involves the appearance of pyroxenes as the dominant ferromagnesian mineral. Diopsidic augite and hypersthene appear in place of hornblende and biotite. The potash liberated from biotite may be taken up in solid solution in the plagioclase or may form a separate orthoclase phase. Quartz is used up extensively in the formation of pyroxene so that it may be lacking or minor. The only important mineralogical difference from the parent rock is the presence of almandite-pyrope solid-solution garnet. There is usually, however, a very marked textural difference. Titanium, which at lower grades forms a separate phase as sphene, is progressively taken back up in the ferromagnesian minerals or in magnetite so that it may not form a separate phase.

Thermal metamorphism of these rocks follows an almost parallel series of mineralogical changes. The one exception is that, because of lower pressure of the thermal environment, almandite often does not appear. Thus metamorphism in these rocks begins by breaking down high-temperature minerals into low-temperature minerals, followed by building back up to high-temperature minerals. Deviations in the end product indicate the extent to which materials have been gained or lost during the process.

METAMORPHISM OF QUARTZO-FELDSPATHIC ROCKS

The quartzo-feldspathic rocks include arkoses, intrusive and extrusive equivalents of the granite family, and granitic gneisses developed in earlier phases of metamorphism. Arkose and silicic extrusives are minor constituents of the geosyncline and thus are involved in regional metamorphism. The plutonic granites and gneisses are also involved where a younger orogeny has followed the same belt as an older orogeny. Again, as in the mafic igneous rocks, the original material is composed of high-temperature minerals, and metamorphism follows a pattern of breakdown to low-temperature minerals and subsequent rebuilding to high-temperature minerals. During the process, marked textural changes occur so that the metamorphic product is similar mineralogically but dissimilar structurally to the parent. The mineralogical changes are indicated in Table 4-8.

The first step is the breakdown of feldspars into their low-temperature equivalents: plagioclase goes to epidote and albite, and potash feldspar goes dominantly to form fine-grained muscovite. Microcline is stable at this stage

Table 4-8.   Regional Metamorphism of Quartzo-feldspathic Rocks

Facies terms (Fyfe, Turner, and Verhoogen)	Greenschist facies		Almandite-amphibolite facies			Granulite facies
	Muscovite-chlorite subfacies	Muscovite-biotite subfacies	Epidote-almandite subfacies	Staurolite-quartz subfacies	Sillimanite-almandite subfacies	
Quartz	Quartz →	→	→	→	→	Quartz
Potash feldspars	(Microcline) Microcline →				→ Orthoclase	Perthite
					Sillimanite →	Sillimanite
Plagioclase	Albite →		→	→ Oligoclase →	Plagioclase → Biotite →	Plagioclase Hypersthene
Biotite-hornblende	Chlorite →	Biotite →	(Almandite) →		Almandite →	Almandite-pyrope
	Muscovite Epidote					

Facies terms (Hietanen)	Greenschist facies		Epidote-amphibolite facies		Amphibolite facies	Granulite facies
	Muscovite-chlorite subfacies	Muscovite-biotite subfacies	Biotite-almandite subfacies	Staurolite subfacies	Sillimanite-orthoclase subfacies	Hypersthene-pyral garnet-sillimanite subfacies

but is often either absent or unidentifiable. The transition from potash feld-spar to muscovite involves a loss of potash, and very commonly, this is lost from the rock. In the case of arkoses this potash may be tied up in the re-generation of clay minerals to muscovite. At any rate the rock usually winds up with some deficiency of potash with respect to alumina. Ferromagnesian minerals of the parent rock are converted to chlorite at the same time. At the next stage microcline usually appears in identifiable grains, and at the same time, chlorite and muscovite react to form biotite. In most rocks, because of the relatively low ferromagnesian content with respect to mus-covite and the high iron to magnesia ratio normal for these rocks, the chlorite is exhausted in this reaction. If not, then almandite will appear at the next step.

At the next stage plagioclase forms from the albite and epidote. Then muscovite breaks down to form orthoclase and sillimanite, biotite begins to break down to yield orthoclase and almandite, and microcline inverts to orthoclase. In the breakdown of muscovite, the earlier loss of potash is expressed in the appearance of sillimanite. In arkoses this deficiency of potash is normal as a result of original clay content, but in granites the amount of sillimanite formed here is a measure of the earlier potash loss. The ultimate stage involves complete dehydration of the rock as biotite breaks down to form hypersthene and feldspar. The alkali feldspar at this stage is commonly a perthite, involving a removal of albite from the plagio-clase and allowing the anorthite content of the remaining separate-phase plagioclase to increase into the andesine range. The lime of the rock is almost entirely in the plagioclase. The resulting hypersthene granite gneisses are the metamorphic charnockites.

METAMORPHISM OF ULTRAMAFIC IGNEOUS ROCKS

Ultramafic igneous rocks are commonly emplaced early in the deformation of geosynclines. They therefore are subjected to metamorphic environ-ments in the later phases and are particularly susceptible to autometa-morphism. In addition, their emplacement into deforming marine sediments subjects the intrusions to an environment of abundant migrating hot solu-tions so that hydrous and carbonate phases are very important products. The fundamental relationships in the system $MgO\text{-}SiO_2\text{-}H_2O$ at high temperatures and pressures are most important and have been investigated by Bowen and Tuttle (1949).* A series of five reactions at increased tem-peratures and with various stabilities is involved, as shown in Fig. 4-20. At the temperature and pressure of curve I, the reaction

Serpentine + brucite $\rightleftharpoons$ forsterite + water vapor

*N. L. Bowen and O. F. Tuttle, The system $MgO\text{-}SiO_2\text{-}H_2O$, *Geol. Soc. America Bull.*, vol. 60, pp. 439–460, 1949.

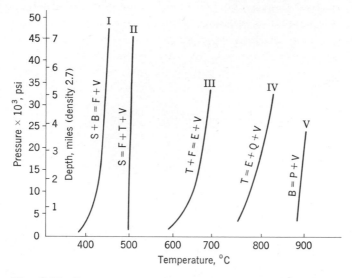

Fig. 4-20. Pressure-temperature curves for reactions in the MgO-SiO₂-H₂O system. The equation on each curve gives the reaction to which that curve refers. Symbols: B–brucite; E–enstatite; F–forsterite; P–periclase; Q–quartz; S–serpentine; T–talc; and V–vapor. (*After N. L. Bowen and O. F. Tuttle, Geol. Soc. Amer. Bull., vol. 60, 1949.*)

is in equilibrium. At any higher temperature, forsterite is stable in association with water vapor. At the environment indicated by curve II, serpentine is in equilibrium with forsterite, talc, and water vapor, but this is the maximum temperature at which serpentine can exist. At temperatures above curve III, enstatite is stable in the presence of water vapor; at any lower temperature, enstatite is not stable but gives rise to talc and forsterite. Curve IV represents maximum conditions for the stability of talc. At higher temperatures it dissociates to form enstatite, quartz, and water vapor. Finally, curve V represents the dissociation of brucite to give periclase and water vapor. Thus serpentine is the diagnostic mineral of lower temperatures, talc of the middle range of temperatures, and enstatite of the upper-middle and high temperatures.

In natural systems these relationships hold, in general, but the addition of iron and alumina in normal proportions changes the absolute temperature and pressure values and even influences the appearance of other mineral phases. The extent to which ferrous iron can substitute for magnesia and ferric iron or alumina can substitute for magnesia plus silica is the controlling factor. In addition, the presence of lime is important. Thus calciferous amphiboles appear, complex chlorites proxy for serpentine, and the orthorhombic amphiboles proxy for talc. The presence of carbon dioxide in the vapor phase may stabilize calcite, dolomite, or magnesite in the rocks.

Special Mafic Igneous Metamorphic Equivalents

Glaucophane schist and eclogite are two metamorphic equivalents of mafic igneous rocks which are quantitatively minor but genetically important. Each represents an abnormally high-pressure environment under which highly sodic ferromagnesian minerals are stable. Glaucophane schist appears to represent a low-temperature ultrahigh-pressure situation. The most characteristic minerals of these rocks are glaucophane, jadeite, and lawsonite. Epidote, muscovite, chlorite, and quartz are other characteristic low-temperature minerals present. The sodic amphibole or sodic pyroxene forms at the expense of albite of the plagioclase. The anorthite molecule goes to form lawsonite, which is almost chemically equivalent to anorthite but is much denser. Sediments associated with the mafic volcanics undergo a similar type of metamorphism, with development of the same critical minerals but in different proportions.

Glaucophane schists are a quantitatively minor rock type in intensely deformed geosynclines and are usually associated with essentially unmetamorphosed rocks or with rocks of the greenschist facies (Coleman and Lee, 1963).* The transition between rock types generally takes place over a short distance. Ernst (1963)† concludes that local differential stress may be important in the development of the glaucophane schist zones. A possible mechanism for the origin of these rocks may be in the concept of creep failure as a mechanism of faulting. Orowan (1960)‡ has demonstrated that faulting at depth cannot be a frictional sliding over a fracture but must be by sudden creep failure by gliding within crystal structures. As stress develops, strain is progressively concentrated into narrow zones of plastic deformation, relieving stress on adjacent rocks. Portions of the geosyncline may be at such a critical depth that this type of failure may take place at relatively low temperature, allowing recrystallization to the glaucophane schist assemblage of minerals. A change in depth conditions due to erosion or gravity sliding could reduce pressure sufficiently to allow frictional sliding on a fracture — that is, faulting — which is normally associated with these rocks. The associated unmetamorphosed or greenschist-facies rocks thus represent those portions relieved of strain by failure in the glaucophane schist zones.

Eclogite, in contrast, represents an environment of very high temperature as well as ultrahigh pressure. Eclogite is dominantly bimineralic,

---

*R. G. Coleman and D. E. Lee, Glaucophane-bearing metamorphic rock types of the Cazadero Area, California, *Jour. Petrology*, vol. 4, pp. 260–301, 1963.
†W. G. Ernst, Petrogenesis of glaucophane schists, *Jour. Petrology*, vol. 4, pp. 1–30, 1963.
‡E. Orowan, Mechanism of seismic faulting, in: D. Griggs and J. Handlin (eds.), Rock deformation, a symposium, *Geol. Soc. America Mem. 78*, pp. 323–345, 1960.

composed of an unusual soda-rich pyroxene called *omphacite* and a lime-rich pyrope garnet. Omphacite is a solid solution of diopside from the pyroxene and jadeite developed from the albite molecule of feldspars. Lime from the feldspar is usually taken up in a grossularite garnet molecule and excess magnesia in a pyrope garnet molecule. Kyanite, enstatite, rutile, and quartz are common accessory minerals. An equation can be written for the reaction:

$$3CaAl_2Si_2O_8 + 2NaAlSi_3O_8 + Mg_2SiO_4 + nCaMg(SiO_3)_2 \longrightarrow$$

$$\underbrace{}_{\text{Labradorite}} \qquad \underset{\text{+ Olivine}}{} \qquad \underset{\text{+ Diopside}}{}$$

$$3CaMg_2Al_2(SiO_4)_3 + 2NaAl(SiO_3)_2 + nCaMg(SiO_3)_2 + 2SiO_2$$

$$\underset{\text{Garnet}}{} \quad + \quad \underbrace{}_{\text{Omphacite}} \quad \underset{\text{+ Quartz}}{}$$

The chemical similarity between eclogite and basalt has been pointed out earlier (Chap. 1).

Eclogites occur in several different environments, each of which is indicative of either very great depth of origin or abnormally high-pressure environments at shallower depths (Coleman et al., 1965).[*] Xenoliths of eclogite, along with other deep-seated rock types, are found in some basalt, ultramafics, and intrusive breccias. Here the xenoliths have apparently been brought up from great depth and have not had time to readjust mineralogically to phases stable at lower pressures. Lensoid masses of eclogite are found in gneisses of the granulite facies. These bodies have been interpreted as fragments of larger masses developed at depth and transported upward by the granitic intrusions into the base of geosynclines. A third occurrence is as bands associated with amphibolites, mica schists, and glaucophane schists in zones of intense deformation. These bodies may have an origin similar to that postulated above for the glaucophane schists. That is, they may represent zones in which failure by creep has been concentrated, but at somewhat greater depth and higher temperature than in the glaucophane schist situation. In many localities eclogites of this environment have been transformed into glaucophane-bearing rocks, apparently as a retrograde-metamorphic effect. However, the glaucophane phase may be the result of continued creep failure at lower temperatures. As mentioned earlier, the transition at the Mohorovicic discontinuity may be one from basaltic or gabbroic rocks above to eclogite below.

[*]R. G. Coleman, D. E. Lee, L. B. Beatly, and W. W. Brannock, *Eclogites* and *eclogites*: their differences and similarities, *Geol. Soc. America Bull.,* vol. 76, pp. 483–508, 1965.

METASOMATISM

Metasomatic Effects Other than Granitization

Metasomatism involves the introduction or removal of any constituents of the rock. In most metamorphic processes some change is involved, but unless the change is marked or involves the obvious introduction of constituents foreign to the normal sequence of minerals, it is overlooked. However, if chemical analyses of a suite of schists derived from one parent formation are compared, there is usually some noticeable change in materials. The effects can be seen in the broad general way by comparison of average compositions of slate, phyllite, and gneiss (Table 4-9). The anhydrous

Table 4-9. Standard Cell of Metamorphic Rocks*

Slate	$K_{4.44}$	$Na_{3.02}$	$Ca_{1.36}$	$Mg_{3.96}$	$Fe_{6.40}$	$Al_{19.95}$	$Si_{56.3}$ O_{160}
Phyllite	$K_{4.72}$	$Na_{3.58}$	$Ca_{1.18}$	$Mg_{3.98}$	$Fe_{5.78}$	$Al_{21.80}$	$Si_{55.5}$ O_{160}
Mica schists	$K_{4.32}$	$Na_{3.40}$	$Ca_{1.87}$	$Mg_{3.69}$	$Fe_{4.99}$	$Al_{18.30}$	$Si_{58.7}$ O_{160}
Muscovite- biotite gneisses	$K_{3.98}$	$Na_{5.36}$	$Ca_{1.88}$	$Mg_{2.39}$	$Fe_{3.68}$	$Al_{17.4}$	$Si_{60.3}$ O_{160}
Plutonic gneisses	$K_{4.88}$	$Na_{6.02}$	$Ca_{1.81}$	$Mg_{1.59}$	$Fe_{2.82}$	$Al_{15.5}$	$Si_{62.3}$ O_{160}

*Analyses taken from A. Poldervaart, The crust of the Earth, *Geol. Soc. America Spec. Paper 62*, table 21:
     Slate: average of 61, line 65, table 21.
     Phyllite: average of 50, line 67, table 21.
     Mica schists: average of 37, line 60, table 21.
     Two mica gneisses: average of 51, line 55, table 21.
     Plutonic gneisses: average of 64, line 51, table 21.

average analyses have been recalculated to a standard cell of a rock containing exactly 160 oxygen ions (Barth, 1952).* The standard cells indicate an increase in total alkali and silica and a simultaneous decrease in ferromagnesian constituents and alumina as metamorphism increases. Although these data represent average values for a limited number of rocks, probably representing considerable variation in initial bulk composition, they do indicate a general change which must be accomplished by metasomatic transfer.

   More obvious effects of metasomatism are seen when volatiles have been introduced into the rock, stabilizing new minerals in appreciable bulk. Tourmalinization of some schist and gneiss is an example. Small amounts of tourmaline are commonly derived from detrital grains, but any appreciable volume must require the introduction of boron, probably as a gas phase. Similarly, the substitution of scapolite for plagioclase involves introduction

*T. F. W. Barth, "Theoretical Petrology," pp. 82–85, Wiley, 1952.

of chlorine and carbon dioxide (Hietanen, 1967).* Chondrodite, topaz, fluorapatite, and fluorite develop in metamorphic rocks into which fluorine has been introduced. Sulfides are often indicative of sulfur introduction; however, many sediments carry sufficient sulfur to produce notable amounts of sulfides. The origin of the metasomatizing solutions or gases can often be found in adjacent intrusive bodies, and this type of alteration is most commonly associated with thermal metamorphism.

Metasomatic Granitization

In Chap. 2 the origin of granite as a product of magmatic processes was discussed. There is evidence, however, that many granites have formed without passing through a magma stage. These rocks have apparently formed by recrystallization and modification of composition in a solid state. The process whereby such granites have been formed is called *metasomatic granitization*. Most petrographers at the present time admit to the efficacy of the process; however, the mechanism by which it takes place and the relative volumetric importance of metasomatic granite and magmatic granite are hotly debated.

The most compelling evidence for granitization is in the nature of contacts at the margins of some granite bodies. Three types of contacts can be recognized. First, are the sharp intrusive contacts between magma and relatively cold rock. Here the magmatic character is evidenced by chilled fine-grained margins, apophyses of the intrusion in the wall rock, and contact breccias composed of stoped xenoliths of the wall rock in a continuous matrix of stringers and dikelets of the granite (Figs. 2-6 and 2-7). Such a contact is unmistakable evidence of a magmatic origin for the granite. A second type of contact is characteristically sharp but lacks evidence of chilling of a melt. The intrusive granite maintains a coarse texture, or is even coarsely porphyritic, right up to its contact with the wall. A pencil line can often be drawn between granite and adjacent metamorphic rock. In such situations it is apparent that the granite has been intruded, but there is considerable question as to the physical nature of the material intruded, that is, whether it was essentially solid or essentially liquid at the time of intrusion. If the rock was emplaced in an essentially solid state, as the contact relationship would suggest, then whether the rock is considered magmatic or metasomatic is simply a question of semantics. If mobility is accepted as the prime requisite of a magma, then these rocks are magmatic. If, however, the igneous rocks are defined as those which have crystallized from a silicate melt, then what is a mass which is emplaced in an essentially solid state? This question will be returned to.

The third contact relationship is the gradational contact in which no

*A. Hietanen, Scapolite in the Belt series in the St. Joe-Clearwater Region, Idaho, *Geol. Soc. America Spec. Paper* 86, 1967.

sharp line of demarcation can be drawn between granite and metamorphic wall rock. Gradual mineralogical and textural changes appear over a broad zone, bounded on one side by obvious metasediments or metavolcanics and on the other by apparently normal nonporphyritic coarse-grained granites (Fig. 4-21). Two opposing interpretations are immediately possible in such situations. One group of granite students would say, "Because all gradations of texture and mineralogy from metamorphic rock to granite are visible, the granite is the end product of the metamorphic process. The granite represents that volume of rock which has been transformed to the ultimate degree." A second group of equally perceptive students of granite would say, "The granite is an intrusive magmatic body which has caused the transformation of the wall rock. Emanations from the magma have worked outward, modifying wall-rock mineralogy and texture until their constituents were exhausted." Both agree that the transition zone has been metasomatically granitized.

Many different types of gradational effects can be seen, depending in part on the nature of the rock undergoing granitization. The transition from greenstone to granite, as it is seen in the Lake Superior region, is most instructive with respect to the final step, yielding uniform-grained granite (Bradley and Lyons, 1953).* In moving from greenstone toward granite, the first effect is the appearance of small ovoid porphyroblasts of alkali feldspar. These increase in size and number until the rock becomes an augen gneiss. The augen consist of large potash feldspar porphyroblasts and granular quartz and are separated by a ferromagnesian-rich matrix of the upper greenschist facies. As the augen grow, they coalesce to form a banded gneiss. The banding is usually very discontinuous and irregular. Some areas show evidence of shear control, with concentration of the more mobile quartzo-feldspathic components along the shears. However, these banded gneisses do not show the continuous uniform banding characteristic of other areas. Large potash feldspar porphyroblasts still dominate the bands. As granitization proceeds, the ferromagnesian bands become more diffuse, quartz and plagioclase become more important in them, and the rock becomes a porphyroblastic gneiss. Through this zone, ferromagnesian constituents gradually decrease, and the gneissosity becomes less distinct until a porphyritic granite is produced (Fig. 4-22). At this stage the rock consists of large ovoid feldspars, separated by a network of finer-grained quartz, feldspar, and ferromagnesian minerals. Finally, the large feldspars break down to form an aggregate of smaller grains. Commonly, zoning or bands of inclusions in the porphyroblasts are preserved in the aggregates without appreciable displacement. A zone in the original porphyroblast

*C. Bradley and E. J. Lyons, A mode of evolution of a granitic texture, in: R. C. Emmons (ed.), Selected petrogenetic relationships of plagioclase, *Geol. Soc. America Mem. 52*, pp. 101–110, 1953.

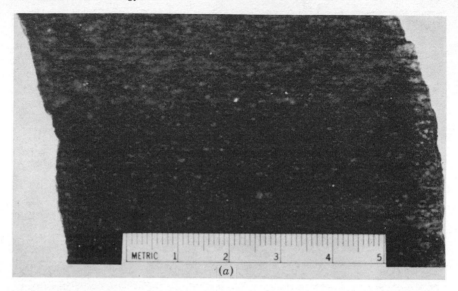

(a)

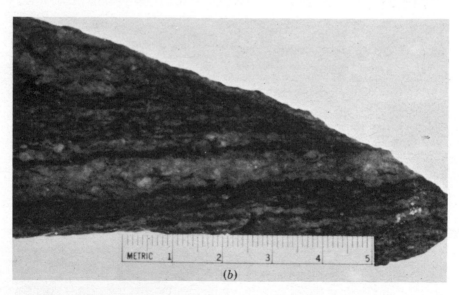

(b)

Fig. 4-21. The transition from sediment to granite gneiss. (*a*), (*b*), and (*c*) are all from one outcrop and illustrate the transformation of an argillaceous unit into gneiss across the Grenville metamorphic front at Coniston, Ontario. (*d*) represents the same type of material ¼ mile south beyond the front. The outcrop from which these specimens were collected is shown in Fig. 3-27.

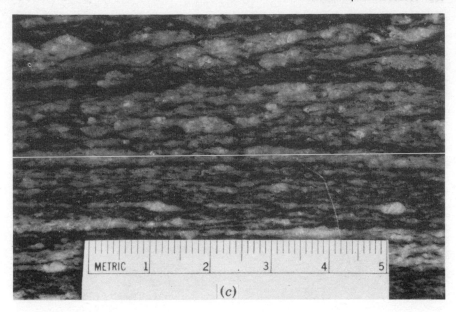

(c)

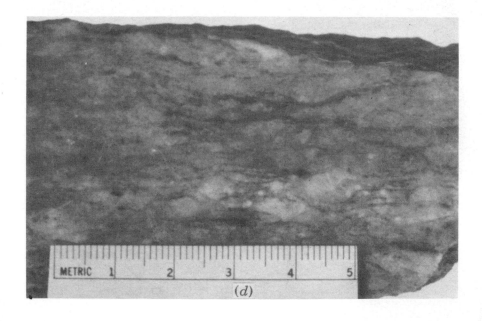

(d)

Fig. 4-22. The transition of gneiss to porphyritic granite. (*a*) A banded gneiss; (*b*) the other end of the same outcrop, showing a coarsely porphyroblastic texture in the left portion of the view; (*c*) a close-up of the porphyroblastic zone in *b*; (*d*) a flow-banded porphyritic granite without gneissic banding typical of outcrops 0.1 mile away. Michigan Highway 95 at Republic, Michigan.

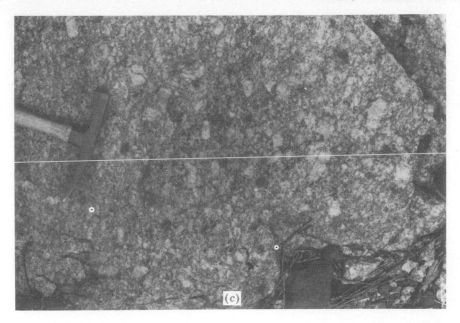

(c)

(d)

may be traced through as many as 10 differently oriented individual small crystals developed from it. Thus a uniform-grained granite is developed. If no movement is involved in this stage, the former porphyritic stage may be preserved through a chainlike network of quartz, feldspar, and biotite, separating ill-defined clusters dominantly of feldspar. Even very slight movement of the mass at this stage will destroy all evidence of the former nature of the rock and produce an isotropic granite. In quarries, zones of isotropic granite can be seen grading into zones or blocks of the network granite, showing that movement has taken place locally.

The mechanism of granitization is also a subject of debate. There are two basic schools of thought: one group advocates the transformation of rocks by ionic diffusion through the crystal structure over long distances and without the assistance of any type of fluid phase (Perrin and Roubault, 1949; Ramberg, 1952),[*] whereas a second group supports transfer of material by a fluid or gaseous phase to the area of reaction (Read, 1957).[†] Diffusion through the crystal structure or along intergranular boundaries over short distances is necessary for most metamorphic processes. The ionic-diffusion, or dry, concept of granitization, however, calls for the ions to be mobilized at depth by orogenic activity and move upward as waves over distances measured in kilometers. There is considerable question as to whether ionic diffusion can operate at a great enough rate to accomplish large-scale transfer in the time available during orogeny. However, transfer by migrating solutions moving through shear channelways, or by diffusion of ions through an intergranular fluid or gaseous phase, is a much faster process. Thus in wet granitization, orogeny or deep-seated intrusion releases fluids charged with liberated ions, particularly alkalis, which penetrate upward and react with rocks to form new materials. The same fluids will remove ions which cannot be tied up by the new minerals. Several sources of fluids can be visualized. First, where sediments are being granitized or where there are abundant associated sediments, the fluids may simply be connate brines liberated by compaction. Second, much water is liberated by the recrystallization of hydrous minerals to give anhydrous metamorphic minerals. In the process water would be reduced by organic material to give carbon monoxide and hydrogen and form a highly reactive gas. Emmons (1955)[‡]

[*]R. Perrin and M. Roubault, On the granite problem, *Jour. Geology*, vol. 57, pp. 257–379, 1949; H. Ramberg, "The Origin of Metamorphic and Metasomatic Rocks," pp. 237–262, University of Chicago Press, 1952.

[†]H. H. Read, "The Granite Controversy," Murby & Co. A collection of eight addresses delivered between 1939 and 1954 which together give an extensive summary of the concepts of granitization. See also a series of studies in the Beartooth Mountains, Montana and Wyoming, Part I: F. D. Eckelmann and A. Poldervaart, The Quad Creek area, *Geol. Soc. America Bull.*, vol. 68, pp. 1225–1262, 1957; Part 3: R. L. Harris, Jr., Gardner Lake area, *Geol. Soc. America Bull.*, vol. 70, pp. 1185–1216, 1959; Part 6: J. R. Butler, Cathedral Peak area, *Geol. Soc. America Bull.*, vol. 77, pp. 45–64, 1966.

[‡]R. C. Emmons and J. Bach, Steel penetration, *Foundry*, April, 1955.

has found that this process is most important in the transfer of silica and iron, where steel is cast in a sand mold. Third, the fluids may come from deeper-seated truly igneous bodies. In the Canadian Shield evidence indicates that the granitized rocks represent only the tops of batholiths (Emmons, 1953)* and that erosion has not cut deeply into the bodies. It is therefore quite possible that truly magmatic bodies may exist at somewhat greater depths and may have caused the granitization. Finally, the fluids may have come from greater depth in the crust or upper mantle. Degasing of the mantle to depths of several hundreds of kilometers is a source of both the required heat and the required material. Convection currents operating in the mantle could transport the minor amounts of alkalis of a thick zone of mantle to the base of the crust and bring about vast changes of bulk composition in the orogenic belt.

The effectiveness of fluid transfer, at least at a late stage of granite history, may be seen over a wide range of scales. Fracturing of the recrystallized rock results in the migration of fluids to the fracture. At a late stage in granitization, these fluids are alkaline, silica-deficient, and ferromagnesian-rich, yielding one type of lamprophyre (Emmons, Reynolds, and Saunders, 1953).† On a small scale these are seen as hairline cracks having a high concentration of biotite. The rock adjacent to the cracks is less ferromagnesian than the average rock at a distance from the crack. Thus a ½-mm-wide biotite-rich fissure may have a zone 1 or 2 cm wide which is impoverished in biotite relative to the rest of the rock. On quarry walls such segregated veinlets may be seen to coalesce into progressively larger dikelets and ultimately into distinct lamprophyre dikes. These represent the drainage of the last fluids. It is interesting to note that in Wisconsin, every commercial granite quarry is cut by lamprophyre dikes. Where there are no dikes, the granite is too rich in ferromagnesian minerals to be of commercial grade.

The granite bodies having sharp contacts but no chilled marginal phases appear in many instances to have been generated from previously granitized materials which have been remobilized in subsequent orogenies. Two mechanisms can be visualized for such remobilization. The first involves plastic flow by recrystallization with a minor, but chemically significant, volume of a volatile phase. These are the parautochthonous granites of Read's (1957)‡ granite series. In the basement sialic rocks underlying a geosynclinal column, temperatures rise until the rocks flow plastically under the pressure of the deformation. Once flow is initiated the mobile material will intrude upward into the root of the geosynclinal sediment pile. Contacts would thus be sharp, because the migrating material was solid in the

*R. C. Emmons, The Argument, in: R. C. Emmons (ed.), *op. cit.*, chap. 9, 1953.
†R. C. Emmons, C. D. Reynolds, and D. F. Saunders, Genetic and radioactivity features of selected lamprophyres, in: R. C. Emmons (ed.), *op. cit.*, chap. 7, 1953.
‡Read, *op. cit.*, pp. 374ff.

same way that a salt dome is solid. Recrystallization through the transporting agent of minor volatiles would result in obliteration of shearing effects but would probably result in preferred orientation of minerals, particularly near the margins of the intrusive. Such preferred orientation, as described by Balk (1937),* is common in these bodies.

The porphyritic character of many such bodies can be accounted for by porphyroblasts and the product of the metasomatic effects of the volatiles late in the process of solid intrusion. These "phenocrysts" commonly contain inclusions of the minerals of the groundmass and therefore cannot owe their large size to early crystallization from a liquid, as is the normal interpretation for phenocrysts. Krauskopf (1957)† has shown that at a pressure of 1,000 atm at 600°C the gas phase in equilibrium with solid but still hot rock would be dominantly water vapor with secondary amounts of other gases, dominantly CO_2, HF, H_2S, HCl, and N_2. Alkali feldspars are quite soluble in supercritical water vapor and would be dissolved from the bulk of the rock (Morey and Chen, 1955).‡ As the mass cooled, these feldspar-bearing solutions, which also contain some excess alkalies and silica, react with the mass of the rock to produce feldspar porphyroblasts by replacement and concretionary growth. Thus they can include minerals of the groundmass, and if they grow while the mass is still mobile, they may develop preferred alinement.

The second process involves the beginning of true melting and produces the anatectic granites (Fig. 4-23). Anatexis apparently involves higher temperatures than those required for plastic flow of the solid and, again, requires the presence of water. At atmospheric pressures granite melts at approximately 1000°C to form an extremely viscous liquid. In the presence of water and under high pressure, the melting temperature is reduced to the order of 650°C, a temperature reached in the upper grades of metamorphism. Thus in the presence of excess water, metamorphism grades into magmatism. As a body of quartzo-feldspathic rock begins to melt, the mobility of the entire mass increases rapidly as a crystal-liquid mush, and intrusion into higher-level rocks follows. However, because the mass is largely solid, there will be sharp contacts developed between intrusion and wall and no evidence of a fine-grained chilled margin. Flow alinement of the crystals is particularly well developed along the margins of the body, and fracturing of crystals is healed by the subsequent crystallization of the melt.

In some partially melted bodies, the melt phase can be separated and

*R. Balk, Structural behavior of igneous rocks, *Geol. Soc. America Mem. 5*, pp. 54–86, 1937.
†K. B. Krauskopf, The heavy metal content of magmatic vapor at 600°C, *Econ. Geology*, vol. 52, pp. 786–808, 1957.
‡G. W. Morey and W. T. Chen, The action of hot water on some feldspars, *Am. Mineralogist*, vol. 40, pp. 996–1000, 1955.

Fig. 4-23. A small anatectic body of granite segregated from a granite gneiss. Georgian Bay area north of Parry Sound, Ontario.

crystallized as stringers and dikes in cooler parts of the body. The composition of such liquid phases will vary with time as successive batches of liquid are separated. Experimental work on the system quartz-orthoclase-albite-anorthite-water, reported by Winkler (1965),* is important in considering the changing composition of the liquid formed. At 2,000 bars water pressure a mixture of quartz-orthoclase-albite has a minimum melting mixture consisting of 34 percent quartz, 40 percent albite, and 26 percent orthoclase. As anorthite is added, the temperature of beginning melting and the composition of the first melt change as follows, according to Winkler. The

Albite: anorthite	Temperature	Quartz	Albite	Orthoclase
∞	670°	34	40	26
7.8	675°	40	38	22
5.2	685°	41	30	29
3.8	695°	43	21	36
1.8	705°	45	15	40

ratio albite to anorthite is given for the entire system — liquid plus solid; it is not given for the liquid, but in all plagioclase melting systems the liquid is always richer in albite than is the solid plagioclase with which it is in equilibrium. Therefore as successive batches of liquid are formed and removed, the anorthite content of the remaining solid is increased. As a result the composition of successive minimum-melting point batches changes to become successively depleted in plagioclase and enriched in orthoclase and quartz. The sequence of liquids would then range from granodiorite or quartz monzonite to granite. This same compositional change may, perhaps, apply on a larger scale where major masses of sialic crust yield successive batches of silicic magma, giving quartz diorite or granodiorite magmas as the early intrusive phase and quartz monzonites and granite magmas as the late intrusive phase. In this connection it is important to remember that the system involved includes not only the quartz-orthoclase-albite-anorthite-H_2O system but also hornblende, a source of anorthite components, and biotite.

Granitization and the Continental Crust — A Hypothesis
The origin of the continental crust has been one of the major mysteries of geology. One concept ascribes the continental masses of granite and granite gneiss to having been derived from a once continuous thin shell rafted together by subcrustal currents. The origin of the granite itself poses a difficult problem: It cannot be accounted for by simple crystallization of quartz and feldspar from the primitive molten mantle material, since the

*H. G. F. Winkler, "Petrogenesis of Metamorphic Rocks," pp. 183ff, Springer-Verlag, 1965.

quartz and feldspar have floated gravitatively to the surface. This would require these minerals to crystallize simultaneously with olivine from a common parent liquid, and the reaction relationships prohibit simultaneous separation of quartz and a magnesian olivine from the same liquid. As an alternative theory it has been suggested that the granite crust developed by repeated foundering and partial remelting of a primitive mafic crust. With each partial remelting, the molten fraction would be somewhat enriched in the low-melting components of the rock, and the high-melting components would sink to the base of the mantle. The melt would move upward gravitationally to enrich a subsequent temporary crust somewhat more in the granitic components. By repeated upward concentration, the final crust was granite when it became stable. This concept assumes that an appreciable portion of the melt from each foundering and partial melting could retain its identity in an environment of strong convective currents and in spite of strong currents produced by foundering of great blocks of crust. It seems that in this environment, dilution by mantle fluid would make such extreme differentiation impossible and at best could produce an intermediate magma for the final crust.

The alternative proposed here is that the initial crust was composed of mafic to intermediate volcanics which subsequently have been granitized selectively under the continents. The granitizing agent called on is highly active volatile material from the upper mantle as a result of degasing of the mantle. During the early Precambrian, degasing was rapid and regional, producing the extensive granitized gneisses and granites of the shields. With time, degasing became restricted to orogenic belts, and therefore the granitized products became more restricted in their areal distribution. By late Precambrian the mantle was depleted in the highly active gases, and water vapor from the mantle resulted in anatexis and dominantly magmatic granites. Thus in this concept the proportion of metasomatic versus magmatic granites changed with time.

The oldest unmetamorphosed rocks of the shield areas are lavas and volcanic agglomerates of mafic to intermediate composition and some sediments derived from them. Granites are consistently intrusive into these rocks or are the product of granitization of them. In many shield areas the oldest rocks are so extensively metamorphosed or metasomatised that the character of the parent rocks is difficult to determine. Still, they can be identified as sediments and volcanics. In the Knifeian series northwest of Lake Superior, where these ancient rocks are least metamorphosed, the volcanics form the base and are intruded by or grade into granite and granite gneiss. Therefore these rocks probably represent the oldest preserved remnants of the crust.

If an initially molten earth is assumed, crystallization would proceed upward from the mantle base by gravitative accumulation of heavy crystals, and the surface layer would remain molten throughout most of the crystallization history. Ultimately, however, a surface crust would begin to develop. This crust would be broken up by convection currents, and portions would founder to be remelted. Some fragments would be rafted together by convection currents and welded by crystallization. Gas streaming through the liquid toward the surface would be trapped under these floating crustal fragments. Some gas would escape around the margins, spattering molten material onto the edges and gradually building the rafts laterally. Other, probably greater, portions of the trapped gas would escape through the rafts along fissures and explosion pipes, bringing molten material to the surface and building up the rafts by layered flows and ash falls. Thus thick floating masses would be built.

Eventually, crystallization would be essentially complete, producing a solid crust having relatively minor amounts of trapped fluid below. This final consolidation would take place when the molten layer became too thin for large-scale convective currents to operate. But two different types of crusts would be present: first, the thickened porous rafts of lava flows and ash beds, and second, a thinner more-dense crust developed over the broad areas which remained fluid the longest. Extreme differentiation of the silicate melt to a granitic residue probably was not important in the process, because at the later stages, crystallization was probably very slow, allowing for complete reaction between interstitial residual fluids and the crystal mush. However, at the late stages, large amounts of volatile materials were undoubtedly trapped in the upper mantle. These volatiles would have encountered two different types of crustal barriers in their upward escape. One was those segments which were to become oceanic segments, the relatively thin compact and dense rocks which would yield to gas pressure by rupture. Here the gases, enriched in the residual components from crystallization, would escape to the surface readily and rapidly, allowing little opportunity for reactions to take place between hot gas and cold rock. The other crustal type was the material which was to become the continental nuclei, the thickened rafted material. This crust, composed of vesicular lava flows, spatter material, and intercalated ash beds, would have been less brittle and therefore less easily fractured. The vast quantity of volatile material rising under such a crust would have been trapped and would have worked its way surfaceward slowly through such material. Thus time would have been available for reactions to occur between hot gas and cold rock.

The volatile phase evolved from the mantle at this early stage would probably consist of hydrogen, water vapor, carbon gases, and halogen gases. A number of highly volatile phases involving rock-forming elements could be formed by reaction between primary silicates and such gases. The silicon hydrides (SiH_4, Si_2H_6, Si_3H_8, and Si_4H_{10}) are highly volatile phases that could form by reactions such as

$$2MgSiO_3 + 4H_2 \rightarrow Mg_2SiO_4 + 2H_2O + SiH_4$$

converting primary orthopyroxene into olivine. Among the alkali compounds containing carbon are two interesting possibilities, the alkali carbonyls ($K_6C_6O_6$ and $Na_6C_6O_6$) and the alkali oxalates ($Na_2C_2O_4$ and KHC_2O_4), which could form by reaction of alkalis dispersed in primary minerals, such as amphiboles and pyroxenes, with carbon monoxide and carbon dioxide gases. All these compounds decompose readily at low temperature at low pressures, the carbonyls decomposing with violent explosion at atmospheric pressure. However, under mantle pressures these compounds theoretically should be stable. In this environment iron and aluminum could form volatile halogen salts ($AlCl_3$, $FeCl_2$, $FeCl_3$), but these volatilize at considerably higher temperatures than the alkali compounds or silica hydrides. There are no calcium or magnesium compounds with high volatility which would be likely to form. Degasing of the mantle therefore could remove large quantities of silica and alkalis upward, leaving a mantle enriched in magnesium and calcium and, to a lesser degree, iron and aluminum.

As such gases rose to shallow depth in the primitive crust, they would encounter the two different types of crust. The unstable alkali carbonyls and oxalates would break down at lower pressure, releasing large quantities of additional gas, and as a result the brittle thin crust of the primitive oceans would rupture and allow rapid escape of the volatile phase. In contrast, the thicker and more permeable rafted continental crust would allow the gas phase to penetrate upward until reduced pressure allowed the decomposition of the unstable phases. Thus alkali would be released at shallow depth in the crust in large quantities and with much available silica in the gas phase. Silica and alkali metasomatism on an extensive scale would result. Iron could be readily mobilized and removed as the carbonyl, as described by Emmons (1955).* Calcium and magnesium could also be readily removed as the soluble bicarbonates. Thus the primitive basalt of the crust would be granitized in its upper portions to yield a crust of granite above and basalt below, resting on mantle materials.

A second stage of granitization would ensue after the first stage of degasing in which volatiles could be released only by deep-seated dislocations in the mantle. The belt of fracturing indicated by intermediate- and deep-focus earthquakes under orogenic belts would become the source of liberated volatiles, and granitization would become less general and limited to distinct orogens. By late Precambrian time the continental crusts were thoroughly modified, and the generation of primary anatectic granite magma became possible. Although degasing of the mantle would not have been complete, the major portion of the excess silica and alkali would have been removed. Therefore, in younger orogenic belts, mantle volatiles supplied heat and energizing agents for mobilization of the crust but did not involve

*R. C. Emmons and J. Bach, Steel penetration, *Foundry*, April, 1955.

extensive metasomatic introduction of material. Remobilized crystal mushes and true magmas of granitic composition become dominant.

The materials removed from the mafic crustal material were probably carried to the surface for the most part. Some would have been tied up in the upper crust in the basification of metamorphics marginal to the metasomatic granites. However, the relatively minor amounts of such rocks in the shield areas will not account for the necessary volume of removed iron and magnesia. Another minor proportion is represented in the alkali-mafic-rich lamprophyres. However, these rocks seem to be much more closely related to the second-stage granitization and younger remobilized rocks. The great Precambrian iron formations may represent iron removed and redeposited as chemical sediments. The iron could have been initially brought to the surface at hot springs. Iron formations are known throughout the Precambrian column but are most abundant following the Algoman Revolution (2.4 billion years ago) and extensive regional granitization. The absence of abundant alumina-rich sediments has always presented a problem in interpreting the iron formations as a product of lateritic weathering of mafic igneous rocks. The abundance of dolomite in the Precambrian and lower Paleozoic sections may be the result of magnesia contributed to the sea by crustal granitization. The greater solubility of magnesia relative to iron in sea water may account for the time lag between iron precipitation and dolomitization.

The author thus accepts several origins for granite bodies and evaluates them as follows:

1. Granite derived by differentiation of mafic magmas: a relatively minor process operative throughout all geologic time.
2. In situ granitization of preexisting volcanics and sediments in the solid state: dominant in the development of the continental crust through mantle degasing; the dominant process in early Precambrian time, becoming less important with time.
3. Remobilized granite without development of much fluid phase: developed by homogenization and mobilization of older granitization products; important in the late Precambrian and Paleozoic.
4. Primary granite magmas: developed by large-scale melting of crustal rocks; a minor process in late Precambrian time, becoming the dominant process in Mesozoic orogenies.
5. Marginal granitization: small-scale metasomatic effects above and marginal to magmatic bodies.

5 IGNEOUS-ROCK PETROGRAPHY

TEXTURE AND STRUCTURE

Texture of igneous rocks refers to those features dependent on size, shape, and arrangement of the component mineral grains. Texture is the result of processes involved before, during, and after crystallization; therefore the textural terms grouped below are roughly in the order in which they develop from the crystallizing melt. Structure in igneous rocks refers to larger-scale features relating polygranular units to each other. Many structural features are not identifiable in hand specimens but are recognizable in the field, i.e., jointing, sheet structure, columnar structure, pillow structure, etc. Several structures are of sufficiently small scale so that they are observable in hand specimens; these are considered here and include vesicular, amygdaloidal, and spherulitic structures. In each case the relationship between physically distinct complex units of the rock is involved, giving a structure rather than a texture. Such structures are considered below along with textures developed in the same general time span of the rock history.

I. Textures and structures inherited from the dominantly liquid phase
  A. Incipient crystals
    1. *Crystallites* (Vogelsang, 1863). The most rudimentary form of crystals, as yet too small to have distinct mineral properties. They are visible in hand specimens of glasses as minute dustlike particles on thin transparent edges (obsidian) and are generally of unidentifiable shape. They are seen microscopically as granules, rods, hairs, and dendritic or feathery objects.
    2. *Microlites* (Vogelsang, 1867). Incipient crystals large enough to have identifiable mineral properties, although still of microscopic size. The aphanitic groundmass of extrusives is usually composed of crystals of microlite size.
    3. *Skeletal crystals* (Fig. 5-1). Incompletely formed crystals in which the faces of maximum growth rate have developed, leaving hollow or hopper-shaped interiors. Normal snowflakes are skeletal crystals.

B. Phenocrysts

Megascopically visible crystals which stand out prominently from a finer-grained groundmass (Iddings, 1889) (Fig. 5-2). In igneous rocks the larger size is usually the result of early slower cooling. Phenocrysts may also represent crystals which were suspended in the liquid and which were never melted from the parent rock during magma generation.

1. Shapes of phenocrysts

   a. *Euhedral crystal* (Cross, Iddings, Pierson, and Washington, 1906) (Fig. 5-3). A grain completely bounded by its own crystal faces. Because early crystals form in a melt and are not in contact with other crystals, their crystal form is nearly perfect. Synonyms are *automorphic crystal* (Rohrbach, 1886) and *idiomorphic crystal* (Rosenbusch, 1887).

   b. *Broken crystals.* Because of movement of the magma, suspended crystals are frequently broken and the fragments separated. Thus a crystal may be euhedral at one end and ragged or fractured at the other end.

   c. *Corroded and embayed crystals* (Figs. 5-4 and 5-21). Because of temperature changes in the magma, a phenocryst may become chemically unstable in contact with the melt and be partially reabsorbed. Thus a euhedral crystal becomes rounded and embayed by reaction.

2. Clustering and orientation of phenocrysts

   a. *Glomeroporphyritic textures* (Judd, 1886) (Fig. 5-5). A clustering together of phenocrysts into clumps. One common cause is the gravitative accumulation of phenocrysts while the magma is standing still and the subsequent breaking up and sweeping out of the crystals in the form of clumps as the magma is again moved.

   b. *Fluidal textures* (Vogelsang, 1867) (Figs. 5-6 and 5-7). An alinement of elements of the rock as a result of liquid flow. Phenocrysts are commonly alined in such a way as to offer the least resistance to movement. Also represented are parallelism of microlites in the groundmass and color banding of the groundmass.

C. Effects of loss of volatiles

1. *Vesiculation.* The development of cavities as a result of freezing-in of gas bubbles escaping from a magma. It is particularly applicable to extrusive rocks, where the resultant rock shows a vesicular structure.

   a. *Vesicle* (Lyell, 1865). A single gas cavity.

   b. *Scoriaceous structure* (Naumann, 1849) (Fig. 5-8). A highly vesicular structure similar to cinder. Gas cavities are very numerous and are roughly equidimensional, separated by thin walls of glass or aphanitic material.

   c. Pumaceous structure (Fig. 5-9). A rock froth, excessively cellular and usually having very thin glass walls separating the gas bubbles. The bubbles are frequently drawn out into elongated tubes due to flow of the froth.

  2. *Amygdaloidal structure* (Figs. 5-10 and 5-11). Rocks in which gas cavities have subsequently been filled with introduced mineral material. An amygdule is a single mineral-filled gas cavity.

D. Other features of the liquid phase

  1. *Xenolith* (Sallas, 1894) (Figs. 2-6 and 2-19). An included fragment of the wall rock in an igneous body.

  2. *Xenocryst* (Sallas, 1894). An included single crystal from the wall rock of an intrusion or one picked up from the floor over which a lava flow moved. Thus quartz-sand grains are often seen in mafic lava flows which are otherwise free of quartz.

  3. *Pillow structure* (Figs. 5-12 and 5-13). Discontinuous spheroidal or ellipsoidal masses of mafic lavas which fit together closely like a pile of squashed marshmallows. This structure is generally interpreted as indicating subaqueous extrusion. A synonym is ellipsoidal lava.

Fig. 5-1. Skeletal crystals of mellilite from slag. Isle Royal Smelter, Houghton, Michigan.

Fig. 5-2. Phenocrysts in an aphanitic groundmass. Euhedral and broken crystals of potash feldspar and plagioclase. Trachyte porphyry, Bannockburn Township, Ontario. (*Ward's Universal Rock Collection.*)

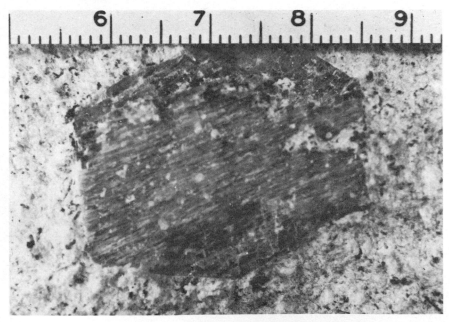

Fig. 5-3. Euhedral crystal of sanidine in an aphanitic groundmass. Trachyte porphyry, Drachenfels, Germany. (*Ward's Universal Rock Collection.*)

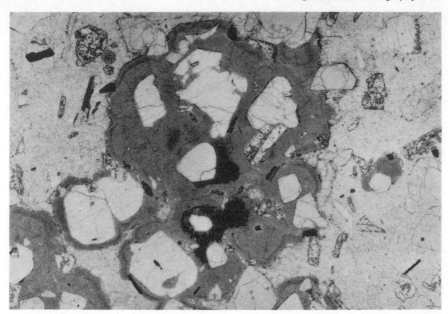

Fig. 5-4. Corroded and embayed phenocrysts in glass. Essentially colorless quartz including glass along embayments, clear rounded sanidine, and clouded euhedral plagioclase and black biotite in a glass dusty with crystallites. Vitrophyre, White Pine County, Nevada. Projection of a thin section.

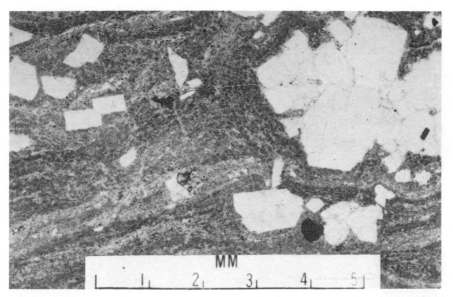

Fig. 5-5. Glomeroporphyritic texture. Clusters of feldspar phenocrysts in an aphanitic groundmass with well-developed flow banding. Rhyolite porphyry, Chaffee County, Colorado. Projection of a thin section. (*Ward's American Rock Collection.*)

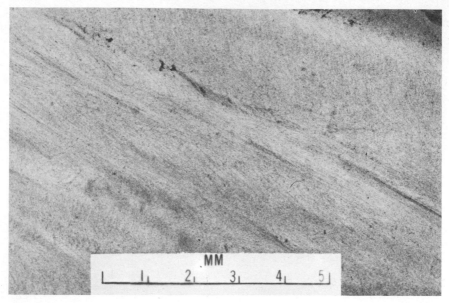

Fig. 5-6. Fluidal texture. Flow banding in obsidian, visible because of variable concentrations of crystallites in the glass. Lake County, Oregon. Projection of a thin section. (*Ward's American Rock Collection.*)

Fig. 5-7. Fluidal texture. Flow banding in an aphanitic rhyolite, visible because of slightly variable texture as well as variable color. Chaffee County, Colorado. (*Ward's Universal Rock Collection.*)

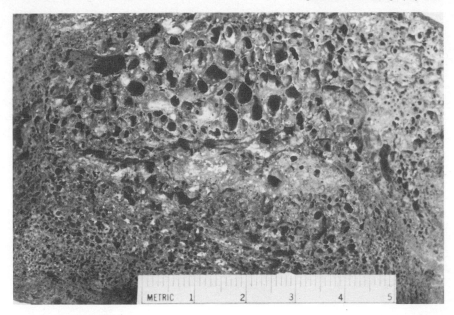

Fig. 5-8. Scoriaceous structure in basalt. The rock can be called scoria. Acco, Idaho. (*Ward's Universal Rock Collection.*)

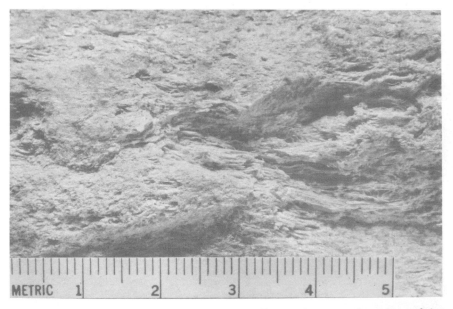

Fig. 5-9. Pumaceous structure. Note the very fibrous character of portions of the specimen due to flow elongation of the gas cavities. Pumice, Lipari Islands, Italy. (*Ward's Universal Rock Collection.*)

Fig. 5-10. Amygdaloidal structure. Gas cavities in the basalt are filled by concentric layers. The minerals include adularia, epidote, calcite, chlorite, quartz, and copper. Keweenawan basalt, Houghton, Michigan.

Fig. 5-11. Flow-alined amygdules in a basalt flow. Keweenawan basalt, South Range, Michigan.

Fig. 5-12. Pillow structure — section of a small pillow. The white band is chalcedony filling the space between two pillows. The chilled margin of a pillow is apparent below the chalcedony. Dark spots in the pillow are amygdules. Hemlock volcanics, Crystal Falls, Michigan.

Fig. 5-13. Pillow structure preserved in basalt metamorphosed to the greenschist facies. Kewatin volcanics, Marquette, Michigan.

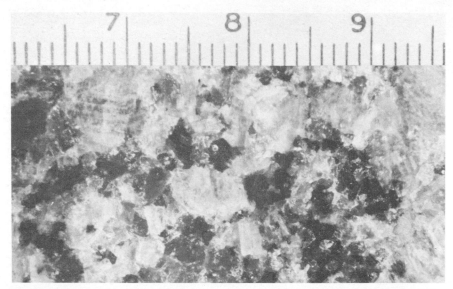

Fig. 5-14. Hypidiomorphic-granular texture in granite. Note the euhedralism of the small white plagioclase and the black biotite crystals, the subhedral to anhedral habit of the larger light-colored potash feldspars, and the anhedral habit of the gray quartz. Scale in mm. St. Cloud, Minnesota. (*Ward's Universal Rock Collection.*)

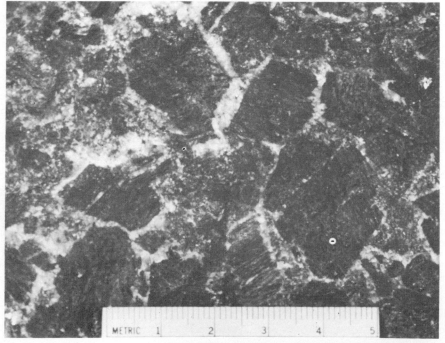

Fig. 5-15. Porphyritic texture. Phenocrysts of hornblende set in a phaneritic ground-mass of quartz and feldspar. Porphyritic diorite near Champion, Michigan.

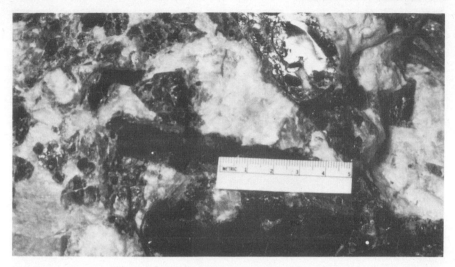

Fig. 5-16. Pegmatitic texture. Large-bladed biotite crystals, smaller white plagioclase and potash feldspar, with still smaller quartz crystals in granite pegmatite. East end of Greater Sacandagan Lake near Northville, New York.

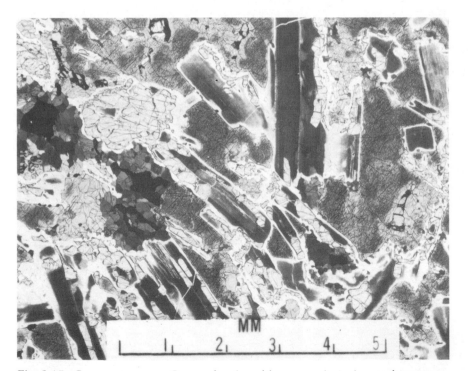

Fig. 5-17. Corona structure. Garnet developed between plagioclase and pyroxene. Elizabethtown, New York. Projection of a thin section. (*Ward's American Rock Collection.*)

Fig. 5-18. Trachytoid texture. Parallel alinement of tabular calcic plagioclase crystals in gabbro. Mellen, Wisconsin.

Fig. 5-19. Trachytic texture. Parallel alinement of sanidine microlites and pheno-crysts in trachyte porphyry. From a small dike near Benton, Arkansas. Projection of a thin section.

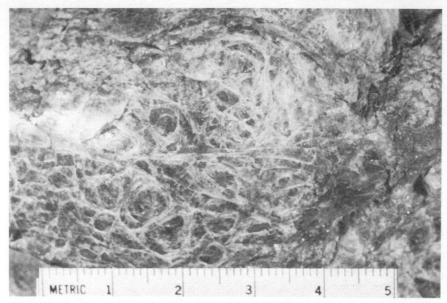

Fig. 5-20. Perlitic structure in glass. Perlite, New Mexico.

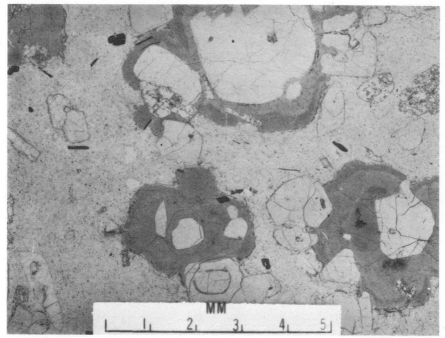

Fig. 5-21. The beginning of devitrification of the glass of a vitrophyre. In the irregular gray patches the glass has begun to crystallize, whereas in the clear areas it has not. Note the embayed quartz phenocrysts with devitrified glass in the embayments. White Pine County, Nevada. Projection of a thin section.

Fig. 5-22. Spherulitic structure. Ideal radial spherulites of mellilite in slag. Isle Royal Smelter, Houghton, Michigan.

Fig. 5-23. Skeletal spherulites in "snowflake" obsidian. Millard County, Utah. (*Ward's Universal Rock Collection*.)

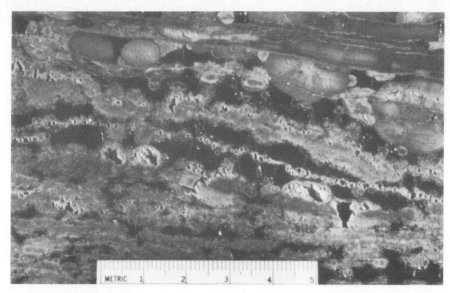

Fig. 5-24. Coalescing spherulites in obsidian. Yellowstone National Park, Wyoming.

Fig. 5-25. Lithophysae. Concentrically developed crystals of trydimite in cavities. Scale in mm. Yellowstone National Park, Wyoming. (*Ward's Universal Rock Collection.*)

Fig. 5-26. Small-scale columnar jointing in slag. Athletic Mining and Smelting smelter, Fort Smith, Arkansas.

Fig. 5-27. Cataclastic texture in anorthosite. The large crystal has been broken and granulated along the fracture and around the edges. Nain, Labrador. (*Ward's Universal Rock Collection.*)

II. Textures developed during the main solidification stage
   A. Degree of crystallinity
      1. *Holohyaline* (CIPW, 1906) (Fig. 5-6). Rocks which are completely glassy and lack megascopic crystals.
      2. *Hypohyaline* (CIPW, 1906) (Fig. 5-4), *hypocrystalline* (Rosenbusch, 1887). Rocks composed of crystals and glass; hypohyaline is dominantly glass, hypocrystalline is dominantly crystal.
      3. *Holocrystalline* (Rosenbusch, 1877) (Fig. 5-14). Rocks composed entirely of crystals.
   B. Shape of individual crystals
      1. *Euhedral* (CIPW, 1906) = *idiomorphic* (Rosenbusch, 1887) = *automorphic* (Rohrbach, 1887) (Fig. 5-3). A crystal completely bounded by its own faces.
      2. *Subhedral* (CIPW, 1906) = *hypidiomorphic* (Rosenbusch, 1887) = *hypautomorphic* (Rohrbach, 1886). A crystal bounded in part by its own faces and in part by surfaces developed through mutual interference of adjacent crystals.
      3. *Anhedral* (CIPW, 1906) = *allotriomorphic* (Rosenbusch, 1887) = *xenomorphic* (Rohrbach, 1886). A crystal not bounded by its own crystal faces but whose form is impressed on it by adjacent crystals.

Based on date of publication, Rohrbach's terms have precedence over both those of Rosenbusch and those of CIPW (Cross, Iddings, Pierson, and Washington). However, the CIPW terms are most commonly used in American literature and are followed in this book.

The shape of the individual crystals is usually indicative of the time of crystallization in igneous rocks. Thus euhedral crystals develop early in the liquid and without mutual interference. As crystals become more numerous and begin to interfere with each other, they become subhedral. Toward the late stages of crystallization, when the only space left for crystal growth is in the interstices between crystals, the new crystals must be anhedral, conforming with the available space.

   C. Rock textures defined by single-crystal shapes
      1. *Panidiomorphic-granular texture* (Rosenbusch, 1887) = *panautomorphic-granular texture* (Rohrbach, 1886). The texture of a rock composed essentially of euhedral crystals.
      2. *Hypidiomorphic-granular texture* (Rosenbusch, 1887) = *hypautomorphic-granular texture* (Rohrbach, 1886) = *granitic texture* (Fig. 5-14). The texture of a rock composed of a mixture of anhedral and either subhedral or euhedral crystals or both.
      3. *Allotriomorphic-granular texture* (Rosenbusch, 1887) = *xenomorphic-granular texture* (Rohrbach, 1886) = *aplitic texture*. The texture of a rock composed entirely of anhedral crystals. Aplitic texture usually implies fine grain size and sugary texture.

D. Rock textures defined by grain size
1. *Aphanitic texture* (Hauy) = *felsitic texture* (Fig. 5-7). A uniform fine-grained texture in which individual crystals are not visible to the unaided eye. Aphanitic texture is the preferable term, in that felsitic implies composition as well as texture.
2. *Phaneritic texture* (Fig. 5-14). A texture in which individual crystals are readily visible to the unaided eye.
   a. *Coarse grained.* Crystals are larger than 5 mm in diameter.
   b. *Medium grained.* Crystals range from 1 to 5 mm in diameter.
   c. *Fine grained.* Crystals are visible but less than 1 mm in diameter.
3. *Porphyritic texture* (Rosenbusch, 1882) (Figs. 5-2 and 5-15). The texture of a rock composed of two distinct grain sizes. It may consist of large crystals in a phaneritic groundmass or large to small crystals in an aphanitic or glassy groundmass. The larger crystals are termed *phenocrysts* (see p. 256). All crystals of one size range are spoken of as belonging to one *generation.* Thus a mineral occurring both as phenocrysts and as part of the groundmass is spoken of as two generations of the mineral.
4. *Pegmatitic texture* (Haidinger, 1845) (Fig. 5-16). The texture of a rock consisting of grains of a wide size range but conspicuously larger than the grain size of the associated parent rock. The term was earlier used by Hauy to indicate an intergrowth of quartz and feldspar in what is now called *graphic granite.*
E. Other textural features
1. *Corona structure, or reaction rim* (Fig. 5-17). A zone of one mineral surrounding another. The rim may be complete or partial, and it may occur only where two specific minerals would be in contact and not where either of those minerals were in contact with any other mineral.
2 Flow effects
   a. *Trachytoid texture* (Fig. 5-18). Parallel alinement of tabular or elongated grains due to movement of the crystal-liquid mush. The term *trachytic texture* (Fig. 5-19) applies specifically to trachytes and requires tabular orthoclase or sanidine.
   b. *Schlieren* (Reyer, 1877). Irregular streaks apparently resulting from a drawing out of partially assimilated xenoliths during flow of a mass. In the granite family of intrusives they are usually ferromagnesian-enriched and appear as dark diffuse streaks. Unassimilated xenoliths are frequently associated with schlieren and have their long dimensions parallel to the flow.
3. *Miarolitic structure* (Rosenbusch, 1887). Small angular gas cavities in phaneritic rocks into which crystals of the rock-forming minerals project.

III. Textures and structures developed after solidification
   A. Postsolidification features of glasses
      1. *Perlitic structure* (Haidinger, 1845) (Fig. 5-20). Concentric shelly cracks that develop in some glasses, probably as a result of contractions. It typically develops in glasses having a notable content of water.
      2. *Devitrification* (Fig. 5-21). The slow process of reorganization of a glass into crystals, usually of very small size. Devitrification frequently takes place first along perlitic cracks and then works out into the body of the glass.
      3. *Spherulitic structure* (Vogelsang, 1872) (Figs. 5-22 to 5-24). Spherical bodies from microscopic size to many feet in diameter which consist of radiating fibers and plates. This structure typically develops in silicic glasses and consists of fibers of feldspar and quartz. In many spherulites, flow structure of the surrounding glass may be traced through the bodies, indicating that they formed after flow ceased. An individual body is called a *spherulite*.
      4. *Lithophysae* (von Richtenhofen, 1860) (Fig. 5-25). Spherulites which consist of concentric shells having hollow interspaces. The cavities are often lined with minute crystals of quartz, tridymite, cristobalite, feldspars, or fayalite. Origin unknown.
   B. Postcrystallization features of crystalline rocks
      1. *Columnar jointing* (Fig. 5-26). The rupture of rock into polygonal prisms, probably due to shrinkage on cooling. Columns are generally perpendicular to the cooling surfaces.
      2. *Cataclastic textures* (Kjerulf, 1885) (Fig. 5-27). Textures developed by the mechanical crushing of rock or mineral grains. Several levels of cataclasis are recognizable in igneous rocks.
         a. Bent grains. These are particularly noticeable as bent mica flakes, curved twin lamella on plagioclase, or curved cleavage surfaces.
         b. Ruptured grains. Ragged anhedral grains set in a matrix of crushed material derived from those grains; it approaches a microbreccia.
         c. Mylonitized rock. Rock reduced to a powdered material.

Some specialized textures and structures found in rocks of only one family will be defined under the rock types to which they are applicable.

CLASSIFICATION

The classification presented in Table 5-1 is based entirely on mineral composition. Three variables are considered. The first, indicated at the top of the table, is silica saturation of sialic minerals, as noted by the presence of

Table 5-1.   Classification of Igneous Rocks Based on Mineral Composition

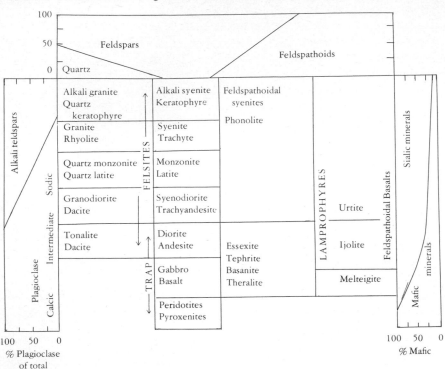

quartz or feldspathoids. In this way four columns are developed: (1) rocks having excess silica as indicated by the presence of quartz, (2) saturated rocks in which the sialic minerals are feldspars with or without minor quartz or feldspathoids, (3) undersaturated rocks in which feldspathoids proxy for feldspars in important amounts, and (4) highly undersaturated rocks in which feldspars are entirely replaced by feldspathoids. The second variable, indicated on the left of the table, is the type and composition of feldspar. The determinants are the ratio of alkali feldspar to plagioclase and the composition of the plagioclase. This gives six horizontal rows which are applicable to the feldspathic rocks. The third variable, indicated on the right of the table, is the ratio of sialic minerals (quartz, feldspar, and feldspathoids) to mafic minerals. This is generally applicable to the entire table but is particularly applicable to those rocks which carry little or no feldspar.

Textural variations are not entered in this table; however, both the phaneritic and aphanitic-porphyritic equivalents are entered in each pigeon hole. In classifying aphanitic-porphyritic rocks, classification must be made on the basis of the character of the phenocrysts, making the assumption that they are a representative sample of the total rock. This is not a valid assumption, but it is the only thing which can be done in mega-scopic work. Two rock terms have been entered which are applicable to nonporphyritic-aphanitic rocks: *felsite*, applicable to the lighter-colored rocks, and *trap*, applicable to the dark-gray or black rocks. The lamprophyres are difficult to tabulate in any manner, but here they are entered with the mafic feldspathoidal rocks. Megascopically, they are characterized by an abundance of ferromagnesian phenocrysts in an aphanitic groundmass.

The remainder of this chapter is a detailed discussion of the following eight major families:

1. Granite-granodiorite
2. Syenite
3. Feldspathoidal syenite
4. Diorite
5. Gabbro
6. Feldspathoidal gabbro
7. Peridotites and pyroxenites
8. Glasses and pyroclastics

It is suggested that once the family to which a rock belongs is identified, the classification and varieties be checked in the discussion of the family.

GRANITE-GRANODIORITE FAMILY

Definition

The granite family includes silica-rich rocks composed of essential quartz and potash feldspar, with or without sodic plagioclase as a separate phase. The main rock terms are defined as follows:

Ratio of potash felspar to plagioclase	Phanerite	Aphanite
9 : 1	Alkali granite	Quartz keratophyre
9 : 1–2 : 1	Granite	Rhyolite
2 : 1–1 : 2	Quartz monzonite	Quartz latite
1 : 2–1 : 9	Granodiorite	Dacite

The main textural varieties are named for granite:

Entirely aphanitic	Felsite
Aphanitic + < 5% phenocrysts	Rhyolite
Aphanitic + 5–50% phenocrysts	Rhyolite porphyry
Aphanitic + > 50% phenocrysts	Granite porphyry
Phaneritic nonporphyritic	Granite
Phaneritic porphyritic	Porphyritic granite

Inasmuch as the potash feldspar to plagioclase ratio cannot be determined in entirely aphanitic rocks, the term *felsite* is applied to this group.

Mineralogy of the Phanerites

Quartz. Essential in all members of the family as a minimum of 10 percent of the rock, averaging 25 to 30 percent.

Color. Usually colorless, milky, or smoky. Color is often variable in single specimens because of transparency of grains and reflections from internal fractures.

Habit. Usually a late mineral to crystallize; therefore it is anhedral but as roughly equidimensional grains. Crystals are commonly clustered into larger aggregates. Less commonly, quartz crystallizes early and will occur as stubby hexagonal dipyramids having dull pitted crystal faces.

Feldspars. Essential in all members of the family as a minimum of 40 percent of the rock but generally 50 to 70 percent.

Potash feldspars. Orthoclase is more common; microcline is less common; both may occur in the same rock and be differentiated by their different colors. Perthitic intergrowths with plagioclase are very common and may be visible megascopically in coarse-grained rocks.

Color. Pink or red and white to gray are the commonest colors; yellowish and greenish tints are less common. Crystals are clouded and nontransparent.

Habit. Crystals are usually late in forming and mutually interfere with each other; they are therefore anhedral in detail. They tend to be blocky, having dimensions near $2a:b:2c$. Carlsbad twinning is common and tends to increase the apparent euhedralism and tabular appearance of the crystals.

Inclusions. Ferromagnesian minerals as dark granules are common and may be distributed in a zonal arrangement. Plagioclase inclusions in perthites may be visible as minute irregular patches or lenses oriented parallel to the (100) face. Quartz and potash feldspars often mutually include each other, indicating simultaneous crystallization.

Alteration. Clay minerals are common, giving a white or yellowish powdery material. Alteration is first noticeable by a loss of luster on the cleavages and then by a loss of hardness.

Plagioclase. A sodic plagioclase which is normally in the oligoclase range of composition. All plagioclase of one generation will have the same composition, but there may be a difference between generations where the rock is porphyritic.

Color. Most commonly white or gray, but it may be greenish, yellowish, or pink. Crystals are clouded and nontransparent.

Habit. Plagioclase commonly crystallizes before quartz or potash feldspars; thus it tends to form euhedral to subhedral crystals of blocky to tabular habit. Dimensions are very commonly $3a:b:3c$. Polysynthetic twinning is characteristic, usually as very fine striations over the entire (001) cleavage. Less commonly, the striations are irregularly distributed and variable in width on the (001) cleavage. Carlsbad twinning is moderately common, combined with polysynthetic twinning.

Inclusions. Ferromagnesian minerals as minute grains are common. Zoning in the crystals may be visible as slight color differences or zones of inclusions (Fig. 2-13). Plagioclase is occasionally mantled by potash feldspar in parallel orientation (Fig. 5-30).

Alteration. As in potash feldspars. Low-grade metamorphism may convert plagioclase to saussurite, a fine-grained mixture of albite and epidote.

Ferromagnesian minerals. Not essential, but they usually make up 5 to 25 percent of rock. In general, ferromagnesian content increases with increase of plagioclase. Biotite is the commonest ferromagnesian mineral in plagioclase-poor rocks and hornblende the commonest in plagioclase-rich rocks. Pyroxenes are rare but are present in some special rock types.

Biotite

Color. Fresh biotite is dark brown to black, occasionally green, and has a splendent luster on the cleavage face. The prismatic zone usually has a dull luster which shows ridges and grooves parallel to the cleavages.

Habit. Euhedral to anhedral flakes. Euhedral crystals are usually hexagonal in outline, having dimensions near $10a:8b:1-5c$. Anhedral crystals usually have a ragged outline but show well-developed (001) faces. Biotite may be uniformly scattered through the rocks in parallel or random orientation. More commonly, it is clustered with quartz or with other ferromagnesian minerals.

Inclusions. Commonly, it is intergrown with muscovite in parallel lamellae. Biotite is included in quartz and potash feldspar and is usually euhedral against them.

Alteration. Most commonly, it is to chlorite, losing the luster of its cleavage surfaces, losing its flexibility, and becoming greenish. This alteration is probably deuteric. Weathering bleaches biotite to a golden luster and color.

Hornblende

Color. Black, having well-developed prismatic cleavage which gives a fibrous appearance.

Habit. Stubby prismatic crystals which show moderately well-developed prism faces but ragged terminations. Dimensions are usually $2a:2b:3-5c$. Hornblende most commonly occurs as clustered grains along with biotite and magnetite.

Inclusions. Hornblende is usually an early mineral; therefore it does not include other minerals but is included in quartz and feldspars.

Alteration. None is usually visible megascopically, although alteration to chlorite can occasionally be recognized.

Accessory minerals

Muscovite. Fairly common in biotite granites, but it is not usual in hornblende-bearing rocks. Some granites have no ferromagnesian minerals and contain up to 10 percent muscovite. Muscovite usually occurs as very irregular colorless or pale yellowish flakes. It is usually a late mineral and may not be primary.

Sodic amphiboles and pyroxenes. May occur as acicular to slender prismatic black to green crystals in alkali granites.

Magnetite and ilmenite. Common accessories which appear as minute grains, plates, or euhedral crystals. They are, in general, much smaller than the silicate minerals and are commonly included in them. Magnetite in particular tends to occur in clusters with biotite. Larger crystals show a bright metallic to submetallic luster, black color, and conchoidal fracture. Alteration to limonite or hematite is common, with a resultant staining of adjacent mineral grains.

Pyrite. A common accessory, usually as irregular rounded blebs.

Zircon. Present in most granites as microscopic rounded or euhedral crystals. Occasionally recognizable megascopically as brown euhedral prisms terminated by the dipyramid.

Apatite. Present in many granites as microscopic hexagonal prismatic crystals; not commonly visible megascopically.

Sphene. Occurs commonly as megascopically recognizable granules or euhedra. Honey-brown color, near-adamantine luster, and wedge-shaped habit are typical. Sphene is particularly common in some granodiorites.

Other accessories not as common as those listed above are:

Allanite	Garnet
Beryl	Monazite
Cassiterite	Sillimanite
Epidote	Topaz
Fluorite	Tourmaline

Mineralogy of the Porphyritic Aphanites

Groundmass. Usually dominates the rock and has quantitatively minor phenocrysts. In some dike rocks (granite porphyry) the phenocrysts dominate, and there is a smaller amount of aphanitic groundmass. Groundmass is usually light-colored: white, grays, tans, brown, or pink. Aporhyolite is commonly dark red. Texture ranges from extremely fine-grained and massive, having a subconchoidal fracture, to fine-grained granular, having a sugary appearance. Flow banding is common in the groundmass, showing up as bands of different color or texture or both (Fig. 5-7). Frequently, flow banding is more noticeable on weathered surfaces than on freshly broken surfaces. Vesicular cavities are common, either rounded or highly irregular, and often are lined with crystal druses.

Phenocrysts. Quartz and feldspars are most common, usually in euhedral, broken, or corroded and rounded forms. Biotite and hornblende are less common, and a pyroxene occasionally occurs. Accessory minerals are rarely visible as phenocrysts.

Quartz. Colorless and smoky varieties are most common. Crystals are usually rounded, corroded, and embayed. The groundmass adjacent to corroded crystals frequently has a noticeable thin zone of slightly different color or texture. Euhedral quartz crystals are usually in the high-temperature form of a hexagonal dipyramid which shows only narrow prism faces, if any.

Feldspar. Transparent colorless sanidine is typical as euhedral tabular crystals frequently mistaken for quartz, but sanidine's perfect cleavage distinguishes them. Nontransparent white, cream, or pink orthoclase crystals are less common. Sodic plagioclase is less common than potash feldspar.

Biotite. The commonest of the ferromagnesian phenocrysts. It usually occurs as thin hexagonal flakes and may be corroded or rounded but not usually ragged as in the phanerites. Black to green, depending on alteration to chlorite. Biotite frequently shows effects of reaction with the magma, developing a zone of magnetite granules embedded in the edge of the crystals.

Hornblende. Not common as phenocrysts. When present it appears as black prismatic crystals in which the axial portion includes the groundmass.

Pyroxene. Rare as phenocrysts. When present it occurs as black or light-green euhedra of stubby prismatic habit and shows well-developed terminations. Commonly altered to cream or greenish pulverulent clay material, retaining the euhedral crystal form.

Vesicle-filling minerals. Several minerals occur characteristically as drusy linings of gas cavities in extrusives of the granite family. These are usually interpreted as representing material concentrated by the gas phase.

Silica minerals (quartz, tridymite, and cristobalite). Quartz is usually as colorless crystals which are too small to determine if they are hexagonal or trigonal. Tridymite is more common as minute pseudohexagonal colorless plates, usually twinned. Cristobalite is less common but is easily recognized as small white to colorless spherical aggregates of intergrown octahedra.

Olivine (fayalite). The iron olivine is stable with silica minerals in this environment and occurs as yellow to brown transparent to translucent plates. An irridescent tarnish is common due to surface alteration.

Topaz. Euhedral prismatic yellow, brown, or colorless crystals are found lining or growing across gas cavities.

Less common crystals in this occurrence are apatite, tourmaline, anatase, and fluorite.

Varieties

Textural and structural varieties

Phanerites

Orbicular granite (Fig. 5-28). Spherical or ellipsoidal masses in a granitic groundmass. Minerals of the masses are arranged in concentric bands and may or may not be the same as the groundmass. Very commonly,

Fig. 5-28. Orbicular structure in granodiorite. The dark masses consist of concentric shells of biotite with secondary amounts of quartz and feldspars. Craftsbury, Vermont. (*Courtesy Dr. R. H. Konig.*)

the orbicules show abnormal concentration of the ferromagnesian minerals.

Rapakivi (Fig. 5-29). A porphyritic granite containing oval phenocrysts of potash feldspar mantled by plagioclase.

Graphic granite (Fig. 5-30). A megascopic intergrowth of parallel skeletal quartz crystals in large potash feldspars; typical of the feldspars in some pegmatites. A granophyre has the same texture on a microscopic scale.

Fig. 5-29. Rapikivi granite. Gray potash feldspar mantled by white plagioclase. Jonesport, Maine.

Aphanites

Aporhyolite. A completely devitrified glass which is dense, nearly non-porphyritic, and usually dark red or brown in color. Flow and other glass textures may be visible microscopically. Flow texture is sometimes visible megascopically on weathered surfaces.

Granophyre. A fine-grained intergrowth of quartz and alkali feldspar. It may form the aphanitic groundmass of rhyolites, or it may occur in the fine-grained phases of small intrusive bodies.

Spherulitic rhyolite (Fig. 5-24). Scattered individual spherulites to aggregates of spherulites without interstitial material.

Fig. 5-30. Graphic granite. An intergrowth of skeletal quartz crystals in potash feldspar, a texture very characteristic of pegmatites. From a granite pegmatite at the east end of Greater Sacandagan Lake near Northville, New York. (*Courtesy Dr. R. H. Konig.*)

Mineralogical varieties

Phanerites

Silexite. Abnormally quartz-rich rock; more than 50 percent quartz.
Alaskite. Granite having no ferromagnesian minerals.
Charnockite. Hypersthene is the ferromagnesian mineral.
Luxullianite. Quartz-tourmaline rock altered from a granite.
Unakite. Quartz-epidote rock altered from a granite.
Granite-family rocks may be named on the basis of the ferromagnesian minerals:

Biotite granite
Hornblende-bearing biotite granite
Biotite-bearing hornblende granite

Hornblende granite

Binary granite = muscovite-biotite granite (in older texts it referred to granite composed of quartz and only one potash feldspar)

Adamellite = quartz monzonite

SYENITE FAMILY

Definition

The syenite family consists of silica-saturated rocks in which the feldspar is in the potash-soda series. Slight oversaturation may be represented by the presence of quartz as less than 10 percent of the rock. Similarly, slight undersaturation may be represented by feldspathoids, usually nepheline, as less than 10 percent of the rock. The main rock terms with respect to evidence of silica content are:

Quartz present < 10% = quartz-syenite

No quartz or feldspathoid present = syenite

Nepheline present < 10% = pulaskite

The main rock terms with respect to feldspar types are:

Ratio of potash feldspar to plagioclase	Phanerite	Aphanite
9:1	Alkali syenite	Keratophyre
9:1–2:1	Syenite	Trachyte
2:1–1:2	Monzonite	Latite
1:2–1:9	Syenodiorite	Trachyandesite

The main textural varieties may be named for syenite:

Entirely aphanitic   Felsite

Aphanitic + < 5% phenocrysts   Trachyte

Aphanitic + 5–50% phenocrysts   Trachyte porphyry

Aphanitic + > 50% phenocrysts   Syenite porphyry

Phaneritic nonporphyritic   Syenite

Phaneritic porphyritic   Porphyritic syenite

Inasmuch as the nonporphyritic rocks cannot be classified according to feldspar type, the term *felsite* is applied to this group, as in the granite family.

Mineralogy of the Phanerites

Feldspars. The dominant mineral of the rock, controlling the texture and color and making up 50 to 100 percent of the rock, usually around 80 percent. Potash feldspars are present in all members of the family.

   Potash feldspars. Orthoclase and perthitic intergrowths of orthoclase and albite are the most common. Microcline is frequent but much less common than in the granite family. Anorthoclase is characteristic of smaller intrusive bodies.

Color. Normally pink or red; may be white or gray with a greenish or bluish cast. Anorthoclase frequently shows a play of colors.

Habit. Usually crystallizes after the ferromagnesian minerals and plagioclase, when plagioclase occurs as a separate phase; therefore potash feldspars are anhedral in detail. Orthoclase, perthites, and microcline are commonly tabular parallel to (010) and have dimensions around $3-4a:b:2-3c$. Well-developed Carlsbad twinning exaggerates the tabular habit and apparent euhedralism. Tablets are commonly parallel, giving a flow alinement. Anorthoclase tends to be blocky and equidimensional.

Inclusions. Grains of ferromagnesian minerals and accessories are commonly included in feldspar. Green needles of aegerine are commonly included in feldspar in the alkali syenites. Perthitic intergrowths are frequently visible megascopically. Mantled crystals with plagioclase cores and potash feldspar rims are occasionally visible.

Alteration. To clay minerals; first noticeable as a loss of luster of the cleavage surfaces, then as a progressive loss of coherence in grains to give a soft powdery material.

Plagioclase. A sodic plagioclase of the oligoclase range is normal as a primary separate phase. Albite occurs in perthites and occasionally appears to have replaced potash feldspars.

Color. Usually white; may be gray due to inclusions.

Habit. An early mineral; therefore it tends to be euhedral and have a tabular habit. Dimensions are around $3a:b:3c$. Polysynthetic twinning appears as very fine striations on (001) cleavage and is usually difficult to see megascopically. As plagioclase becomes more abundant in the monzonites, striations become more evident. Replacement albite is frequently so weakly striated as to be indistinguishable from potash feldspar.

Inclusions. Commonly includes the dark minerals; often included by potash feldspars.

Alteration. As in the potash feldspars.

Ferromagnesian minerals. Not essential, but it is usually present as 10 to 30 percent of the rock. Biotite and hornblende are the most common, but pyroxene is frequently important.

Biotite

Color. Dark brown to black, having a high luster on cleavage flakes when fresh. Frequently altered to chlorite, at which time it assumes a greenish color and loses the luster of the cleavage.

Habit. Often the first primary mineral to form; therefore it appears as euhedral tabular hexagonal plates and has dimensions around $4a:3b:$

1–2*c*. Commonly, the euhedra are corroded, resulting in anhedral rounded flakes. Biotite is often clustered with hornblende and surrounded by smaller prisms of hornblende.
Inclusions. Seldom includes any megascopically visible grains; often included in feldspars.
Alteration. To chlorite.

Hornblende. Less common than biotite when plagioclase is minor, hornblende becomes more common as plagioclase content increases.
Color. Black and lustrous when fresh, but it becomes duller and greenish as it is altered to chlorite. Prismatic cleavage gives it a fibrous appearance.
Habit. Frequently an early mineral as subhedral prismatic crystals with well-developed prism faces but poor terminations; dimensions are around $2a : 3b : 5{-}7c$. In other syenites, hornblende is contemporaneous with feldspar, forming anhedral equidimensional grains. It is commonly clustered with biotite.
Inclusions. May include feldspar and other late minerals along the *c* axis of prismatic crystals.
Alteration. Not usually noticeable, but alteration to chlorite may be seen.

Pyroxene. Augite and sodic pyroxenes occur in some syenites. Sodic pyroxenes are characteristic of alkali syenites, as are sodic amphiboles; however, sodic amphiboles are not megascopically distinguishable from hornblende.
Color. Augite is usually as black grains; sodic pyroxene is distinguishable only as green needles of aegerine.
Habit. Augite is as stubby prismatic crystals or equidimensional anhedral grains; aegerine is as acicular needles, singly or in radiating tufts at the margins of augite or along cleavage planes in feldspars.

Accessory minerals

Quartz. Present in quartz syenites as the last mineral to crystallize; therefore it is anhedral and interstitial and the grain shape depends on the interstices. Usually colorless to milky.

Nepheline. Present as an accessory in some syenites. If present as more than 10 percent, the rock belongs to the nepheline syenite family.
Color. Usually gray, pink, or white; occasionally yellow or bluish; greasy or waxy luster. Frequently difficult to distinguish from quartz, but luster and lack of marked conchoidal fracture are distinctive; weathering characteristics are most useful in distinguishing nepheline from quartz. It is also difficult to distinguish from sections of feldspar in which cleavage does not show. The absence of systematic cracking in nepheline aids here.

Habit. Late and anhedral; usually interstitial between feldspars, therefore taking the shape of interstices.

Alteration. Usually the earliest mineral to alter completely, yielding clay minerals. Therefore on a weathered surface, if interstitial areas between feldspars are filled with clay or are open cavities, nepheline may be indicated. Interstitial quartz would not weather out this way, so a weathered surface may distinguish quartz from nepheline.

Minor accessory minerals

Sphene. Fairly common as megascopically visible crystals or granules. Honey-yellow color, adamantine or near adamantine luster, and wedge-shaped habit are characteristic.

Magnetite. Usually present as minute granules clustered with ferromagnesian minerals or included in the feldspars.

Corundum. Occurs occasionally in syenites but more commonly in feldspathoidal syenites. Euhedral hexagonal prismatic crystals, brown to gray in color and of large size, are characteristic.

Minor accessories, usually of microscopic size, include apatite, zircon, allanite, and epidote.

Mineralogy of the Porphyritic Aphanites

Groundmass. Usually dominates the rock; phenocrysts are minor. Usually light-colored: white, gray, pink, yellowish, or greenish. It is generally microcrystalline and visibly granular (Fig. 5-3) and is not often dense, showing a subconchoidal fracture, as are many rhyolites. Microscopically, the groundmass is frequently composed of numerous parallel feldspar tablets, giving a trachytic texture (Fig. 5-19). This is noticeable megascopically as differences in luster or granularity in different directions. In slightly coarser material, the rock tends to split parallel to the (010) cleavage. Such surfaces have a less-lustrous surface, and grains appear rounded and equidimensional. Surfaces at right angles have a brighter luster and numerous parallel lathlike cleavages. Flow textures are common in finer-grained groundmass as bands of slightly contrasting colors or textures. Small angular to round gas cavities are common and may be lined by a druse of minute crystals. Granular groundmasses are particularly susceptible to weathering to argillaceous material, which is usually iron-stained.

Phenocrysts. Feldspars are by far the most abundant, and potash feldspars are the commonest and largest. Ferromagnesian minerals are not abundant as phenocrysts, but biotite, hornblende, and pyroxene may occur. Flow alinement of phenocrysts is common as parallel or subparallel alinement of tabular or prismatic crystals.

Feldspars. Clear colorless sanidine or white, yellowish, or pinkish orthoclase are the most common. Crystals are usually euhedral but may be broken or corroded and embayed. Sanidine is typically tabular parallel to 010, whereas orthoclase crystals are often elongated along the *a* axis. Carlsbad twinning is common, and zoning is often seen as contrasting colors or alined inclusions. Plagioclase is less common than the potash feldspars and is generally white or pink oligoclase. Polysynthetic twinning is often poorly developed or is present in only a portion of the crystal. Zoning is common in plagioclase and is visible as cores with mantles of different shades of pink. Weathering of the groundmass often frees euhedral unaltered feldspar phenocrysts.

Ferromagnesian phenocrysts
Biotite. Most common; occurs as thin hexagonal plates or round corroded flakes.
Hornblende. Less common; may occur alone or with biotite. Usually hornblende occurs as prismatic crystals without terminations and includes groundmass along the axis of the crystal.
Pyroxene. Most commonly occurs as minute needles of aegerine which, when small enough and abundant, give a green color to the groundmass. Occasionally, euhedral prismatic crystals of diopsidic-augite or aegerine-augite are visible.

Mineralogical varieties. Syenites may be named on the basis of the dominant ferromagnesian mineral:

Hornblende syenite
Augite syenite
Biotite syenite

Special names are based on the proportions of ferromagnesian minerals:

Nordmarkite. Albite-perthite quartz syenite having no ferromagnesian minerals.
Yogoite. Equal feldspar and ferromagnesian minerals.
Shonkinite. Ferromagnesian minerals are dominant over feldspars.

Syenite lamprophyres are dike rocks having ferromagnesian phenocrysts in an aphanitic groundmass:

Minette. Biotite phenocrysts.
Vogesite. Hornblende phenocrysts.

Other terms:

Laurvikite. Perthite-augite syenite; usually dark with a play of colors on feldspar.
Orthophyre. Orthoclase porphyry (obsolete).

FELDSPATHOIDAL SYENITE FAMILY

Definition

The feldspathoidal syenite family consists of alkali-rich rocks which are strongly undersaturated with respect to silica, as represented by the presence of at least 10 percent feldspathoids. This is a quantitatively minor group of rocks, but it has a wide variety of mineral assemblages and contains many rare minerals. A classification based on the ratio of potash feldspar to plagioclase is not necessary here, since the plagioclase in the more common members of the family is albite, which is difficult to distinguish megascopically from potash feldspars. A classification on the basis of feldspathoids present and on the mode of occurrence is most important, as follows:

Plutonic rocks (deep-seated or larger intrusives)
    Dominant feldspathoid

Nepheline	Nepheline syenite
Sodalite group	Sodalite syenite
Analcite	Analcite syenite
Leucite	Leucite syenite (very rare)

Hypabyssal rock (shallow or small intrusives)

Dominant feldspathoid	Texture	
Nepheline	Granular	Nepheline syenite aplite
	Porphyritic	Nepheline syenite porphyry
	Tinguaitic	Tinguaite
Leucite	Porphyritic	Leucite syenite porphyry
	Tinguaitic	Leucite tinguaite
Sodalite group	Tinguaitic	Sodalite tinguaite
		Noselite tinguaite

Extrusive rocks
    Without phenocrysts (felsite)
    With phenocrysts of feldspar plus feldspathoid

Nepheline	Phonolite
Nepheline + sodalite	Sodalite phonolite
Nepheline + leucite	Leucite phonolite
Leucite (no nepheline)	Leucite trachyte
Sodalite (no nepheline)	Sodalite trachyte

The tinguaitic texture is limited to these rocks and consists of a fine-grained granular aggregate of feldspar and feldspathoids having abundant randomly oriented needles of aegerine.

Mineralogy of the Phanerites

Feldspars. Essential to rocks of this family and must compose a minimum of 10 percent of the light-colored minerals. Most commonly, feldspar composes 40 to 80 percent of the rock, averaging around 60 percent.

Members of the potash-soda series of feldspars are most common primarily as perthites and anorthoclase. Albite as a distinct primary phase is less common. Sodic plagioclase in the oligoclase to andesine range occurs in some rare members of the family as a separate phase.

Color. Most commonly some shade of gray, from light to dark. Shades of green or red are not uncommon. When as a separate phase rather than in perthites, albite is often white. Crystals are clouded with minute inclusions of primary minerals or alteration products, and therefore are nontransparent.

Habit. May be either tabular parallel to (010) or equidimensional. Tabular crystals are often alined in a trachytoid texture so that the texture and luster of the rock differ in different directions. When broken parallel to flow planes, feldspars appear as oval or rhombic cleavages showing a dull luster. When broken at right angles, they appear as bright rectangular cleavages. Dimensions are around $4a:b:4c$. Crystals are anhedral in detail but show Carlsbad twinning, which increases the apparent euhedralism. Equidimensional crystals tend to be blocky but anhedral in detail of grain borders.

Inclusions. Minute grains and needles of the ferromagnesian minerals are commonly included. Perthitic inclusion of albite in potash feldspar can be recognized megascopically in coarser rocks. Albite may also mantle perthite or potash feldspar as a very thin rim.

Alteration. To clay minerals, but usually after alteration of nepheline. Orthoclase seems to be the least stable feldspar and albite the most stable.

Feldspathoids. Nepheline is by far the most important. Leucite occurs only very rarely in phaneritic rocks in some unusual dike rocks. Nepheline may be accompanied by the sodalite group (sodalite, noselite, hauynite) or by analcite. Cancrinite may be either primary or secondary after nepheline. The feldspathoids must make up at least 10 percent of the light-colored minerals, but usually compose 15 to 25 percent and may be as high as 40 percent.

Nepheline

Color. White, gray, yellow, greenish, or pink; most commonly white or grays; usually shows a greasy luster. Luster, lack of good conchoidal fracture, poor hexagonal prismatic cleavage, association, and weathering character distinguish it from quartz.

Habit. Occasionally as euhedral hexagonal prismatic crystals which may be stubby, giving hexagonal tablets, or elongated. More commonly, as anhedral grains mutually interfering with feldspar; often interstitial between feldspars as the last mineral to crystallize, therefore having the shape of the interstices.

Inclusions. Minute inclusions of earlier minerals, frequently in a zonal-growth distribution. These may appear megascopically as zones of slightly different color or luster. Acicular aegerine needles often cut through nepheline randomly, especially when the nepheline is late and interstitial.

Alteration. Deuteric or hydrothermal alteration to finely fibrous zeolites or to scaly yellow cancrinite is common. Weathering to clay minerals seems to be more rapid in nepheline than in feldspars; therefore, weathered surfaces of nepheline-bearing rocks are pitted, having feldspars standing out in relief, whereas nepheline is altered to clay.

Sodalite group. The different minerals can be differentiated only by chemical tests and are usually indistinguishable megascopically.

Color. Characteristically blue; less commonly white, gray, or pink.

Habit. Euhedral dodecahedral crystals showing good dodecahedral cleavages. Sodalite may also be late and interstitial between feldspars and nepheline, in which case it is often colorless, white, or gray and very difficult to recognize megascopically.

Inclusions. Common in euhedra with regular distribution through the crystal, giving zones or patches of different hue in the crystal. Noselite and hauynite often have dark rims due to inclusions.

Ferromagnesian minerals. Pyroxenes of the diopsidic augite-aegerine series are the most characteristic ferromagnesian minerals. Dark sodic amphiboles are frequently associated. Biotite is less common. Ferromagnesian minerals range from 5 to 30 percent of the rock, averaging around 15 percent.

Pyroxene. Dark euhedral stubby prismatic crystals or equidimensional anhedral grains of augite or aegerine-augite are characteristic. Associated with these are single needles or radiating tufts of green aegerine. The aegerine may appear as needles growing out of the aegerine-augite crystals, as tufts filling interstices, or as single crystals or clusters along cleavage in feldspar or cutting through nepheline. Many varieties of pyroxene can be found and be differentiated microscopically, but only these two can be differentiated megascopically.

Sodic amphiboles. Megascopically, they would be called hornblende. They appear as black lustrous prismatic crystals. Some may have a greenish or brownish cast. Typically, they are intimately associated with the pyroxenes in clusters.

Biotite. Brown euhedral hexagonal tablets are occasionally present. Small irregular biotite flakes often surround the other ferromagnesian minerals. Large crystals may be corroded and have a darker hardened rim.

Accessory minerals

Cancrinite. Typical of some syenites and may be either primary or secondary. Characteristically canary yellow but may be white, gray, or pink, in which case it is not readily identifiable.

Sphene. Common as honey-brown wedge-shaped euhedral or anhedral grains. Adamantine luster and hardness aid in differentiation.

Other common familiar accessory minerals present are magnetite, ilmenite, garnet, zircon, tourmaline, corundum, and fluorite. Calcite may be present as a primary mineral.

Rare accessory minerals include a wide variety of red and yellow titanium and zirconium silicates plus abundant rare earth metals.

Mineralogy of the Porphyritic Aphanites

Groundmass. Generally dark greenish or grayish colors; less frequently light tan or buff colors. Dense with greasy luster and platy to flat conchoidal fracture. Usually the groundmass dominates the rock, containing only a minor amount of phenocrysts. Flow structures may be prominent through zones of contrasting color or texture or as alinement of tabular crystals either in the groundmass or as phenocrysts.

Phenocrysts. Soda-potash feldspars and nepheline are commonest. Euhedra of leucite are less common. Sodic pyroxene and amphibole are the normal ferromagnesian minerals as phenocrysts.

Feldspar. Soda orthoclase or soda sanadine are commonest. Crystals are euhedral and either tabular parallel to (010) or prismatic and elongated on the a axis. Calsbad twinning is common. Crystals are usually white to gray but may be colorless and transparent. Alteration to fibrous zeolites is more common than alteration to clays.

Nepheline. Identifiable megascopically only when present as phenocrysts. Typical habit is as stubby hexagonal prismatic euhedra which are colorless, white, or gray. Deuteric alteration gives rise to zeolites; weathering gives clays. Nepheline or zeolites may be detected in the groundmass by their reaction with acid to form a silica gel. If a flat or sawed surface is treated with HCl for a few seconds it will be discolored, and the silica gel can be rubbed off the dried surface with one's thumb. The gel forms rubbery rolls resembling eraser crumbs.

Leucite. Gray trapezohedrons occur in some rare rocks. Crystals are euhedral or corroded and commonly contain irregular inclusions of the groundmass. A zonal structure can be seen in many larger crystals.

Ferromagnesian phenocrysts. Small prisms of sodic pyroxene are commonest, and even they are rarely visible megascopically. Crystals are usually stubby prisms having well-developed faces. Acicular green aegerine crystals may occasionally be seen with the augite. Hornblende occasionally occurs as minute black prisms.

Accessory minerals. Not usually visible megascopically; but sphene, apatite, and magnetite may be found. Biotite and olivine are rare accessories.

Varieties

Phanerites
Malignite. Over 50 percent ferromagnesian minerals.
Ditroite. Nepheline syenite having a granular texture.
Foyaite. Nepheline syenite having a trachytoid texture.
Litchfieldite. Gneissoid albite-nepheline syenite; often contains appreciable cancrinite.

Aphanites
Wyomingite. Phlogopite phenocrysts in a leucite groundmass. Some lamprophyres belong in the nepheline syenite family but, because nepheline is not identifiable megascopically, are discussed with the gabbro family.

DIORITE FAMILY

Definition

The diorite family consists of intermediate rocks in which plagioclase in the andesine range dominates the sialic minerals. Oversaturation with respect to silica may be indicated by the presence of quartz (quartz diorite = tonalite). Slight undersaturation may be represented by the presence of minor feldspathoids. With increased feldspathoids, diorites grade into the essexite-theralite family. The main textural varieties are named for diorite:

Entirely aphanitic	Felsite or trap
Aphanitic + < 5% phenocrysts	Andesite
Aphanitic + 5–50% phenocrysts	Andesite porphyry
Aphanitic + > 50% phenocrysts	Diorite porphyry
Phaneritic nonporphyritic	Diorite
Phaneritic porphyritic	Porphyritic diorite

Equivalent terms for more than 10 per cent quartz are:
Aphanitic = dacite
Phaneritic = tonalite

Inasmuch as nonporphyritic aphanites cannot be classified according to feldspar type, the term *felsite* is applied unless the rock is very dark gray to black, in which case it may be called *trap*.

Mineralogy of the Phanerites

Feldspars. Plagioclase usually dominates the rock, composing 40 to 80 percent of the rock, averaging 60 to 70 percent. Because the remainder is dominantly dark ferromagnesian minerals, the light and dark minerals appear subequal in proportions, and the rock assumes a gray or salt and pepper appearance. The normal tendency is to greatly overestimate the percentage of dark minerals. Potash feldspars are accessory minerals.

Plagioclase. Generally anhedral, randomly oriented, subequidimensional grains.

Color. White to gray, becoming yellowish, greenish, or red on alteration. Crystals are almost always clouded with inclusions, therefore are seldom transparent.

Habit. Anhedral and roughly equidimensional to tabular parallel to (010); dimensions are then around $2-3a:b:2-3c$. Apparent euhedralism is enhanced by polysynthetic twinning; the straight lines of the twin lamella are more readily seen than the irregular or sutured boundaries of the grains. Polysynthetic twinning is prominent in fresh unaltered grains. It may be either fine uniform lamellae distributed uniformly over the (001) faces or of variable width and distributed irregularly. Alteration of feldspar destroys the twinning so that in many rocks it is difficult to see.

Inclusions. May be zonally arranged, especially with abundant minute inclusions in the core of the crystal and none in the rim. Minute inclusions of ferromagnesian minerals are particularly common. Thin rims of potash feldspar over plagioclase are occasionally seen.

Alteration. Deuteric alteration of feldspar in diorite is common. Alteration to saussurite, a fine-grained mixture of albite and epidote, is more common than in the granite family but less common than in the gabbro family. Feldspars are converted to a dense greenish mass, losing cleavage and twinning. Alteration to sericite or kaolinite is common and may be zonal, having an altered core and an unaltered rim. Chlorite may develop in feldspar as a result of chloritization of ferromagnesian minerals. Similarly, iron staining may be developed as ferromagnesian minerals alter. Weathering alters feldspars to clays.

Potash feldspars. Compose less than 10 percent of the feldspar. If greater than 10 percent, the rock is a granodiorite or syenodiorite, depending on the presence or absence of quartz. Therefore potash feldspar is an accessory mineral. When present it crystallizes after plagioclase and before or with any quartz. It appears as mantles on plagioclase, as interstitial grains, or as interstitial micrographic intergrowths with quartz. In the last habit it is often pink and has a peculiar fine granular texture.

Quartz. An accessory in diorites but essential in tonalities, where it composes 10 to 30 percent of the rock. Usually the last mineral to crystallize and is therefore interstitial between feldspars and ferromagnesian minerals. Usually colorless and therefore takes on the color of its background.

Ferromagnesian minerals. Usually compose 10 to 50 percent of the rock, averaging around 30 percent. Because of their dark color in a lighter groundmass, their percentage is often overestimated. Hornblende is the characteristic ferromagnesian mineral, but biotite and pyroxene are common.

Hornblende
Color. Usually dark green to black. Fresh and unaltered it is normally black, showing lustrous cleavage surfaces; altered it becomes green, showing a dull luster or fibrous character.
Habit. Usually prismatic and more or less elongated along the *c* axis. Usually euhedral against feldspar but shows poor or ragged terminations. Commonly clustered with other ferromagnesian grains, where it is anhedral against pyroxene and euhedral or anhedral against biotite. A fibrous dark-green hornblende, uralite, is a common deuteric-alteration product of pyroxene and is therefore intimately intergrown with pyroxene.
Inclusions. Clusters of hornblende crystals may include corroded rounded grains of pyroxene in a reaction relationship. Similarly, biotite flakes may coat hornblende. Accessory minerals (magnetite, apatite, sphene, etc.) may be associated in the clusters of hornblende.
Alteration. To chlorite and to iron ores as magnetite or iron staining of adjacent minerals. Chlorite penetrates into and replaces hornblende as a fine-grained scaly aggregate.

Biotite.
Often the dominant ferromagnesian mineral in tonalites.
Color. Black to dark brown in unaltered material, becoming dull and green on alteration to chlorite. Biotite may be leached to a light golden brown by weathering.
Habit. Usually thin hexagonal tablets euhedral against feldspar and anhedral against the other ferromagnesian minerals. The flakes are commonly clustered with hornblende or are rosette clusters around magnetite.
Inclusion. Magnetite is commonly included in a cluster of small biotite flakes. Minute crystals of minor accessories (zircon, sphene, etc.) are common inclusions but are rarely visible megascopically.
Alteration. To chlorite, selectively along certain leaves of the biotite crystal, peripherally around the crystal, or completely. Leaching of biotite removes iron and produces golden-brown flakes.

Pyroxenes.
Both orthorhombic and monoclinic pyroxene are found in diorites, but monoclinic pyroxene is the more common.

Color. Usually dark green to black granular diopside or augite; platy greenish-black diallage is less common. Orthorhombic pyroxenes are more fibrous in appearance and are greenish-black hypersthene or brown bronzite.

Habit. Usually the earliest mineral to crystallize and therefore is euhedral or rounded and corroded. Diopside and augite tend to form roughly equidimensional grains which are anhedral due to corrosion reaction to form hornblende. Diallage is often platy parallel to (100) and may appear as stellate crystal intergrowths. Hypersthene tends to be prismatic and elongated parallel to the *c* axis. A fibrous appearance is typical due to three cleavages in one zone.

Inclusions. None usually visible in pyroxene; but pyroxene is commonly included in hornblende-biotite clusters.

Alteration. To uralite, a fibrous greenish amphibole in parallel crystallographic orientation. Less commonly to chlorite.

Accessory minerals

Magnetite, titaniferous magnetite, or *ilmenite* are usually the only megascopically recognizable accessories. Magnetite usually occurs as small anhedral granules particularly associated with the ferromagnesian minerals. Minute octahedrons may occasionally be found. Ilmenite tends to occur as platy crystals.

Sphene may be visible as small honey-brown wedges having an adamantine luster.

Apatite, zircon, and *garnet* are common accessories, but usually as microscopic grains or crystals.

Pyrite is common as irregular masses.

Mineralogy of the Porphyritic Aphanites

Groundmass. Highly variable in proportion to the phenocrysts, but it usually dominates the rock. Glassy, dense microcrystalline, and fine granular textures are all characteristic. Color varies from light to dark gray, red, brown, green, and almost black. Gas cavities are common as either rounded vesicles or intergranular pores in the coarser groundmasses. Flow textures are often seen as bands of slightly different color or texture or both. Flow alinement of phenocrysts is also common, in that elongated prismatic hornblende or tabular feldspar crystals may be more or less perfectly alined.

Phenocrysts. Plagioclase in the oligoclase to andesine range is abundant in association with any of the ferromagnesian minerals. Quartz is present in dacites as phenocrysts.

Feldspar

Color. Usually white; may be completely colorless and glassy. Alteration results in yellowish, greenish, or red colors. Occasionally, two different types of plagioclase may be recognizable as phenocrysts.

Habit. Euhedral tabular crystals, often broken. Clustering of small crystals into intergrown aggregates is common. Polysynthetic twinning is highly variable. Commonly, the twin lamellae are of irregular width and are unevenly distributed over the (001) cleavage so that portions of the crystal are highly twinned and other portions are untwinned; this is characteristic of andesine compositions in these rocks. Very fine or very coarse uniform twinning is also seen. Zoning is very common and may be seen megascopically as contrasting color tints, alteration characteristics, primary inclusions, or twinning characteristics. Corrosion of crystals may be seen in the zoning, in that a euhedral clear zone may mantle a rounded corroded core. Similarly, broken crystals may be mantled by euhedral outer zones.

Inclusions. Minute dustlike particles are common, discoloring the crystal, but recognizable mineral inclusions are rare.

Alteration. Deuteric or hydrothermal alteration to saussurite, an albite-epidote mixture, is common, giving a dense fine-grained greenish aggregate. Chloritization occurs when the ferromagnesian minerals are extensively altered hydrothermally. Weathering produces a chalky white to yellow or red clay.

Quartz. Essential in the dacites. Occurs as colorless or white to gray hexagonal dipyramidal euhedra or extensively corroded crystals.

Biotite

Color. Black or brown; may be almost red in transmitted light through thin flakes. Becomes green and dull-lustered on alteration.

Habit. Hexagonal euhedral tablets, often corroded and embayed. Corrosion often makes the peripheries of the flakes darker-colored, harder, and granular appearing.

Alteration. To chlorite; becomes dull-lustered, greenish, and loses its highly micaceous cleavage.

Hornblende. The characteristic ferromagnesian mineral of the phenocrysts of andesites.

Color. Black and lustrous, having well-developed cleavages.

Habit. Elongated prismatic forms having poorly developed terminations. Crystals commonly include groundmass material along the *c* crystallographic axis (Fig. 5-31). Occasionally shows effects of reaction with the magma, in that the outer portion of the crystals is converted into a very fine-grained aggregate of augite and magnetite (Fig. 5-32). This causes chips of the material to adhere to a magnetized knife blade and destroys the cleavage of the surface layers.

Alteration. Usually fresh, but it may be altered to chlorite or iron-stained clay minerals.

Fig. 5-31. Hornblende phenocrysts in andesite porphyry. Note the well-developed prismatic habit but the poor terminations and inclusions of groundmass along the axis of the crystals. Mt. Shasta, California. (*Ward's Universal Rock Collection.*)

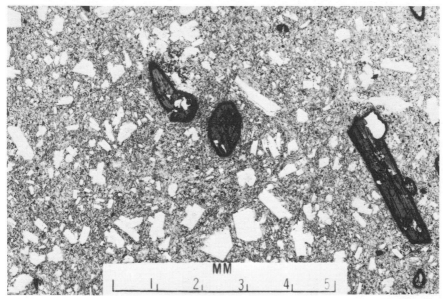

Fig. 5-32. Reaction rims on hornblende. Projection of a thin section of the same rock as Fig. 5-31, showing the dark granular rims on the hornblende. The rims consist of fine grains of augite and magnetite. The white phenocrysts are plagioclase. Mt. Shasta, California. (*Ward's American Rock Collection.*)

Pyroxene. The dominant ferromagnesian mineral in the groundmass. It occurs commonly as phenocrysts of augite or hypersthene or both. They cannot be differentiated easily unless the crystals are large enough so that the (100) cleavage of hypersthene is visible.

Color. Very dark green to black, having a duller surface texture and luster than hornblende.

Habit. Usually stubby prismatic crystals which show a well-developed prism zone and terminal faces. Rarely shows any corrosion effects.

Alteration. Hypersthene alters readily to serpentine. Augite is less susceptible to alteration but will alter to chlorite, iron-stained clays, etc.

Olivine. Minor in andesines and absent in dacites. Occurs in pyroxene-bearing rocks but is not normally found with biotite.

Color. Yellow-green glassy granules when unaltered, but alteration develops rusty red, black, irridescent, and dull-green colors.

Habit. Usually as rounded corroded granules, often intimately associated with or mantled by pyroxene. May occur as euhedral crystals tabular parallel to (100) and somewhat elongated on *c;* thus dimensions are around *a* : *2b* : *3c.*

Alteration. Several types of alteration are seen. Slight oxidation results in darkening of the crystals, often with the development of an irridescent tarnish. Further oxidation develops a rust-red color, but without appreciable alteration. Complete oxidation and leaching result in a powdery limonite residue. Hydrothermal alteration results in the development of serpentine or a scaly copper-red interleaving of serpentine and hematite. Serpentinization may be complete or partial along the periphery of the grain and internal cracks, leaving isolated granules of unaltered olivine.

Accessory minerals. Rarely visible megascopically. Hexagonal prisms of apatite, octahedrons of magnetite, wedges of sphene, or irregular granules and blobs of sulfides may be recognizable.

Varieties

Textural and structural varieties

Orbicular diorite, or *tonalite*, consists of spherical or ellipsoidal masses in a phaneritic groundmass. Minerals in the orbicules are usually arranged concentrically and may be the same as or different from those in the groundmass.

Andesite tuff and *breccia* are particularly common as accumulations of pyroclastic materials.

Mineralogical varieties. These are named on the basis of the ferromagnesian mineral:

Hornblende diorite = normal diorite

Biotite diorite = hornblende + biotite
Pyroxene diorite = hornblende + pyroxene
Hornblende-bearing biotite-diorite = biotite predominates over horn-
blende, etc.

Similarly:

Hornblende andesite = normal andesite
Pyroxene andesite = pyroxene proxies for hornblende
Pyroxene-bearing andesite = pyroxene plus hornblende in pheno-
crysts, etc.

Special names are based on the proportion of ferromagnesian minerals:

Leucodiorite. Very little ferromagnesian minerals.
Meladiorite. More than 50 percent ferromagnesian minerals.
Trondhjemite. Leucodiorite consisting essentially of quartz and
plagioclase.
Anorthosite (in part). More than 90 percent plagioclase in the oligoclase-andesine range. Indistinguishable megascopically from anorthosites of the gabbro family.

Some lamprophyres belong in the diorite family, but since the only recognizable minerals are the ferromagnesian phenocrysts, they will be discussed
with the gabbro family.

GABBRO FAMILY

Definition

Mafic igneous rocks composed essentially of plagioclase in the labradorite-
bytownite range plus pyroxene. The rocks are essentially saturated with silica. Slight undersaturation is expressed by the appearance of olivine. Slight
oversaturation is expressed by development of late-forming quartz if the
excess silica cannot be accommodated by the variable composition of feldspar and pyroxene. The main classification is thus based on the presence or
absence of quartz or olivine or both:

Quartz gabbro and quartz basalt
Gabbro and basalt
Olivine gabbro and olivine basalt

The main textural varieties may be subdivided for gabbro-basalt:

Entirely aphanitic   Trap
Aphanitic + < 5% phenocrysts   Basalt
Aphanitic + 5–50% phenocrysts   Basalt porphyry
Aphanitic + > 50% phenocrysts   Gabbro porphyry
Phaneritic nonporphyritic   Gabbro
Phaneritic porphyritic   Porphyritic gabbro

Special textural and rock terms are applied to dike rocks, small intrusions, and the central masses of thick flows, where the texture is phaneritic but not typically gabbroic. The textural terms are:

Diabasic texture (Figs. 5-33 and 5-34). Composed of tabular plagioclase with smaller interstitial granular pyroxene crystals.

Ophitic texture (Fig. 5-35). Small tabular plagioclase embedded in larger anhedral pyroxene crystals.

The resulting rock terms are:

Diabase. A nonporphyritic phaneritic rock having a diabasic texture. Dolerite is an almost synonymous term no longer used in American geology but retained by the British.

Ophite. A nonporphyritic phaneritic rock having an ophitic texture.

The following types of terms may be derived and are commonly used:

Olivine diabase
Quartz diabase
Olivine diabase porphyry, etc.

The term *trap* is applied to any nonporphyritic aphanitic rock which is very

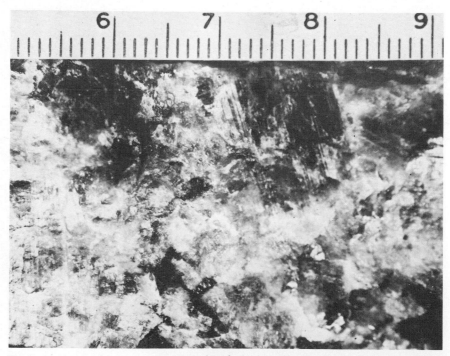

Fig. 5-33. Diabasic texture. Large subhedral plagioclase crystals containing anhedral interstitial augite (dark, particularly in bottom center). Split Rock, Minnesota. (*Ward's Universal Rock Collection.*)

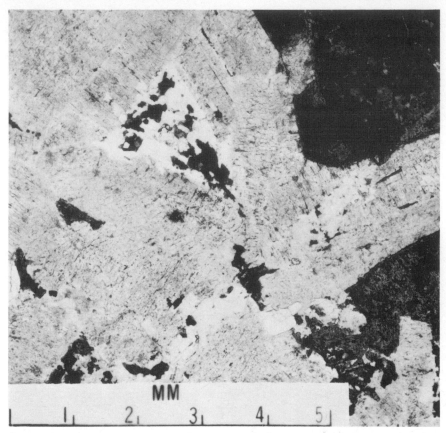

Fig. 5-34. Diabasic texture. Anhedral augite (dark) interstitial between subhedral plagioclase laths. The white interstitial material is a later sodic plagioclase developed as a result of assimilative reaction. The diabase was a xenolith in a younger granite. Mellen, Wisconsin.

dark gray or black, regardless of composition. It is thus equivalent to the term *felsite* as used for lighter-colored nonporphyritic aphanitic rocks.

Mineralogy of the Phanerites

Plagioclase in the labradorite, less commonly bytownite, range and pyroxene in subequal amounts compose 80 to 100 percent of the rocks. Olivine, biotite, iron ores, and apatite are the common accessories. Many large gabbro bodies have a layered structure with alternating layers impoverished in one constituent and enriched in another; thus feldspar-rich anorthositic gabbro may alternate with feldspar-poor melagabbro. The layers vary from a fraction of an inch to many feet in thickness.

Plagioclase

Color. Normally dark gray to black due to swarms of microscopic inclusions; less commonly light gray or green and glassy. Alteration produces opaque white, yellowish, green, or red materials.

Fig. 5-35. Ophitic texture. The bright area is a reflection from a single cleavage of a large augite crystal. The dark areas are tabular subhedral plagioclase crystals. Wichita Mountains, Oklahoma.

Habit. Generally tabular parallel to (010) and has dimensions around $5a:b:5c$ but may be more nearly equidimensional. Tabular crystals often tend to be oriented with (010) subparallel throughout the rock, giving a flow structure. Equidimensional crystals are more commonly randomly oriented. Crystals are anhedral in detail of contacts, but well-developed polysynthetic twinning often gives the appearance of straight-sided euhedralism. The (001) cleavages tend to be elongate or lath shaped, with well-defined polysynthetic-twinning lamellae running parallel to the lengths of the laths; (010) cleavages are un-twinned, duller in luster, and tend to be ovoid or rounded rhombus shapes. Albite twinning is usually coarse relative to the thickness of the crystal and very well-defined due to the large (001) $\wedge$ (010) angle. Carlsbad twinning is commonly associated with albite twinning. In many anorthosites and some gabbros the plagioclase shows a more or less well-developed cataclastic texture of highly angular fragmental plagioclase crystals set in a crushed groundmass of plagioclase. Zoning in plagioclase is not commonly visible megascopically.

Inclusions. Swarms of microscopic inclusions give plagioclase its dark color but are not visible individually. Inclusions of pyroxene and olivine crystals may be visible.

Alteration. Saussurite, a mixture of albite and epidote, is a common alteration product, particularly as a result of low-grade metamorphism. It is compact fine-grained, greenish or white, and lacks any feldspar cleavage or twinning. Weathering produces kaolinitic clays.

Pyroxene. Augite is the normal pyroxene of gabbros. It may be proxied for by hypersthene, in part or completely, in which case the rock is called norite rather than gabbro.

Color. Essentially black. Augite generally has a dull luster, and the cleavage is usually too poorly developed to be recognized. Some titaniferous varieties show a submetallic luster and conchoidal fracture and look very much like magnetite. This effect is probably due to abundant microscopic inclusions of rutile. Hypersthene shows much better cleavage, which results in a fibrous appearance. Alteration of pyroxene to uralite often results in a dark-green color.

Habit. Anhedral prismatic crystals are the rule, showing mutual inter-ference between pyroxene crystals and between pyroxene and plagio-clase. Hypersthene tends to be somewhat more euhedral than augite and appears as slightly elongated stout crystals fibrous parallel to the length. In gabbros with marked parallelism of tabular plagioclase, aug-ite tends to be interstitial between the feldspars, but the two minerals mutually interfere with each other.

Inclusions. Hypersthene commonly, and augite less commonly, include rounded granules of olivine in a reaction relationship. Pyroxene is often surrounded by secondary amphibole in parallel crystallographic orientation. Pyroxene may be surrounded by granular primary hornblende or flaky biotite.

Alteration. Deuteric or metamorphic alteration produces uralite from pyroxene. Uralite consists of fibers of hornblende or actinolite intergrown into the pyroxene in parallel crystallographic orientation. Uralitization may be partial or complete but usually develops from the surface of the pyroxene inward so that the habit, and sometimes even the cleavage, of the pyroxene is preserved. Chlorite, serpentine, and epidote with iron oxides are other common alteration products.

Accessory minerals

Olivine. Essential in olivine gabbros and in olivine norites. It averages 15 percent or less in most rocks in which it occurs, but it may be the only ferromagnesian mineral and have a higher percentage.

Color. Basically a glassy yellowish green. Alteration produces a variety of effects, including irridescent surfaces, black, red, and green colors.

Habit. A very early mineral; therefore grains are either euhedral or corroded. Euhedral crystals are most commonly prismatic, somewhat flattened parallel to (100), elongated on the c axis, and have dimensions around $a:2b:3-4c$. The magma reacts with olivine to produce pyroxene, resulting in rounded granules of olivine often included in pyroxene.

Alteration. Slight oxidation produces an irridescent tarnish on the olivine crystals and darkens the color. Further oxidation develops a rust-red color, but without loss of other physical properties. Complete oxidation and leaching leave a limonite residue. Deuteric alteration converts olivine to serpentine, first along the periphery and then along cracks through the crystal, leaving residual grains of unaltered olivine. The serpentine is usually dark-colored due to minute magnetite granules which developed simultaneously. Ultimately, the crystal is replaced completely by serpentine, but the serpentine which develops in the body of the crystal away from the cracks is free of magnetite and is therefore somewhat lighter-colored.

Hornblende. Usually minor; may proxy for augite and be primary, most commonly as uralite, or secondary after pyroxene.

Color. Primary hornblende is normally very dark to black. Secondary uralite is usually greenish to greenish black.

Habit. Primary hornblende is usually late when it is a minor constituent and occurs as thin mantles on pyroxene or as anhedral crystals associated with biotite. When hornblende proxies for pyroxene and is a major constituent, the rock would probably be classified as a diorite on megascopic examination. Secondary amphibole, or uralite, develops as a fibrous replacement of augite, as described previously.

Inclusions. Primary hornblende is often clustered with biotite and magnetite and may have pyroxene in the center of the cluster.

Alteration. To chlorite.

Biotite. More common than primary hornblende, but usually in minor amounts. Typically, it occurs as irregular flakes intergrown with the other ferromagnesian minerals as rosettes of flakes around iron ores or, less commonly, as discrete hexagonal plates.

Quartz. When present, it is interstitial between the plagioclase crystals. It may occur as discrete grains or be intergrown with potash feldspar in a granophyric texture. Skim-milk-blue quartz occurs as distinct grains in some norites.

Minor accessory minerals

Magnetite, titaniferous magnetite and ilmenite. Common in gabbros as black submetallic granules, plates, and large irregular grains. Often associated with the ferromagnesian minerals, particularly biotite.

Chromite. Common in some gabbros as black granules and octahedra and in some of the layered bodies may compose the bulk of certain layers.

Pyrite, pyrrhotite, and chalcopyrite. Common in some gabbros and norites as irregular intergrown masses which represent an immiscible sulfide melt. In some floored bodies the sulfides accumulate along or near the floor as economic ore deposits.

Apatite. Commonly present as megascopically recognizable yellow hexagonal prismatic needles which may be mistaken for olivine on the basis of color, but their greatly elongated hexagonal shape is characteristic. Minute apatite crystals occur in most gabbros.

Garnet. Often appears as red-brown granular material along the contacts between plagioclase and pyroxene. It may be a primary reaction product between the two minerals, but it probably often represents metamorphism of the rock.

Other accessories, rarely visible megascopically, are other spinel minerals as minute green or brown octahedra, graphite, rutile, titanite, etc.

Mineralogy of the Diabases and Ophites
Plagioclase and pyroxene dominate, with olivine, magnetite, apatite, and granophyric intergrowths of quartz and orthoclase as common accessories.

Plagioclase. Consistently lath-shaped and usually dark-colored. When small the crystals appear as narrow lath-shaped cleavages, showing two to several twin lamellae but generally a relatively small number. The crystals are usually randomly oriented. Although the (001) cleavages are rectangular, the borders are irregular and the crystals are anhedral in detail.

Pyroxene. Normally augite or the low-calcium monoclinic pyroxene pigeonite. These are not distinguishable megascopically. In rocks with a diabasic texture, the pyroxenes occur as small irregular or rounded granules interstitial between the plagioclase tablets. In this occurrence they are difficult to identify megascopically. They appear as small dull-black to greenish granules without notable cleavage. In ophitic-textured rocks the pyroxenes occur as relatively large anhedral grains enclosing the smaller tablets of plagioclase. In the Keweenawan volcanics of Michigan, the pyroxene increases in size at a rate of about 1 mm per 10 ft from the cooling surface. In thick flows and sills the pyroxene may exceed 1 cm in diameter. In these coarser sizes they appear as irregular equidimensional polygonal blocks, and the poor pyroxene prismatic cleavage can be recognized. Minute laths of plagioclase are randomly scattered through the pyroxene cleavages. Weathering often brings out the ophitic texture so that the surface of the weathered rock will be lumpy and pitted. Some pyroxenes will be so oriented that they weather more readily than others, developing the pits and leaving the less readily weathered pyroxenes standing out as lumps.

Granophyric intergrowth of very fine-grained quartz and potash feldspar. Characteristic of portions of many sills. The granophyre is a late residual liquid from differentiation and is interstitial between the feldspars of diabasic-textured rocks. It is often pink or red but may be white or gray, in which case it is difficult to recognize megascopically.

Accessory minerals. Magnetite, olivine, and apatite are commonly recognizable accessories and have the same characteristics as in gabbros.

Mineralogy of the Porphyritic Aphanites

Groundmass. Usually dominates the rock and has quantitatively minor phenocrysts. Commonly vesicular or amygdaloidal and contains gas cavities ranging from scattered small bubbles to extremely abundant and large cavities having thin fine-grained to glassy walls, as in scoria. Color is normally dark gray to black, but it may become brown or rust red due to oxidation. Scoriaceous surfaces of flows are particularly susceptible to oxidation. The texture of the groundmass is usually visibly granular, and individual minute crystals may be visible but undeterminable as to mineral. The groundmass often includes much brown to opaque-black glass, but this is masked by crystalline material to such an extent that the rocks seldom appear glassy. Flow structures are not usually visible in the groundmass, although phenocrysts may be alined or vesicles may be drawn out into tubes by flow. The groundmass weathers to a deep-red clay residue. Where glass is prominent, the groundmass is often more susceptible to weathering than are the phenocrysts.

Phenocrysts. In basalt and basalt porphyries, plagioclase or olivine or both are commonest; pyroxene and hornblende are less common. A group of dike rocks having a dark aphanitic groundmass and ferromagnesian phenocrysts is known as the *lamprophyres*.

Plagioclase. Labradorite to bytownite phenocrysts are common, ranging from minute crystals to groups several centimeters in dimensions.

Color. Often green, white, gray, or colorless; sometimes discolored pink or brownish by infiltration of iron oxides. When the crystals are small they usually appear very dark because of their background of dark material and their transparency.

Habit. Tabular euhedra are characteristic but are frequently broken or clustered into glomeroporphyritic textures; tabular on (010), having dimensions $5-10a:b:4-8c$. Twinning on the albite law usually results in relatively few, yet sharply defined, lamellae. Carlsbad twinning is common. Zoning may be recognizable megascopically because of zonal distribution of inclusions or zonal alteration.

Alteration. The green color is frequently due to incipient chloritization or saussuritization. Typical weathering is to kaolinite, more or less iron-stained.

Olivine. Commonest recognizable ferromagnesian phenocryst. May be the only phenocrystic mineral but is usually associated with plagioclase phenocrysts.

Color. Basically glassy yellow or green granules having a good conchoidal fracture. Alteration may produce irridescent tarnish or red, black, or green colors.

Habit. Prismatic euhedra are common, but most crystals have been rounded by magmatic corrosion. Basalts often contain clusters of olivine granules having almost no associated minerals in the clots (Fig. 5-36). They range from clusters of a few grains up to masses several inches in diameter. These apparently represent either gravitatively accumulated crystals swept off the magma-chamber floor by extrusion or xenoliths of deep-seated origin.

Alteration. The first stage of oxidation produces an irridescent surface tarnish. Further oxidation produces a rust-red discoloration. Complete alteration produces a limonite residue. Under other conditions a red alteration product known as *iddingsite* or a similar-appearing intergrowth of serpentine and hematite may develop. Serpentinization of olivine is not as common as in the coarse-grained rocks, probably because of the more ready escape of water vapor from extrusive rocks, but it is known.

Pyroxene. Usually augite, common as phenocrysts but often overlooked megascopically because of its similarity in appearance to the groundmass.

Color. Black.

Habit. Stubby prismatic crystals, often tabular parallel to (100). Dull luster and poor cleavage make it easily overlooked; however, on weathered surfaces the pyroxenes may stand out in relief.

Fig. 5-36. Olivine nodule in basalt. The nodule consists of 1- to 3-mm granules of olivine containing an occasional bright-green grain of chrome diopside and black grains of a spinel-group mineral. San Carlos, Gila County, Arizona.

Alteration. Usually fresh, except in old or metamorphosed rocks. Here chloritization plus development of iron oxides is common.

Hornblende. Black "basaltic hornblende" (also called lamprobolite and oxyhornblende) is a common phenocryst. Crystals are elongated jet-black prismatic units. Corrosion and alteration to a fine-grained mixture of augite and magnetite are common.

Vesicle-filling minerals. The minerals which fill the gas cavities of amygdaloids are usually low-temperature hydrothermal minerals. Portions of their constituents may have been derived from the primary minerals by the action of hot gases. Other constituents have been introduced from outside the rock by hydrothermal solutions or groundwater. The amygdule filling may be composed of only one mineral, but more commonly, it is composed of several minerals in concentric layers. The most common minerals are:

Zeolites
Chlorites
Rhombohedral carbonates

Epidote
Quartz, chalcedony, opal, etc.
Potash feldspars (adularia)
Kaolin, hematite, etc.
Ore minerals (copper in Michigan)
Many rare minerals

The lamprophyres. A group of aphanitic porphyritic dike rocks having phenocrysts of ferromagnesian minerals set in a dark alkali-rich groundmass. The groundmass usually contains abundant second-generation ferromagnesian minerals in a matrix of soda or potash-rich minerals or both, such as feldspars, nepheline, melilite, leucite, analcite, or sodalite-group minerals. Many of these are extensively altered to fibrous zeolites. The classification of the lamprophyres is based on the type of phenocrysts and the groundmass minerals. Therefore, megascopically they cannot be accurately classified and should be called biotite lamprophyre, hornblende-augite lamprophyre, etc. Ferromagnesian-phenocryst combinations may include one, two, or three of olivine, augite, hornblende, and biotite, except that olivine and biotite are not normally associated.

Biotite. Very common as euhedral hexagonal tablets, often several centimeters in diameter. The crystals may be irregularly corroded and often show a dark reaction rim around the periphery of the crystal. Crystals usually are randomly oriented but may show some flow alinement and bending of crystals. Biotite is very dark brown with black rims; it weathers to a pale brown.

Hornblende. Common as large stubby prismatic euhedral crystals or as elongated prismatic euhedra. Color is very dark brown to black. Corrosion effects and dark granular reaction rims are common. The crystals often weather out in relief.

Pyroxene. Usually augite, titaniferous augite, or sodic augite, which are indistinguishable megascopically. The pyroxenes are usually smaller than biotite or hornblende and are difficult to recognize because of their dark color and poor cleavage, which makes them merge into the groundmass. Usually as prismatic euhedra or somewhat tabular parallel to (100). The crystals commonly weather out in relief. Partial alteration to serpentine is common.

Olivine. Usually as small euhedra more or less completely serpentinized. The crystals have a characteristic prismatic habit terminated by a dipyramid. Serpentinization gives a green or iron-stained brown color, whereas fresh material is glassy and yellow-green.

Many lamprophyres contain calcite, feldspathoids, or zeolites in the groundmass and are therefore subject to reaction with HCl. The feldspathoids and zeolites leave a silica gel on reaction with acid. This can be detected by rubbing a dried, acid-treated, flat surface with one's thumb. The silica gel rubs off like rubber-eraser crumbs.

Varieties

Textural and structural varieties

Phanerites

Orbicular gabbro. Consists of spherical or ellipsoidal masses in a gabbro groundmass. The minerals of the orbicules are arranged concentrically.

Pegmatitic gabbro. Local patches of pegmatitic texture and gabbro composition.

Gneissoid gabbro. Metamorphically reconstituted gabbro.

Aphanites

Scoria. Scoriaceous basalt (Fig. 5-8).

Vesicular basalt.

Amygdaloidal basalt (Fig. 5-10).

Glomeroporphyritic basalt.

Ellipsoidal basalt = pillow lava (Fig. 5-13). Lava flow broken up into close-fitting ellipsoidal masses, presumably because of subaqueous extrusion.

Mineralogical varieties

Phanerites

Anorthosite. Dominantly plagioclase with less than 10 percent other minerals.

Norite. Gabbro with orthorhombic pyroxene in place of or associated with monoclinic pyroxene.

Troctolite. Gabbro with olivine the dominant ferromagnesian mineral.

Cumberlandite. Magnetite-rich olivine gabbro.

Aphanites

Spilite. Albite basalt; albitization may be primary or metasomatic.

Tholeiite. In the modern sense, a silica-saturated or oversaturated basalt. Usually a chemical analysis is required to differentiate it from an alkaline olivine basalt. Tholeiites are usually olivine-free or olivine-poor and differentiate toward a granophyric residue.

Oceanite. Olivine-rich basalt.

Ankaramite. Pyroxene-rich olivine basalt.

Lamprophyres. The more important lamprophyres are listed below for the sake of familiarity with the terms.

Name	Groundmass	Phenocrysts
Minette	Orthoclase	Biotite
Vogesite	Orthoclase	Hornblende and/or pyroxene
Kersantite	Sodic plagioclase	Biotite
Spessartite	Sodic plagioclase	Hornblende and/or pyroxene
Odinite	Calcic plagioclase	Hornblende and/or pyroxene
Camptonite	Nepheline and plagioclase	Hornblende and/or pyroxene
Monchiquite	Zeolite and analcite	Olivine, pyroxene, hornblende
Ouachitite	Analcite and zeolites	Biotite ± hornblende and pyroxene
Alnoite	Melilite	Olivine, hornblende, and/or pyroxene

FELDSPATHOIDAL GABBRO FAMILY

Definition

The feldspathoidal gabbro and basalt family consists of ferromagnesian-rich rocks in which feldspathoids proxy in part or completely for feldspar. The feldspar is a calcic plagioclase, usually in the labradorite range, with or without potash feldspars. Feldspathoids present are dominantly nepheline, leucite, analcite, or melilite and may have minor associated sodalite groups minerals. The rocks are quantitatively very minor. They occur as dikes or differentiated sills or as lava flows, but rarely in large bodies. Thus no direct equivalents from one crystallization environment to another can usually be recognized. The feldspathoids, with the exception of leucite, are late-forming minerals distributed interstitially in the groundmass and are therefore not usually recognizable megascopically. As a result these rocks would commonly be indistinguishable from normal gabbros and basalts. The following material merely gives a basic classification applicable to those rocks in which feldspathoids may be visible.

Varieties

Phanerites with calcic plagioclase

Essexites:	Calcic plagioclase	25%
	Potash feldspar	10%
	Nepheline	10%
	Augite and hornblende	40%
	± Analcite, olivine, biotite, apatite, ores	
	Dark gray to black, granitic texture	
Theralite:	Calcic plagioclase	35–15%
	Nepheline	5–10%
	Augite and olivine	50–70%
	± Biotite, amphibole, apatite, ores	
	Spotted, dark gray, granitic texture	

Teschenite:	Calcic plagioclase	30–10%
	Analcite	15–10%
	Hornblende and augite	50–75%

± Olivine, biotite, orthoclase, apatite, ores
Gray to dark gray, porphyritic texture with phenocrysts of augite and labradorite

Phanerites without feldspar

Urtite:	Nepheline	70–80%
	Hornblende or pyroxene	25–15%

± Other feldspathoids, apatite, ores
Ferromagnesian minerals are highly sodic
Medium grained, light colored

Ijolite:	Nepheline	50–70%
	Pyroxene	45–25%

±Biotite, other feldspathoids, apatite, titanite, ores
Mottled gray, granitic texture

Melteigite:	Nepheline	24–45%
	Ferromagnesian, pyroxene, ±biotite or horn-blende	70–50%

± Garnet, apatite, titanite, ores
Dark or mottled dark and light colors, granular texture
Nepheline is visible in this series as gray, brown, or pink anhedral material; greasy luster, poor conchoidal fracture, and lack of good cleavage distinguish it
Uncompahgrite = melilite ijolite
Fergusite = leucite ijolite
Missourite = leucite melteigite

Aphanites with plagioclase

Tephrite = plagioclase + feldspathoid + pyroxene
  Nepheline tephrite. The feldspathoid is nepheline.
  Leucite tephrite. The feldspathoid is leucite.
Basanite = plagioclase + feldspathoid + pyroxene + olivine
  Nepheline basanite. The feldspathoid is nepheline.
  Leucite basanite. The feldspathoid is leucite.
Tephrites and basanites are ash-gray, dark-gray, or black rocks with phenocrysts of plagioclase and pyroxene. Olivine phenocrysts are present in basanites. Nepheline is usually limited to the aphanitic groundmass and is therefore unidentifiable. It occasionally occurs as minute hexagonal prisms that are recognizable. Leucite usually appears as phenocrysts as well as in the groundmass and forms

euhedral trapezohedrons of light-gray color. Dark inclusions, often systematically arranged, are normal in leucite.

Aphanites without plagioclase

Feldspathoidites. Composed of feldspathoid and ferromagnesian minerals but without olivine.

> Leucitite = leucite + pyroxene
> Nephelinite = nepheline + pyroxene
> Melilitite = melilite + pyroxene

Feldspathoidal basalts. Composed of feldspathoid and ferromagnesian minerals including essential olivine.

> Nepheline basalt = nepheline + olivine and pyroxene
> Leucite basalt = leucite + olivine and pyroxene
> Melilite basalt = melilite + olivine and pyroxene

Phenocrysts. Usually the ferromagnesian minerals. Pyroxene and biotite are common in all the rocks, and olivine is common in the feldspathoidal basalts.

Nepheline and leucite occur as phenocrysts in those rocks without olivine but not in the feldspathoidal basalts. Members of the sodalite group may also appear as phenocrysts. Melilite does not usually appear as phenocrysts.

Groundmass. Ash gray to black, dense to granular, and basaltic in appearance. It is composed of microcrystalline feldspathoid and ferromagnesian material, with fine granular ore minerals and occasionally some glass.

PERIDOTITES AND PYROXENITES

Definition

Ultramafic igneous rocks composed dominantly of ferromagnesian minerals or ore minerals and containing less than 10 percent feldspar. The rocks occur as intrusive bodies, and no equivalent extrusives are known. They occur as gravitative accumulations of early crystals along or near the floors of mafic sills, as distinct bands in layered gabbros, or as distinct homogeneous intrusive bodies. The intrusive bodies may be the result of mobilization of gravitatively accumulated masses or may be intrusions of largely crystalline primary magmas. Some ultramafic masses are metamorphic or metasomatic in origin.

Classification

Classification is based broadly on the proportions of the various ferromagnesian minerals. The fundamental subdivision is based on the ratio of olivine to the other silicates; further subdivision is based on the ratio of pyroxene to hornblende and biotite. The commoner rock types are listed below, according to this twofold subdivision:

More than 90% olivine
 Dunite
50 to 90% olivine
 Olivinites
   Olivine + mica = mica olivinite
   Olivine + mica + less pyroxene = kimberlite
   Olivine + monoclinic pyroxene = wehrlite
   Olivine + monoclinic and orthorhombic pyroxene = lherzolite
10 to 50% olivine
 Peridotites
   Hornblende + olivine = cortlandtite
   Orthorhombic pyroxene + olivine = saxonite
   Orthorhombic pyroxene + olivine + ore minerals = harzburgite
Less than 10% olivine
 Dominantly biotite = glimmerite
 Dominantly hornblende = amphibolite or hornblendite
 Dominantly pyroxene = pyroxenite
  Pyroxenites may be named on the basis of the type of pyroxene, such as diopsidite, bronzitite, hypersthenite, etc. Websterite is a mixture of hypersthene and augite.
  Intermediate rocks are best expressed as a compound name, such as hornblende-bearing pyroxenite, biotite pyroxenite, etc.

Texture

A uniform granular texture composed of rounded or polygonal equidimensional anhedra is typical of these rocks, particularly when one mineral greatly predominates. Porphyritic textures are less common. A cataclastic texture is common in some pyroxenites, consisting of large rounded or highly angular anhedra in a matrix of finer-grained material identical to the large crystals. This texture gives the appearance of having been produced by the intrusion of a coarsely crystalline mass containing very little fluid and a resultant grinding of grains against each other as the mass moved. Brecciated textures are common in kimberlites, which contain abundant fragments of other rock types.

Mineralogy

Olivine

Color. Glassy-green granules are typical of the fresh material. Serpentinization is common, producing dull-lustered green to greenish-black compact aggregates.

Habit. Usually roughly equidimensional rounded or polygonal granules. Where olivine dominates the rock, the grains are of uniform size. Where olivine is quantitatively minor, it may occur either as rounded euhedral phenocrysts in a typical prismatic habit or as rounded granules included in larger crystals of pyroxene or hornblende.

Alteration. Serpentinization of olivine is usually present to some degree except in the purest dunite. Serpentinization can take place only at temperatures below 500°C in the presence of water vapor. It begins along the margins of crystals and along cracks through the crystals and is usually accompanied by the formation of minute granules of magnetite. As a result of the magnetite, this early formed serpentine is often dark green. Residual granules of olivine are left from the uncracked portion of the crystal. Subsequently, the residual olivine is serpentinized, forming an amorphous material of lighter color. Occasionally, the resultant mesh-work texture in serpentine may be recognized megascopically and indicate an olivine source. Talc is a less common alteration product from olivine, developing at temperatures above 500° in the presence of water vapor. It generally occurs as white to pale-green, coarse, flaky aggregates between the olivine grains rather than as a replacement of single olivine grains.

Pyroxene. Both orthorhombic and monoclinic pyroxenes are common, the orthorhombic varieties being the most usual.

Color. Black and various shades of green or brown. The orthorhombic pyroxenes tend to have a fibrous appearance due to cleavages. Green or brown colors are characteristic of enstatite. A variety with a metalloid brownish-bronze reflection off the (100) cleavage is common and is called *bronzite.* Hypersthene is less common in these rocks and tends to be greenish black to black. Augite is the characteristic monoclinic pyroxene and is generally greenish black to black but lacks the good cleavage and fibrous character of hypersthene. Some augite shows a well-developed parting and a metalloid luster on the (100) face and is called *diallage.* Small granules of bright-green chrome-diopside occur in some dunites.

Habit. Usually in anhedral grains, roughly equidimensional and of uniform size. Mutual interference of growing grains or rupture of grain margins by cataclasis prevents euhedral habits. In the olivinites, pyroxenes may appear as phenocrysts either because of their inherent larger size than olivine or because of selective alteration of olivine to a fine-grained product. In the peridotites, pyroxene commonly forms large grains including smaller granules of olivine.

Alteration. Serpentinization is common, particularly in the orthorhombic pyroxenes. The serpentine developed from pyroxene is often quite distinct from that developed from olivine, in that it has a platy structure more or less preserving the structure of the pyroxene. This material is known as *bastite* and lacks the veined mottling of olivine replacement. Uralite and chlorite are the common alteration products of augite.

Hornblende. An accessory mineral in most ultramafic rocks, except cortlandtite and amphibolite. Uralite developed from augite is fairly common.

Color. Dark green, brown, or black, usually showing excellent cleavage, although inclusions of granules of olivine or pyroxene or alteration may mask the cleavage.

Habit. Roughly equidimensional or elongated prisms are typical. In both, the crystals are anhedral in detail but may appear straight-sided because of cleavage. Prisms in igneous amphibolites tend to be randomly oriented; those in metamorphic amphibolites are parallel, with either a linear or a planar structure. Linear alinement has all the *c* crystallographic axes in subparallel alinement, whereas planar structure has *c* crystallographic axes in one plane but randomly oriented within that plane. Alinement is seen in some apparently igneous amphibolites in small dikes having alinement parallel to the walls. In cortlandtite, large crystals of hornblende include numerous granules and flakes of olivine, pyroxene, magnetite, and ores.

Alteration. To chlorite. Fibrous green amphibole (uralite) is an alteration product from pyroxene in parallel crystallographic orientation to the original pyroxene. Rocks in which the amphibole is uralite should be called *pyroxenite* rather than amphibolite or hornblendite.

Mica. Phlogopite is the common mica of ultramafic rocks due to its characteristic high magnesia content. Biotite may be present in the more iron-rich rocks. The two minerals cannot be distinguished megascopically and should be named on the basis of high magnesium or high iron content.

Color. Brown lustrous flakes, except where very fine-grained; it may then be brown or green, depending on the extent of alteration.

Habit. When recognizable megascopically, micas occur as scattered irregular flakes. Some peridotites have a groundmass of very fine-grained mica not identifiable megascopically.

Alteration. To chlorite.

Accessory minerals

Feldspars. May be present in some ultramafic rocks transitional to gabbro. The feldspar is usually calcic plagioclase and is late and interstitial between the ferromagnesian minerals.

Spinel minerals. Magnetite and chromite are particularly common accessory minerals of the ultramafic rocks, frequently attaining sufficient proportions to constitute an ore. The minerals may be disseminated uniformly through the rock but more frequently occur concentrated in certain bands. Some bands become essentially monomineralic magnetite or chromite, and the resultant rocks may be called *magnetitite* or *chromitite*. Both typically appear as small black octahedra or rounded granules. Spinel, hercynite, pleonaste, and picotite are other common accessory spinels.

Garnet. Pyrope is a common accessory in olivinites and peridotites. It usually occurs as rounded crystals without faces, often surrounded by a granular zone of radially arranged pyroxene or hornblende and other minerals. The material is fine-grained and not identifiable megascopically, but the rim is very noticeable.

Others

The titanium minerals ilmenite, rutile, and perovskite are present but minor.

Apatite is a fairly consistent accessory, especially in magnetite-rich rocks.

Pyrrhotite and *pyrite* are minor.

Carbonates (calcite, dolomite, and magnesite) are common in veins and are disseminated through serpentinized rocks.

GLASSES AND PYROCLASTICS

The glassy rocks are those extrusives which have cooled so rapidly that crystallization of the fluid phase is impossible. Most commonly, glasses are in the silicic to intermediate range of compositions, but unless they contain phenocrysts, they cannot be placed in a rock family. Chemical analysis is required for accurate placement, although a moderately good identification can be made by a determination of the index of refraction and the specific gravity. Mafic igneous magmas usually crystallize so rapidly that residual glass is masked by microcrystalline material, and the resulting rock does not have a glassy appearance even though glass may compose an appreciable percentage. The glasses, as a result, are classified megascopically on appearance and texture as follows.

Glasses containing few or no phenocrysts
    Obsidian. Massive; vitreous luster (Fig. 5-6).
    Pitchstone. Massive; resinous luster.
    Perlite. Spheroidally cracked (Fig. 5-20).
    Pumice. Cellular, glass froth (Fig. 5-9).
    Tachylite. Massive basalt glass.
Glasses containing phenocrysts
    Vitrophyre (Figs. 5-4 and 5-21).

Glasses containing few or no phenocrysts

Obsidian

Color. Normally black; may be brown or red or colorless. Different colors are usually intermixed in streaks or bands or contorted in a marble-cake effect. Black obsidian often has transparent bands which when viewed along thin edges can be seen to contain minute dustlike crystallites. The opaque bands contain a much higher concentration of these crystallites. The color of obsidian depends on size, concentration, and mineral composition of these crystallites. The luster of obsidian is highly vitreous.

Structure. Massive and compact, with a very well-developed conchoidal fracture. Spherulites are common in obsidian, ranging in size from that of fine shot to several centimeters in diameter (Fig. 5-23). They show a radial platy structure and often show concentric structure. They consist of fine plates of feldspars and quartz or tridymite and often include complex spherical aggregates of cristobalite and platy crystals of fayalite. The margins of spherulites may be sharp and abrupt, or they may be highly irregular, feathery, and obscure. Spherulites may be randomly scattered through the glass or concentrated along flow lines (Fig. 5-24). They may become so abundant as to coalesce into distinct bands. Lithophysae may also occur in obsidian and consist of spherical masses of concentric shells separated by thin void zones (Fig. 5-25). The solid shells are finely crystalline minute crystals of quartz; tridymite, feldspar, fayalite, and topaz may coat the shells, developing euhedral faces into the voids.

Composition. The majority of obsidians have compositions equivalent to granites and rhyolites. Characteristically, they contain less than 1 percent water. Thin splinters fuse in a bunsen burner to give a colorless bubbly glass which cannot be refused.

Pitchstone

Color. Brown, green, gray to almost black, with a characteristic pitchy or resinous luster. The colors may be uniform, streaked, or marbled. The glass is translucent to transparent on thin edges. Crystallites are abundant in pitchstones.

Structure. Massive, with well-developed conchoidal fracture as in obsidian, but generally the fracture surfaces do not extend over as large an area as in obsidian. This is probably because of the more extensive cracking which is common in pitchstones.

Composition. As obsidian, but usually with more water. Most pitchstones contain 5 to 8 percent water. On heating to red heat in a bunsen burner, pitchstone becomes white and opaque, increases in volume several times, and cracks extensively. Chips as much as 2 mm thick are affected throughout, whereas only the thinnest edges of obsidian are affected.

Perlite

Color. Gray, blue-gray, or light brown, with a pearly luster. Translucent on thin edges but often not transparent due to incipient crystallization of the glass.

Structure. Concentrically cracked spheroidal structure. Spheroids range from a few millimeters to several centimeters in diameter and may be separated by a matrix without visible cracking or may be close-packed with polygonal outlines. Cracking may extend to the center of the spheroids, but often an uncracked ball of glass remains in the center. These may be colorless, smoky, or black and obsidianlike. In the southwestern United States they are called *Apache tears* (Fig. 5-37). They have been described under the name marekanite.

Composition. As obsidian, but with a water content consistently between 2 and 4 percent. Shrinkage cracking may be dependent on the critical water content. Behaves like pitchstone in heating to red heat in a bunsen burner.

Pumice

Color. White to light gray, yellowish, or brownish. A dull vitreous to silky luster is characteristic of the rock mass due to gas cavities, but the glass itself has a highly vitreous luster.

Structure. Highly cellular; a rock froth. Gas cavities range from minute bubbles to cavities several millimeters in maximum dimensions. The cavities may be spherical, but they are more commonly drawn out into tubes by flow of the glass. The gas cavities make up the major percentage of the bulk and are separated by paper-thin walls of glass. Most pumice is sufficiently porous to float in water, and the cavities are completely isolated so that permeability is nil and the pores cannot become

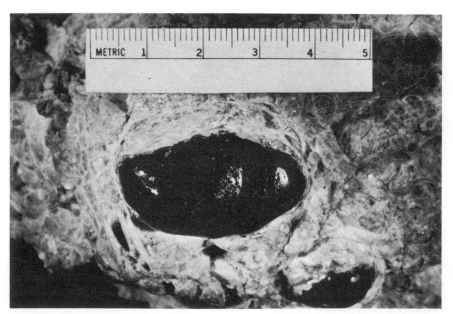

Fig. 5-37. "Apache tears." New Mexico.

filled with water. Pumaceous rocks may be produced by a welding together of hot glass fragments in an ash flow.

Composition. As obsidian, usually with a low water content in the glass.

Tachylite

Color. Brown or black, vitreous luster, usually clouded with crystallites and microlites.

Structure. Massive or scoriaceous. Basalt glasses are rare, occurring at the chilled margins of dikes or lava flows. Usually, cooling is not rapid enough to prevent a large degree of crystallization. As a result basalt glass is usually interstitial between crystals and is not recognizable megascopically. Hairlike threads of basalt glass are formed by bursting of bubbles in lava lakes, spinning a thread. These hairs, known as *Pele's hair,* accumulate down wind in quiet areas.

Composition. Basaltic.

Glasses containing phenocrysts

Vitrophyre. A porphyritic glass. The glass may have the characters of obsidian pitchstone, or perlite.

Phenocrysts are most commonly colorless transparent sanidine and plagioclase, high-temperature quartz, biotite, and less commonly, hornblende and pyroxene. They are usually euhedral but may be broken or corroded. Vitrophyres may be classified on the basis of the phenocrysts.

Alteration of glasses. Because of their chemical instability, glasses are readily altered and are rarely seen in older rocks. Devitrification is the commonest alteration effect and involves gradual crystallization to minute grains of quartz and feldspar and results in a stony aphanitic texture and loss of glassiness. In perlite, devitrification is initiated along the contraction cracks so that the glassy luster is lost and the material becomes white and earthy appearing. After complete devitrification the only evidence of the original glassy nature of the rock is in relict spherulites, lithophysae, and flow banding.

Pyroclastics. Pyroclastic rock consists of fragmental material erupted from volcanoes and accumulated as essentially solid materials. Fragments are classified on the basis of size (Fig. 5-38). The size designations and the corresponding terms for consolidated accumulations are:

Size	Fragment	Consolidated
> 32 mm	Bombs (liquid)	Agglomerate
	Blocks (solid)	Volcanic breccia
4–32 mm	Lapilli	Lapilli tuff
< 4 mm	Ash	Tuff

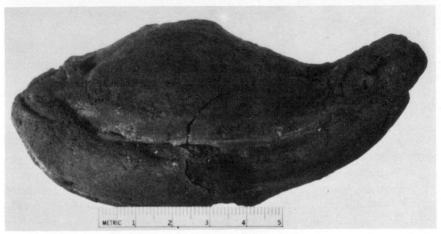

Fig. 5-38. Volcanic bomb. Dish Hill Crater, San Bernardino County, California.

The fragments are classified on the basis of their origin as:

Juvenile. Fresh magmatic ejected material.
Accessory. Solid volcanic rock derived from the conduit or crater walls.
Accidental. Any rock type derived from the subvolcanic basement.

The smaller-sized materials of a juvenile nature may be classified on the basis of their composition as:

Vitric ash. Composed of glass fragments.
Crystal ash. Composed of single crystals or crystal fragments; crystals are often partially coated by glass.
Lithic ash. Composed of rock fragments, usually of accessory or accidental origin.

Pyroclastic materials may fall in basins of sedimentary accumulation or may be washed into such basins by streams so that all gradations between pyroclastics and sediments are possible.

Volcanic breccias. Brecciated texture with angular to rounded fragments of extrusive igneous rocks plus fragments of accidental materials set in a groundmass of fine ash is characteristic. The breccia is named for the dominant extrusive rock type, such as rhyolite breccia or andesite breccia. The visible rock fragments have the lithic character of any lava type. The groundmass is usually fine-grained and is recrystallized to such an extent that its original character is obscured.

Vitric tuffs. These are composed essentially of glass fragments and minor crystals and rock fragments. The glass is in the form of shards, fragments of bubble walls. The fragments are thin plates, Y-shaped, curved, or pumaceous. All are shapes which could be derived from the explosive rupture of highly vesicular glass. The shards are usually a fraction of a millimeter to a few millimeters in size and may not be recognizable megascopically. The glass is highly susceptible to alteration by devitrification, alteration to clay of a bentonitic type, or silicification to a chertlike rock called *porcelanite*. Basaltic vitric tuffs alter to palagonite, having a brown sandy appearance with crystals of basaltic minerals. Palagonite is a yellow to orange amorphous material which coats basaltic fragments or fills the interstices between the fragments. Zeolites and carbonates are also abundant in the interstices between fragments altered to palagonite.

Vitric tuffs erupted in the glowing ash clouds give rise to welded tuffs, or *ignimbrite* (Fig. 5-39). In the hot-gas-charged clouds, the glass fragments remain soft until they settle out, then as a result of the weight of the accumulating material, the shards are compressed and welded to form a solid mass. Subsequent alteration or devitrification will completely destroy any pyroclastic texture, and the rock becomes indistinguishable megascopically from lava flows. When unaltered, however, vitric tuffs may retain pyroclastic texture and grade upward into ash beds which have not been welded.

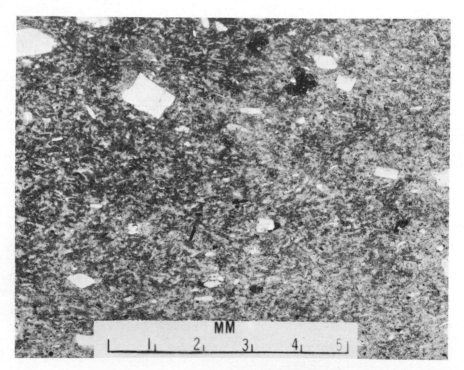

Fig. 5-39. Welded vitric tuff — ignimbrite. Broken phenocrysts of quartz and feldspar are set in a groundmass which shows abundant remnant volcanic shards. Castle Rock, Douglas County, Colorado. Projection of a thin section.

Crystal tuffs. These are composed of crystals and have lesser amounts of rock fragments and glass. Probably no rock composed largely of crystals exists. The crystals are derived from single phenocrysts in the magma, clusters of phenocrysts, floating crust material on lava lakes, or from cold wall rock of the conduit or crater. Characteristically, they are broken, cracked, or completely fragmented. In general, crystals of accessory or accidental origin are much more intensely fragmented than are crystals of juvenile origin. Films and clots of glass often adhere to the crystals. Distribution of crystals in the tuff may be nonuniform. In comparing local areas of a few square feet in a porphyritic lava flow, the proportion of different minerals as phenocrysts is approximately constant. In contrast, in crystal tuffs there are commonly wide variations in proportions of mineral species because of the explosive eruption of clots of crystals.

The groundmass of crystal tuffs is composed of glass shards and very fine dust. Commonly, the presence of shards is the best direct evidence of pyroclastic origin. As the proportion of matrix increases, the crystal tuffs grade into vitric tuffs and no clear line of distinction can be picked.

Lithic tuffs. These are composed of fragments of cold solid rock of accessory or accidental origin, including deep-seated rocks, earlier lava flows and ashes from the conduit wall, and solidified crust over the conduit developed during quiescent periods. Thus the texture and rock type represented are widely variant. These fragments are set in a matrix of glass shards, pumice fragments, and fine dust (Fig. 5-40). The lithic fragments often show effects of hydrothermal alteration or thermal metamorphism.

Fig. 5-40. Lithic tuff. The large fragments are weathered andesite set in a fine-grained matrix of volcanic ash. Gaffey, Colorado. (*Ward's Universal Rock Collection.*)

Partial fusion of feldspar and quartz in granite or feldspathic-sandstone fragments is common. Contact metasomatic products from limestone and dolomite are found in great variety in the fragments from Monte Vesuvius. Fragments from the conduit and crater walls of volcanoes often show effects of modification by superheated steam prior to eruption. In general, the fragments in lithic tuffs are coarser than those in crystal or vitric tuffs. This is in part a result of difficulty of recognizing rock fragments in the finer-grained materials.

6 SEDIMENTARY-ROCK PETROGRAPHY

TEXTURE

Texture in sedimentary rocks deals with the size, shape, surface characteristics, and arrangements of individual grains of the rock. A megascopic examination permits the use of only qualitative or descriptive terms. However, some statistical definitions are given below, inasmuch as the medium- and coarse-grained materials are often sized by mechanical means or direct measurements. For detailed techniques, reference should be made to standard texts on sedimentary petrography.

SIZE OF PARTICLES

Sedimentary materials are classified on the basis of size, as indicated in Table 6-1. The sizes, in millimeters, entered in the table are the limiting sizes of each class.

Statistical Size Measurement of Single Fragments

In megascopic studies the following definitions are applicable to rudites, less readily applicable to arenites, and not applicable to lutites.

1. *Nominal diameter.* The diameter of a sphere whose volume is equal to that of the fragment. It is determined by measuring the volume of water displaced by the fragment.
2. *Average diameter.* The arithmetic average of the maximum, intermediate, and minimum diameters of a fragment measured at right angles to each other but not necessarily passing through a common point.
3. *Intermediate diameter.* The diameter of a fragment measured at right angles to both the maximum and minimum dimensions.

Size Distribution in a Bulk Sample

The distribution of particles by size in a bulk sample is determined by direct measuring of many fragments or by screening of rudites, screening of arenites, or determining the settling velocity of lutites in water. The data are plotted graphically, and five statistically important sizes for the bulk sample are read from the graph:

Table 6-1. Size Terms for Clastic Particles

Size		Descriptive terms			
mm	*φ**	*Common terms*	*Greek derived*	*Latin derived*	*Megascopic criteria*
512	−9	Boulders			
256	−8	10 in.			
128	−7	Cobbles			
64	−6	2½ in.	Psephites	Rudites	Size determined by direct measurement of grains or by estimate
32	−5	Very coarse			
16	−4	Coarse (PEBBLES)			
8	−3	Medium			
	−2	Fine			
4		³⁄₁₆ in.			
2	−1	Granules			
1	0	Very Coarse			
½	+1	Coarse			
¼	+2	Medium (SAND)	Psammites	Arenites	¹⁄₁₆ mm is the finest size in which single grains are visible in the rock
⅛	+3	Fine			
¹⁄₁₆	+4	Very fine			
¹⁄₃₂	+5				Silt-size individual grains are not visible in rock, but the sediment feels gritty between the teeth
¹⁄₆₄	+6	Silt			
¹⁄₁₂₈	+7		Pelites	Lutites	
¹⁄₂₅₆	+8				
¹⁄₅₁₂	+9	Clay			Sediment feels pasty between the teeth
¹⁄₁₀₂₄	+10				

*The size measure ϕ is equal to the negative logarithm to the base 2 of the size in millimeters. Thus 1 mm = 0ϕ, ½ mm = +1ϕ, ¼ mm = +2ϕ, etc.

1. *Maximum size.* The size of the largest fragment, either extrapolated from the curve or measured directly.
2. *Third quartile (Q_3).* That specific size which is smaller than 25 percent of the sample and larger than 75 percent of the sample, as determined on the curve.
3. *Median diameter (Md).* That specific size which is smaller than 50 percent of the sample and larger than 50 percent of the sample, as determined on the curve.
4. *First quartile (Q_1).* That specific size which is smaller than 75 percent of the sample and larger than 25 percent of the sample, as determined on the curve.
5. *Mode, or Modal diameter.* That size which is most frequent in the size distribution of the sample, i.e., the most abundant, therefore the most typical, particle in the sediment. This size can be estimated megascopically for rudites and arenites.

Sorting. Sorting is a measure of the size range in a sediment and is defined quantitatively by the equation $So = \sqrt{Q_3/Q_1}$. In megascopic examination of sediments, qualitative terms can be applied. The following definitions are applied, depending on the number of ϕ classes included in 90 percent of the material, minor admixtures, or coarser and finer material omitted:

Size range, ϕ units	Descriptive term
Less than 2 ϕ classes	Well sorted
2–3 ϕ classes	Moderately well sorted
3–6 ϕ classes	Moderately sorted
6–10 ϕ classes	Poorly sorted
More than 10 ϕ classes	Very poorly sorted

These terms may be readily applied to rudites and arenites on megascopic examination.

Skewness. Skewness is the measure of the asymmetry in the distribution of sizes in a sample and is defined quantitatively by the equation $Sk = Q_1 Q_3 / Md^2$. In an ideal distribution there are both coarse and fine admixtures to the main bulk of the sample. However, in most rudites and arenites this is not the case. Usually such sediments either have a main bulk of coarse material, falling in one to three ϕ classes with an important admixture of finer material, or have a main bulk of fine material with an important admixture of coarser material. The first is said to be *negatively skewed (Sk* less than 1.0), and the second is said to be *positively skewed (Sk* greater than 1.0). The significance of skewness is unclear, but markedly skewed sediments probably indicate two distinct sources for the sediment or two distinct modes of transportation of the sediment.

SHAPE OF PARTICLES

In describing the shape of particles, two distinct properties must be distinguished: sphericity and roundness. Sphericity is dependent dominantly on the original shape of the clastic particle, and the effect of abrasion in modifying the shape is minor. Roundness, in contrast, is a measure of the modification of the original shape by abrasion during transportation.

Sphericity. Sphericity is the extent to which individual particles approach a sphere in shape. Descriptive terms are best applied to sand and larger-sized particles on the basis of their maximum, minimum, and intermediate diameters. Four fundamental shapes can be recognized:

1. *Equant.* All three diameters are subequal.
2. *Tabular.* Maximum and intermediate diameters are subequal, but the minimum diameter is much smaller.
3. *Prolate.* Intermediate and minimum diameters are subequal, but the maximum diameter is much greater.
4. *Bladed.* Maximum, minimum, and intermediate diameters are all markedly unequal.

Special terms may be applied to fragments of distinct geometrical shapes, particularly in unabraded or slightly abraded pebbles whose shapes have been inherited from bedding or jointing of the parent rock. Such terms are prismatic, pyramidal, tetrahedral, rhomboidal, etc.

Quantitatively, sphericity is defined as the ratio of the nominal diameter to the maximum diameter. A perfect sphere has a ratio of 1; any other shape has a ratio of less than 1.

Roundness. Roundness is a measure of the extent of abrasion, causing rounding of edges and corners of fragments. Roundness is independent of sphericity. The following five qualitative terms can be applied from megascopic examination:

1. *Angular.* Fragments show little or no effects of abrasion. The edges and corners between faces are sharp and angular, and the numerous secondary corners seen in a profile are sharp.
2. *Subangular.* Fragments show definite effects of abrasion. The fragments still have their original shape; edges and corners are rounded, but the original faces are unworn. Secondary corners are less abundant and rounded.
3. *Subrounded.* Fragments show considerable wear. The edges and corners are rounded to smooth curves, and original faces are considerably reduced. Secondary corners are much rounded and reduced in numbers.

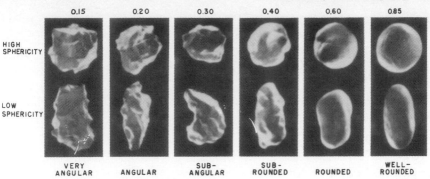

Fig. 6-1. Comparison chart for visual estimation of roundness. (*From AGI Data Sheet 7, GeoTimes, 1958, Am. Geol. Inst.*)

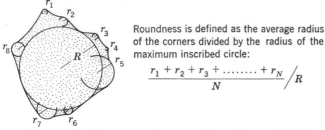

Roundness is defined as the average radius of the corners divided by the radius of the maximum inscribed circle:

$$\frac{r_1 + r_2 + r_3 + \ldots\ldots + r_N}{N}\bigg/ R$$

Fig. 6-2. Method for measuring roundness. (*From AGI Data Sheet 7, GeoTimes, 1958, Am. Geol. Inst.*)

4. *Rounded.* Most surfaces show effects of wear. Edges and corners are reduced to broad curves, but some flat surfaces remain of the original faces. Secondary corners are not noticeable.

5. *Well rounded.* The surfaces of fragments are entirely abraded, and no remnants of the original faces are left. The entire fragment is bounded by broad curved surfaces. No secondary corners remain.

The determination of roundness for a megascopic description is generally made by comparison of the grains with a standard chart, such as that given on the AGI Data Sheet #7 (Fig. 6-1).

Quantitatively, roundness is defined as the ratio of the average radius of curvature of the corners to the radius of the maximum inscribed circle. It is determined by measurements on the outline of a grain, as shown in Fig. 6-2.

SURFACE TEXTURES OF CLASTIC GRAINS

The surface texture of grains is in part indicative of the environment of transportation of the material. It should be used with caution, however, because several different environments can apparently produce the same surface effects. This is particularly true in the cases of polishing and frosting of sand grains. Characteristic surface textures are described below for gravel and sand-sized materials.

Surface Textures of Granules and Larger Sizes

Dull, or mat, surface (Fig. 3-6). Fragments exhibit a lusterless finish having low reflectivity. The surface may be either smooth or rough, but it is marked by innumerable minute scratches and pits. Normal stream transportation produces such a surface.

Impact scars (Fig. 3-30). Conical or crescentic cracks which extend into the rock and which were produced by impact with other pebbles at high velocity. The maximum diameter of the scars is usually about $\frac{1}{10}$ the maximum diameter of the pebble. Usually such scars develop in brittle massive rocks, such as vein quartz, in mountain torrents.

Striated surfaces (Fig. 3-8). Parallel, divergent, or random sets of scratches on flat surfaces of pebbles. These are characteristically developed by glacial abrasion.

Faceted surfaces (Fig. 3-7). Flat smooth surfaces which are produced by abrasion. Glacial abrasion usually produces flat striated surfaces. Wind abrasion by sand produces rounded to concave surfaces, more or less highly polished, which intersect along sharp edges. Wind-abraded pebbles are known as *ventifacts.*

Polished surfaces. Smooth, flat, or curved surfaces which have a high reflectivity. Some small pebbles and granules may be polished in stream or beach environments, but most polished stones represent special environments. Four origins of polish can be recognized:

1. *Desert varnish.* A polish or glaze developed on pebbles in a desert region, presumably by water evaporating on the surface and leaving a film of dissolved salts, particularly iron or manganese oxide or silica. Desert varnish is frequently brown to black, and on a freshly broken surface, it can be seen to discolor the stone to a depth of only a few millimeters.
2. *Ventifacts.* Wind-polished stones from desert regions. A surface polish, together with facets, is developed by wind-driven sand and silt. Only hard materials normally take a polish; softer materials become faceted but show a mat finish.
3. *Gastroliths.* Highly polished stomach stones of ancient reptiles, frequently found in clusters in Mesozoic sediments. The polish presumably is developed by abrasion and chemical action in the gizzards of reptiles.
4. *Compaction polish.* A polished surface developed on pebbles and

boulders during compaction of a clay matrix in boulder clays. Slickensides are often developed in the adjacent clay during the compaction.

Crystal faces. Minute crystal faces are developed on pebbles by secondary enlargement. Quartz pebbles in secondarily enlarged quartz sand are particularly affected. The surface will show minute triangular faces which are parallel over broad areas of the pebbles. The faces are developed in crystallographic continuity with the crystals of the pebble, and they aid in identifying the internal texture of the pebble. Chemical etching can produce the same effect in the form of minute parallel etch pits.

Surface Textures in Sand-sized Grains

Impact frosting. A chipping of the surface of grains by impact, giving a surface texture like frosted glass. Sandblast of dry grains in an eolean transportation system is the dominant cause. Frosting is limited to protuberances and convex surfaces and does not extend into cavities. Individual chips are relatively large but otherwise similar to impact scars on pebbles.

Diagenetic frosting. Quartz-sand grains are commonly pitted during cementation, especially by calcite, and may result in a frosted surface. The pitting often does not affect the entire surface but is patchy. It may extend into cavities and crevices in the grain.

Solution frosting. Prolonged transportation in undersaturated solutions will cause frosting by direct solution of grain material. Such frosting affects all parts of the grains and extends into pits and concavities on the surface.

Desert chemical frosting. Frosting of sand grains in deserts often extends into concavities and furrows and cannot be the result of impact. The solvent action of dew is thought to be the active agent, dissolving quartz in salt-charged moisture and reprecipitating the silica upon evaporation.

Coatings. A variety of coatings can give a frosted effect to sand grains. Electron micrographs often show surfaces coated with minute clay crystals and other minerals.

Polished surfaces. These are smooth lustrous surfaces which have a high reflectivity and which retain the transparency of the mineral grains. Polished surfaces are apparently of subaqueous origin and are the result of chemical processes rather than abrasion. Polished surfaces extend to the bottom of cavities and furrows in the grain, mitigating against abrasion. Direct solution in undersaturated solutions or chemical precipitation from oversaturated solutions produces the polish.

Crystal faces. Sand grains may be secondarily enlarged during diagenesis to produce megascopically visible crystal faces. In extremely porous sands or scattered grains in limestone, quartz may develop as perfect euhedra.

ARRANGEMENT OF CLASTIC GRAINS

Isotropic versus Anisotropic Fabric

Coarse clastic grains normally depart markedly from spherical shapes; therefore, a preferred orientation of grains is normal when they are deposited by current action having a consistent direction of flow. Either variable directions of current flow or no appreciable current result in random orientation of grains and an isotropic fabric. The orientation of tabular and bladed pebbles is often recognizable megascopically and is an aid in current-direction determinations. Stream-deposited pebbles dip upstream and give an imbricate structure. Similarly, beach-deposited materials dip toward the water as a result of wave action.

Orientation of flaky clay particles may also be random or preferred. The slow accumulation of single clay flakes out of standing water generally results in random orientation and an isotropic fabric. In a hand specimen these materials break with a blocky fracture and in random directions. Such sediments are described below as mudstones and claystones. Lake-bed clays and the underclays of the Pennsylvanian coals are examples. Compaction of the clays usually results in the development of preferred orientation wherein the clay flakes parallel the bedding. This commonly results in the development of fissility, and the rock is properly termed a *shale*.

Packing of Grains

Packing is the arrangement of particles whereby they are supported in the gravitational field by contact with their neighboring particles. Theoretical packing patterns are developed by considering uniform-sized spheres. The possible ideal-packing patterns are illustrated in Fig. 6-3. The most-open packing (Fig. 6-3a) leaves 47.64 percent voids, and the most-close packing (Fig. 6-3f) leaves 25.95 percent voids. In natural sediments packing varies widely from these ideal values, because grains are neither spherical nor uniform in size. In normal sediments deposited under agitated conditions, packing is as close as possible. An open packing develops by settling of fine-grained materials out of quiet water. Burial by other sediments generally results in compaction and closer packing. Sedimentary rocks, particularly sandstones, may develop an open packing during diagenesis by replacement, by forcing apart of grains by growing cement crystals, or by selective solution of unstable sand-sized grains. This last method is particularly common where sand-sized fragments of fossils were dissolved during diagenesis.

Porosity. Porosity is defined as the percentage of pore space in the total volume of rock. Typical ranges of porosity of sediments and sedimentary rocks are given in Table 1-1. Porosity may be either original or secondary in origin. Original porosity is developed at the time of deposition and depends on size, size range, and shape of the particles. Size is important in

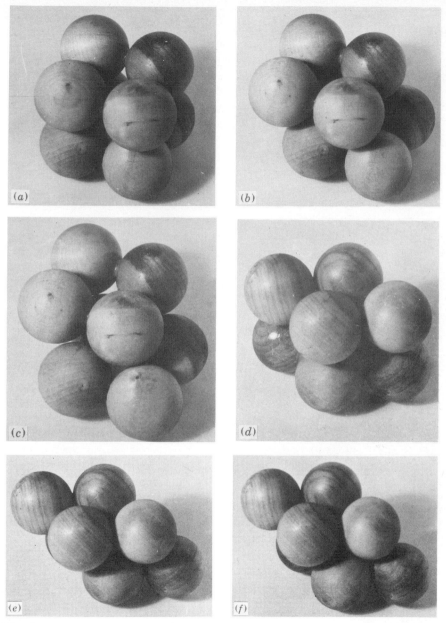

Fig. 6-3. Packing patterns for spheres. The two basic patterns are the square and the 60° rhombus. Each pattern has three different arrangements, with each sphere touching either one, two, or three spheres in the underlying layer. The pattern in (a) is the most open, having eight spheres at the corner of a cube. The pattern in (f) is the closest possible, having eight spheres on the corners of a rhombohedron.

only the lutites, where settling velocity out of water or air is so low that grains strike the bottom at velocity too low to fall into stable positions. In well-sorted arenites and rudites, porosity is independent of size. Size range is important in these sediments, because the smaller sizes can occupy the interstices between the larger fragments. A simplified system can be visualized by considering the packing of marbles in a box of baseballs.

Secondary porosity develops during diagenesis or subsequent fracturing and will either increase or decrease porosity. Compaction and cementation generally result in a reduction of porosity. Recrystallization of argillaceous or calcareous materials can either increase or decrease porosity, depending on the extent of recrystallization and the possibility of change from one mineral to another during the process. Metasomatism may likewise result in either increased or decreased porosity. For example, dolomitization of reef limestones often results in a high porosity, whereas dolomitization of bedded limestones usually results in nonporous rocks.

Porosity must be determined experimentally, but an estimate can be made based on the density of the rock. Table 1-1 gives this relationship for some sedimentary rocks.

Permeability. This is the property of the rocks which allows the transmission of fluids. It is dependent on the size of the channelways which connect the pore spaces of the rock. Permeability is dependent on size of grain and sorting of the sediment. Most minerals are wetted by water, which means that each grain is coated with a film of adsorbed water. The films are of approximately the same thickness for all grain sizes. Therefore, in fine-grained sediments the channelways between pores are plugged by this fixed-water film and no water can move through the rock. As grain size increases, the size of the channelways increases relative to the thickness of the fixed-water film. As a result effective size of the channelways and permeability increase with increase of grain size. In poorly sorted sediments the channelways between coarser grains are plugged by the finer grains, and therefore, permeability is controlled by the size of the finer-grained material. Permeability may be increased or decreased by diagenetic processes.

Permeability can be estimated megascopically by the behavior of drops of water on the surface of a specimen. In impermeable rocks, drops of water will sit on the surface without being adsorbed. In permeable rocks, the drops will disappear into the rock. Degrees of permeability can be estimated by the rate at which the drops disappear.

TEXTURES IN CHEMICAL SEDIMENTS

Descriptive terms applicable to chemically precipitated sediments, including evaporites, are borrowed from igneous and metamorphic petrology. Grain size and nature of grain contacts are the dominant features. Special terminology has been developed for limestones and will be defined with the classification and descriptions of those rocks.

Macrocrystalline equigranular texture (Fig. 6-4). The texture of phaneritic rocks having visible crystalline texture and crystals of generally uniform size. The rocks may be termed *coarse*, *medium*, or *fine-grained*, as in the igneous rocks. The grain boundaries are usually straight or smooth curves, giving polygonal outlines to each grain. Such a relationship is called a *mosaic texture*. Other grains may have highly irregular and interlocking boundaries, giving a *sutured texture*. Individual grains vary in shape, being equidimensional, fibrous, or platy to sheaflike aggregates (Fig. 6-5).

Porphyroblastic texture (Fig. 6-6). A texture characterized by two distinct grain sizes. This texture may be the result of primary crystallization, recrystallization, or cataclastic modification. The porphyroblasts and groundmass may be the same mineral or two different minerals. Either porphyroblasts or matrix may dominate the rock. A *mortar texture* consists of rounded or highly irregular porphyroblasts in a fine granular matrix of the same mineral and is apparently the product of mechanical crushing.

Microcrystalline and cryptocrystalline textures. The textures of aphanitic rocks lacking visible crystals. The two terms differ in having microscopic or submicroscopic grains, respectively. Microcrystalline rocks

Fig. 6-4. Macrocrystalline equigranular texture. Rock salt composed of coarsely crystalline halite. Houston, Texas.

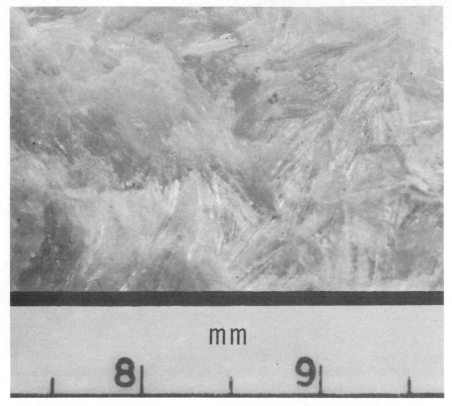

Fig. 6-5. Macrocrystalline platy texture. Radiating bundles of thin plates of anhydrite. Lockport, New York. (*Ward's Universal Rock Collection.*)

generally are visibly granular or have an earthy luster, whereas crypto-crystalline rocks have a porcelaneous to glassy luster.

Oolitic and pisolitic textures. These contain spherical or slightly ellipsoidal bodies showing radial or concentric structures or both. They are accretionary bodies, usually of radiating fibrous crystalline material, which may have a nucleus of a sand or silt grain of quartz or a fossil fragment, or they may have no visible nucleus. Four types of bodies can be recognized:

1. *Oolith* (Fig. 3-47). Size ranges from 0.25 to 2.00 mm, most commonly 0.5 to 1.0 mm. A rock composed of ooliths is an oolite; its texture is oolitic.
2. *Pisolith* (Fig. 6-7). Larger than 2.00 mm in dimensions.
3. *Spastolith.* A deformed oolith.
4. *Pseudoolith.* Rounded grains of fine-grained materials, often associated

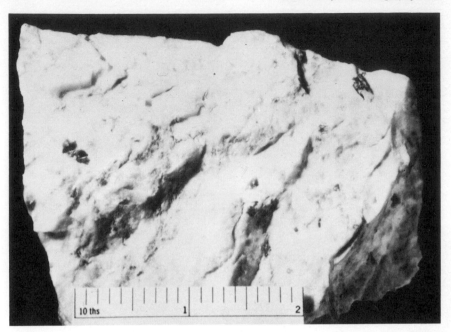

Fig. 6-6. Porphyroblastic texture. Larger crystals of gypsum in microcrystalline gypsum (alabaster). Oklahoma.

with ooliths but without concentric structure. Pseudooliths are generally larger than their associated ooliths.

Colloidal textures. These are exhibited as botryoidal or reniform surfaces with or without concentric banding and radial fibrous texture. Many chemical sediments show evidence of having been originally deposited as a colloidal gel. Subsequent dehydration of the gel produces an amorphous mass of hardened material. Subsequent crystallization at first produces a fibrous texture having crystalline fibers perpendicular to the surface of the mass. Complete crystallization yields a microcrystalline or cryptocrystalline mass having a mosaic texture. Shrinkage cracks may develop in the mass during dessication and be preserved or filled with other minerals.

STRUCTURE

Structure of sedimentary rocks deals with the larger features of the rock and is usually seen best in the field. Primary structures are related to the bedding and are dominantly of mechanical origin. Secondary structures are, in general, related to diagenetic changes in the rocks and are of a chemical origin.

Fig. 6-7. Pisolites. Large concentrically zoned pisoliths of bauxite in a matrix of fine-grained structureless bauxite. Bauxite, Arkansas.

PRIMARY STRUCTURES (MECHANICAL)

Bedding

Bedding is the most characteristic feature of sedimentary rocks. Two distinct properties of sedimentary rocks are often confused in the term bedding: (1) the actual stratification, or layering, which is a primary depositional property of the rock, and (2) the splitting characteristics, or physical parting, of the rock along planes parallel to stratification, which may be the result of primary depositional features or may be secondary in origin due to compaction, groundwater activity, weathering, etc. Some basic, but not uniformly accepted, definitions are listed below, following, in general, the terminology of McKee and Weir (1953).*

Stratum. A distinct sedimentational unit; the term has no thickness connotation. It may be homogeneous throughout or may grade in texture or composition from bottom to top, but it is separated by an abrupt change in character from the overlying and underlying strata.

Bed. A stratum; the term usually connotes a thickness of more than 1 cm.

*E. D. McKee and G. W. Weir, Terminology for stratification and cross-stratification in sedimentary rocks, *Geol. Soc. America Bull.*, vol. 64, pp. 381–389, 1953.

Lamina. A stratum; the term usually connotes a thickness of less than 1 cm.

Descriptive Terminology of Stratification and Splitting Properties

Unit thickness		Descriptive terminology	
Approximate	Defined	Stratification	Splitting
> 4 ft	> 120 cm	Very thick bedded	Massive parting
4 ft–2 ft	120–60 cm	Thick bedded	Blocky parting
2 ft–2 in.	60–4 cm	Thin bedded	Slabby parting
2 in.–$\frac{1}{2}$ in.	5–1 cm	Very thin bedded	Flaggy parting
	1 cm–2 mm	Laminated	Shaly or platy parting
	< 2 mm	Thinly laminated	Papery parting

Massively bedded is synonymous with thick bedded or very thick bedded but is not equivalent to massive parting. Many laminated shales, when fresh and unweathered, break out in massive-parting blocks independently of the stratification. It is only after weathering that the bedding-plane parting (fissility) develops. Similarly, laminated sandstones and siltstones may show flaggy or slabby parting.

Visual Evidence of Stratification

Stratification is apparent in outcrops and hand specimens at three distinct levels of visual evidence. The first and most obvious is the contact between strata of markedly different lithic character, such as abrupt contacts between sandstone and shale or between limestone and shale. Such bedding surfaces represent an abrupt change in the sedimentation and may represent an interval of nondeposition.

The second level of observational evidence is a distinct physical separation of adjacent strata whose lithic character is similar but not necessarily identical. The physical separation of the strata develops as a result of secondary processes and may follow stratification and be controlled by it, or it may merely be almost parallel to the stratification. The opening between strata is the result of groundwater solution, weathering, or, possibly, unloading of the rocks due to erosion. Where it follows stratification, it usually reflects some textural or compositional difference between the strata. These are usually the same features which control the splitting character of the rocks. Some characteristics which cause bed separation and splitting are:

A. The presence of thin clay seams along stratification surfaces of limestones and sandstones; in limestones these may be residual clays from differential groundwater solution along selected bedding surfaces.

B. A change in porosity at stratification surfaces because of
   1. Increase in argillaceous content
   2. Degree of cementation
   3. Degree of recrystallization in limestones
   4. Change in type of clastic particles in clastic limestone
C. Compositional changes
   1. Mica-rich lamellae in sandstone
   2. Concentration of metastable and unstable accessory minerals in sandstones, particularly magnetite
   3. Concentration of unstable accretionary bodies along stratification surfaces, such as siderite pellets or pyrite nodules
   4. Thin gypsum partings of primary origin or secondary origin due to oxidation of pyrite
D. Preferred orientation of flaky particles parallel to stratification
   1. Compacted and recrystallized clay minerals in shales
   2. Claystone flakes concentrated along stratification surfaces
E. Concentration of organic debris along stratification surfaces during intervals of very slow deposition
   1. Thin bone beds
   2. Thin layers of fecal pellets
   3. Seams of lignitic or carbonaceous material
   4. Reworking of the surface of a stratum by worms and other scavenging animals
F. Mechanical rupture along or close to stratification surfaces
   1. Earth-tide flexure (?)
   2. Vertical isostatic movements
   3. Jointing and deformation
   4. Unloading of overburden by erosion

   The third level of evidence of bedding is seen in laminated or massively bedded units, where there is no separation of strata or marked contrast of lithic type. Some of these features make laminations visible in fine-grained clastic sediments, and others make recognition of bedding direction possible in thick units. Each represents a temporary change in sediment supply, current changes, rate of sediment accumulation, or other environmental change. The effects are generally textural or compositional.

A. Parallelism of clastic units in sediments
   1. Parallel mica flakes scattered uniformly through sandstone
   2. Parallelism of broken or abraded fossil fragments: shell fragments, fenestellate bryozoan chips, single crinoid stem plates, trilobite pygidia, etc.
   3. Edgewise conglomerate, intraformational conglomerate, and clay flakes

    4. Zones of imbricate pebbles or boulders in otherwise randomly oriented conglomerate

B. Grain-size gradations

    1. Graded bedding involving upward decrease in grain size in fine clastic sediments or upward increase in proportion of fine clastics in a stratum

    2. Alternations of silt- and clay-sized materials, as in varves

    3. Change in size of clastic particles in limestone, as upward decrease or increase in size of crinoidal debris or ooliths

    4. Cut-and-fill structures in conglomerate, with a textural change at the base of channels

    5. Sandy lenses in conglomerate

    6. Local coarsening of sandstone along surfaces of nondeposition owing to winnowing of fines from an accumulated sand followed by deposition of an identical sand

    7. Silty films and lenses in uniform shale, often resulting in only a local granular appearance

C. Geopetal structures — structures showing the horizontal direction at the time of deposition. They are generally recognizable as partially or completely filled cavities, such as the interiors of articulated bivalves or coiled shells. The lower part of the cavity is commonly filled with clastic debris and the upper part is either empty or filled with coarse secondary crystalline material. The contact represents the original horizontal surface.

D. Concentration of minor constituents in certain strata

    1. Heavy-mineral concentrations in sandstone as a result of wave, wind, or current action

    2. Glauconitic zones

    3. Layers and lenses of scattered quartz sand in limestone

    4. Variations of clay content

    5. Feldspathic zones in sandstone

    6. Ashy beds

    7. Variations of organic content in shale as carbon films or hydrocarbons

E. Secondary effects

    1. Variations in cementation, such as carbonate-cemented lenses in generally silica-cemented sandstones

    2. Concentration of concretionary bodies in zones

    3. Color changes due to original differences in clay content, oxidation state of iron, concentration of heavy minerals, etc.

    4. Secondary color changes due to weathering of iron minerals, alteration of feldspar to clay, etc.

    5. Differential weathering, giving honeycombed or washboard surfaces

False Bedding

Many processes involved in the geologic history of sediments produce features which can be confused with stratification. The distinction between such features and true stratification usually depends on a recognition of the criteria of true stratification. Deformation and weathering phenomena are the commonest origins of false bedding, although some may be primary depositional features.

A. Deformation features and associated metamorphic features giving the appearance of bedding
   1. Jointing of massively bedded strata
   2. Recrystallization of fault gouges and breccias in small outcrops of massively bedded strata
   3. Cleavage development in shales
   4. Metamorphic differentiation, causing mineral segregation along fracture planes
   5. Deformation of clastic units, pebbles, fossils, ooliths, etc., into parallel lensoid units not parallel to stratification
   6. Hydrothermal mud injections along fractures; they may contain "xenoliths," giving a pseudoconglomerate
B. Weathering features giving false bedding
   1. Spalling parallel to the surface due to unloading or frost action
   2. Case-hardening by cementation along fractures
   3. Color-banding due to oxidation or enrichment in iron oxides parallel to fractures, sometimes giving liesegang structure
C. Occasional primary sedimentation features giving false bedding
   1. Mica, heavy minerals, or coarser sand grains often accumulating in the troughs of ripple marks; as a series of rippled thin lamellae accumulate, this will result in textural or compositional banding across the stratification
   2. Large clastic dikes or fissure fillings in a limited outcrop not recognizable as such
   3. Vertical burrows and root tubes giving an impression of bedding which is at right angles to the stratification
   4. Large-scale cross bedding, as in large deltaic deposits, and initial dip of sediments, as in reef-flank deposits, representing true stratification but misleading relative to regional stratification

Graded Bedding

Graded beds are sedimentation units in which grain size decreases from the bottom of the unit to the top. Two general types occur: the first, and commonest, has fine-grained material throughout the unit, but an important admixture of coarser material occurs at the bottom and becomes finer upward through the unit; the second type has fine-grained material only at the top and a uniform size increase toward the bottom. The first type appears to represent settling from a turbidity current which has come to rest, whereas the second appears to represent the waning of current velocity over an interval of time.

Turbidity currents consist of a thoroughly mixed slurry of sand-, silt-, and clay-sized particles dispersed in water. When the current stops flowing, all particles begin to settle to the bottom. Settling velocity depends on size: large sizes settle faster than smaller sizes. Therefore, while sand from throughout the slurry is settling, silt and clay from the bottom of the slurry settle and accumulate together. Subsequently, silt from throughout the slurry settles to the bottom and is accompanied by more clay-sized material. The last material to accumulate is the finest clay sizes. The resulting bed will consist of sand, silt, and clay at the base, grading upward into silt and clay, and finally into clay alone. Sequences of beds accumulated in this manner show a sharp contact between the sandy base of one layer and the clayey top of the underlying layer (Fig. 3-32). This type of bedding is characteristic of the graywacke assemblage of sandstone and shales.

Either a waning current over a long period of time or the seasonal introduction of sediments results in the second type of graded bedding. Each current velocity can deposit only one limited range of grain sizes. Coarser sizes are not brought into the area, and finer sizes are removed. Therefore, as a current decreases gradually in an area, the size of grain deposited decreases (Fig. 3-34). Similarly, as a current increases in competence, the size of deposited material may increase, giving an inverted graded bedding. Similar grading is produced by seasonal supply of sediments, especially to lakes. The coarse material represents rainy seasons, whereas the finer material represents dry seasons or winter freezes. This produces one type of varved silt-clay deposit.

Current Bedding

The activity of currents in introducing sediments is often indicated by the internal structure of strata and by the surface features of bedding surfaces. Many of these features can be used to interpret the direction of current movement and are therefore important in reconstructing the paleogeography of the area. Included in these features are cross bedding, ripple mark, and sole markings.

Cross bedding (Fig. 3-36). This consists of thin lamellae to very thick beds deposited as inclined units because of the introduction of sediments from one direction by currents. The inclined surface on which cross-bedded strata are deposited is developed by sedimentary processes, either by scouring of channels or troughs or by deposition as terracelike deposits. Terrace building is most commonly deltaic, but similar effects may be developed by dune migration. The three basic shapes of cross-bedded bodies are tabular, lenticular or troughlike, and wedgelike.

The inclined strata tend to be planar or curved and are generally concave upward so that the base of the lamellae becomes tangent to the surface on which the cross-bedded unit rests. Ideally, the lamellae also become tangent at the upper surface, but commonly the upper portion has been removed by erosion before deposition of the next unit. Undeformed units have a depositional slope most commonly between 12 and 36°, averaging between 16 and 20°. There are reported examples having dips up to 50°.

A wide range of size is known, from cross-bedded units a few millimeters thick to units whose thickness exceeds 1,000 feet. In northwest Arkansas, a sandstone-siltstone-shale formation dips 5 to 8° to the south, whereas the underlying clastic limestone dips less than 1° to the south. The sandstone-siltstone-shale sequence is more than 1,800 feet thick and represents giant foreset beds filling into a basin. Such relationships lead to considerable confusion in structural and stratigraphic interpretation. Most commonly, cross-bedded units range from a few inches to a few feet in thickness, and the individual inclined strata are proportionately smaller.

Ripple mark (Figs. 6-8 to 6-10). This consists of a series of parallel to subparallel ridges and troughs developed on the surface of accumulating

Fig. 6-8. Ripple mark on Baraboo quartzite of Precambrian age. Devils Lake State Park, Wisconsin.

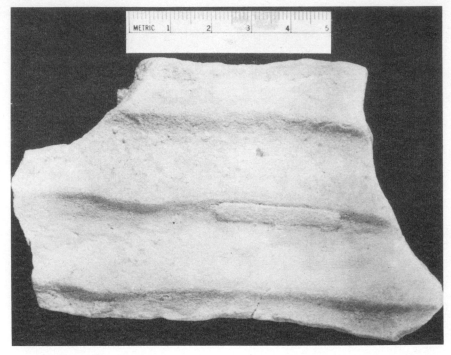

Fig. 6-9. Oscillation ripple mark in siltstone. Clarksville, Arkansas.

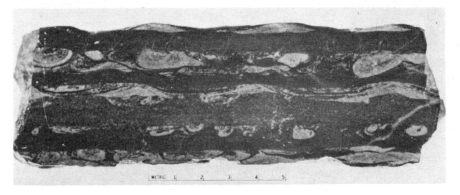

Fig. 6-10. Cross section of ripple mark and load casts. The rock is a green gray-wacke. Sedimentary features are preserved in the pink arkose. Fern Creek formation of Archean(?) age, Sturgeon River Falls, east of Iron Mountain, Michigan.

granular sediments owing to the motions of the depositing medium. It may be symmetrical or asymmetrical in cross section and vary in wavelength from a few centimeters to several tens of feet. The small-scale ripple mark is commonest.

Current ripple mark is small scale, having a wavelength from 2 to 10 cm. It is distinctly asymmetrical, and the steep side is in the direction the current moved. Ideally, the ripples develop as parallel ridges perpendicular to the current direction. Most, however, are curved and irregular or anastomosing due to changes in current and irregularities of the sediment floor. Subaqueous current ripple mark may be developed in any depth of water but is more common in shallow water. It has an amplitude to wavelength ratio of 1 : 4 to 1 : 10. Heavy-mineral grains or platy fragments tend to concentrate in the troughs or on the steeper down-current side. Eolean current ripple mark are more variable in their directions than the subaqueous variety and are much shallower. An amplitude to wavelength ratio of 1 : 15 to 1 : 50 is characteristic.

Oscillation ripples are also small scale and are generated by wave action without appreciable current. These ripples have sharply pointed ridges and rounded troughs and are symmetrical. They develop only in shallow water, where wave action can reach bottom and where there is no interference on the bottom by plants. In general, their wavelength increases with water depth.

Interference ripples consist of two sets of ripples developed on the same bedding surface at any angle to each other. Current direction and wave direction are commonly at an angle to each other, resulting in the development of a sediment surface consisting of a series of oval depressions separated by small ripple ridges.

At high water velocities, large wavelike ripples develop in the sediments. These may be symmetrical or asymmetrical and may migrate with the current or against it, depending on the nature of the sediment and the actual current velocity. At high water velocities, the surface of a sand bed is set in motion to a considerable depth and develops ephemeral sand waves. If the velocity decreases, these waves become stabilized and give rise to *metaripples*, which are asymmetrical and of large wavelength. Tidal currents around reefs and banks develop waves in calcareous sediments which, when preserved, are called *para-ripples.* These may be symmetrical or asymmetrical and often show a concentration of coarse organic detritus in the troughs. Their wavelengths are generally a few feet.

Sole markings (Fig. 6-11). As high-velocity currents, particularly turbidity currents, pass over unconsolidated sediments, they cut shallow grooves and gouge out pits. These may then be filled with sediment as the current ceases to flow, and they are preserved as irregularities on the underside of the overlying stratum. They are best seen as sandstone or siltstone fillings into shale. The structures are parallel to the direction of flow.

Fig. 6-11. Sole markings—prod casts on the base of a sandstone unit. Sand has filled into grooves dug on the top of an underlying shale. The grooves were presumably cut by tools carried by a turbidity current. Stanley formation of Mississippian age, north of Hot Springs, Arkansas.

Soft-sediment Structures

After deposition but prior to burial and lithification the sediments may have their surfaces marked by a number of processes. They may also be deformed by movement of the soft material, usually as a downslope creep or a rapid slump. Such features are indicative of the depositional or the immediate postdepositional environment.

Mud cracks (Figs. 6-12 and 6-13). Desiccation of fine-grained sediments deposited under water but subsequently exposed to the atmosphere results in the development of wedge-shaped cracks extending down into the sediment and making polygonal patterns on the stratification surface. The polygons range in size from a few millimeters to a meter in diameter. Size is dependent on the thickness of the bed, the amount and type of colloidal material, and the rate and extent of desiccation. Mud cracks develop on tidal flats as well as in lakes and on flood plains, but they are probably most commonly preserved in the continental environment, where loose sediment can be drifted into the cracks before the clay flakes formed by cracking can be moved.

Crystal impressions (Fig. 6-14). Ice and salt crystals forming in the surface layers of fine-grained water-saturated sediments leave impressions on the stratification surfaces which may be preserved by filling with the next layer of sediment. Ice crystals usually yield elongate wedge-shaped, feathery, or dendritic structures. Salt crystals give polygonal or hopper-shaped depressions.

Swash and rill marks. Waves washing up on unconsolidated sediments leave small arcuate ridges known as *swash marks* at the point of maximum advance. The outflow of the waves or tides results in the development of small branching channels or tear-drop-shaped depressions around larger objects on the beach. Both types of depressions are elongated down the

Fig. 6-12. Desiccation cracks in weakly consolidated sediment. The sediment is a lake-bed accumulation of volcanic ash. Florissant, Colorado.

Fig. 6-13. Mud cracks in argillaceous siltstone. Note the two scales of polygonal cracking.

beach toward the strand line and are known as *rill marks*. Swash and rill marks tend to be destroyed by subsequent wave action but may be preserved.

Pits and mounds. These are small features on the surface of unconsolidated sediments made by the impact of falling rain drops, falling hail, spray, or by gas bubbles trapped on the surface of sediments or rising through the sediments. Impact pits usually have a raised rim completely or partially around them. Gas bubbles trapped on the surface of sediments do not leave such rims but produce shallow round depressions. Gas rising through the sediments builds up small mounds having craters in the center. Larger pits are developed by springs rising through unconsolidated material. The rising current destroys small-scale stratification and concentrates the coarser materials by flushing out the fine-grained materials. Pits and mounds are also developed by burrowing organisms.

Fig. 6-14. Salt-crystal impressions in fine-grained sandstone. Salt crystals grew in the underlying sediment and subsequently dissolved out, leaving impressions. These were filled by sand before the impressions could be destroyed by wave action. De Queen limestone of Cretaceous age, De Queen, Arkansas.

Mud balls and armored mud balls. Semiconsolidated mud masses may be broken loose from a bed by current action or by partial desiccation and be rolled along the bottom to become rounded or cigar-shaped bodies. Similarly, larger clastic fragments may be rolled through soft mud and collect a coating of mud. Such bodies are preserved as mud balls. Occasionally, mud balls roll through coarse sand and fine gravel and become coated with embedded grains of coarse material. These are armored mud balls. Structures similar to mud balls are also developed as concretions.

Load casts (Figs. 6-10 and 6-15). Massive sediments, such as sands, deposited on soft unconsolidated sediments, such as mud, sink into the mud unevenly because of inequalities of loading and subsequent compaction of the mud. This produces a highly irregular contact involving complicated

Fig. 6-15. Load casts. The base of the lowest sand unit of a turbidity-flow deposit. The heavy sand has settled differentially into the underlying soft mud. Jackfork formation of Mississippian age, Aly, Arkansas.

infolding of the sand into the mud. These load casts are usually on a scale of a few millimeters to centimeters of interpenetration.

Slump structures (Fig. 3-36). Water-saturated sediments deposited on a sloping floor may slump downhill as a unit. Three typical resultant structures have been described in Chap. 3 under Deposition of Clastic Particles. These are sand wedges having an intricate contorted stratification and cutting into shale, contorted bedding in one set of strata in between simple-bedded overlying and underlying strata, and pull-apart structures. Slump or compaction also results in intricate small-scale faulting.

Flame structures. These are tongues of semiconsolidated sediment drawn up in a flowing mass of mud. These structures are most commonly seen at the base of turbidity-flow deposits.

SECONDARY STRUCTURES (CHEMICAL)
Accretionary Bodies

Accretionary bodies are regular to highly irregular masses which occur in sedimentary rocks and which show evidence of growth in the semiconsolidated or consolidated sediment by the chemical segregation of minor constituents of the rock (Fig. 6-16). Some may have developed at the time of deposition by the accumulation of gelatinous materials into clots and globs. Most, however, are the product of diagenetic processes. The time relationship between deposition and segregation can be determined in some bodies. Accretionary bodies are generally lumped under the terms *concretions* or

Fig. 6-16. Accretionary body in place in shale. Note the continuity of the shale beds over and under the body, indicating the diagenetic origin of the body. Note also the cracking of the body and the subsequent filling with calcite. Calcareous concretion in black shale, Fayetteville formation of Mississippian Age, Fayetteville, Arkansas.

nodules, except for special forms such as geodes and septaria. Concretions consist of pure or dominantly pure single-mineral segregations which appear to have made room for themselves by replacing or forcing aside the surrounding sediment. Nodules have formed by a filling of voids in the sediments and the incorporation of the sedimentary materials in the body.

Chert nodules. These are essentially pure microcrystalline quartz generally found in limestone. They may contain carbonate fossil fragments, finely disseminated carbonates, or carbonaceous material. Their shape varies from lenticular or spheroidal to highly irregular tuberous bodies, which may be widely scattered or be so close as to coalesce, forming continuous irregular beds. Most commonly they are distributed in a way which is somehow related to the stratification of the host limestone, either along stratification surfaces or along the central portions of individual strata. The chert masses may be completely structureless, concentrically banded, or irregularly color-mottled. Many have a soft white rind on the outside of the mass, and some have a soft irregularly shaped powdery center. The chert varies in color from white to black or brown but is most often gray.

Clay ironstones (Fig. 6-17). These are bodies consisting dominantly of microcrystalline siderite plus admixed clay in various proportions. These bodies are most common in black shale. The structures are most commonly flattened parallel to the bedding of the shale and are discoidal to irregularly rounded. They often show evidence of having been collapsed, giving the shape of a wheel with a bulbous hub cap. Cracking of the surface and brecciation of the center usually accompany the collapse. This structure indicates desiccation of a lenticular colloidal mass after an outer shell had hardened. The unaltered interior of clay ironstone concretions is composed of black microcrystalline material, but oxidation usually discolors the surface brown or yellow and often results in concentric spalling of surface layers. When oxidation is complete, the centers of the bodies may be hollow or may contain loose powdery material (Fig. 6-18). Occasionally, layers may be separated so that the interior is loose and rattles in a hardened oxidized crust.

Septaria and melikaria (Fig. 6-19). These are large nodules characterized by sets of roughly radial and concentric cracks in their centers. The cracks become thinner toward the surface, and the structure gives the effect of

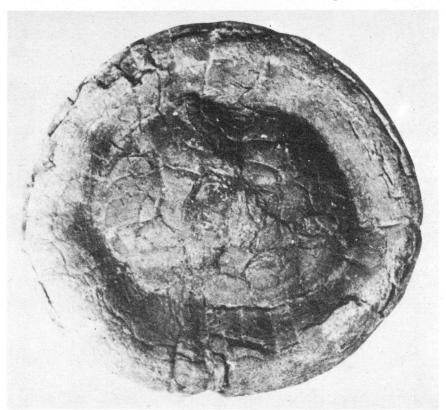

Fig. 6-17. Clay-ironstone concretion. Typical external cracking as a result of desiccation of the original colloid is evident. A cross section would show extensive brecciation and veining by calcite. Fayetteville formation of Mississippian age, Fayetteville, Arkansas.

Fig. 6-18. Liesegang weathering of clay ironstone. Weathering has been controlled by cracks, probably joints, in the original concretion; iron migrates to the cracks and leaves a soft clay core. Savanna formation of Pennsylvanian age, Petit Jean State Park, Morrilton, Arkansas.

Fig. 6-19. Septarian concretions and melikaria weathered from the concretions. Imo formation of Mississippian age near Leslie, Arkansas.

desiccation of a colloidal mass. The cracks are commonly filled with mineral material, usually calcite, introduced later by groundwater. Clay ironstones commonly show septarian structure. Melikaria are boxwork patterns of material filling the septarian cracks left after weathering and removal of the original accretionary body.

Rosettes and spherulites. These are radiating macrocrystalline bodies of discoidal or spherical shape, consisting essentially of one mineral. The best examples are rosettes of pyrite or marcasite associated with coal or carbonaceous shale and spherulites of siderite and chalcedony. Pyrite also occurs as single crystals or crystal clusters in many sediments without the radial structure. Barite rosettes, or desert roses, consist of bladed coarsely crystalline barite as a cement in sand (Fig. 6-20). The crystals are crudely radial in orientation and include much sand.

Geodes. These are hollow subspherical bodies having a rind of chalcedony and an interior lined with crystals. They are interpreted as having grown by radial expansion of a cavity lined with chalcedony. The growth of the lining crystals usually occurred after the cavity had completed its growth. The lining crystals are most commonly quartz but may also include carbonates and sulfides. Hydrocarbons may be present in the cavity.

Calcareous concretions (Fig. 3-45). These are varied bodies formed by the localization of carbonate cement in sediments. The masses are composed dominantly of the host sediment, with calcite filling the pore spaces. Discoidal to bizarre-shaped bodies occur in shale, often containing a nucleus of organic material. Apparently, the presence of the organic matter caused local concentration of the cement. Bedding commonly can be traced through the concretion, and impressions or carbon films left from plants are found along

Fig. 6-20. Barite rosettes. Large barite crystals developed as cement in sand. Norman, Oklahoma.

one bedding plane. In sandstone containing a paucity of carbonate available for cementation, the carbonate may accumulate to form spherical masses of coarsely crystalline cement with much included sand. The remaining sand may be left essentially uncemented or a different type of cement may be introduced later. Bedding of the sandstone can often be traced through the bodies. Occasionally, the large calcite crystals are euhedral single crystals or clusters of randomly oriented crystals, giving so-called sand-calcite crystals (Fig. 3-44). These usually have a rounded scalenohedral habit. Similar spherical masses form as a result of local silica cementation, but the silica is usually microcrystalline or amorphous rather than coarsely crystalline (Fig. 6-21).

Solution Structures

Stylolites. These are highly irregular junctures between adjacent units of soluble rock, consisting of columns and depressions and usually having striated sides. The columns appear to have developed parallel to the direction of pressure and are commonly perpendicular or nearly perpendicular to the stratification. Stylolitic seams may be parallel or may bifurcate, forming a coarsely braided pattern. Accumulations of insoluble residues, such as clay or iron or manganese oxides, are common along the seams. Evidence of interpenetrations are common in the truncation of fossils, pebbles, ooliths, etc., at the edge of the columns, and these should be sought. Stylolites are

Fig. 6-21. Spherical mass of silica-cemented sandstone. The remainder of the sand was cemented with calcite.

most common in calcareous rocks but are also found in sandstone, quartzite, and chert. They are unknown in highly argillaceous rocks.

Cone-in-cone (Fig. 6-22). These are calcareous structures in shales and are commonly associated with concretions. They consist of nested cones, usually having the cone axis perpendicular to the bedding and the base of the cone up. The sides of the cones are usually ribbed and are marked by concentric ridges and grooves. Microscopically, the cones consist of fibrous crystals of calcite or siderite, which are subparallel to the cone axis. The structures indicate growth against a confining (load) pressure. Where cone-in-cone has developed on both sides of a concretion, the cone bases are up on top and down on bottom of the concretion, indicating outward growth of crystals against pressure.

Solution sinks and fissures. Ancient karst topographies have been preserved on all scales through burial by younger sediments, from large sinks to narrow fissures and irregular lumpy solution surfaces. The time of solution and the question of whether it occurred under a cover of sediment or at the base of the soil can often be determined by the stratification of the solution-cavity fill. Modern solution of limestones under a cover of fissile shale often results in the draping of the shale over knobs on the limestone surface. Solution can be so extensive that limestone units several feet thick are completely missing at the surface but are encountered in hillsides when excavation cuts back 10 to 20 ft. It is then important to distinguish between modern and ancient solution effects.

Fig. 6-22. Cone-in-cone structure in limestone.

DESCRIPTIONS OF ROCKS

Sedimentary rocks are discussed here in the eight major readily recognized groups. No attempt is made to develop an overall classification because of the complexity of origin and the gradation between fundamental rock types. Common variations within the major groups are indicated and broad classification terms are given. Discussion of the features follows a suggested outline for the detailed observation of the rocks.

CONGLOMERATES AND BRECCIAS

Definition

These are the clastic sedimentary rocks in which fragments larger than 2 mm in diameter compose an important portion of the whole. There is no set limit on the minimum percentage of the coarse-sized material which must be present to classify a rock as conglomerate or breccia. Most commonly 30 to 50 percent of pebble- or larger-sized materials are required, although many rocks having as little as 10 percent gravel have been considered conglomerates. Generally, rocks with less than 30 percent coarse material should be considered pebbly sandstone, boulder claystone, conglomeratic limestone, etc.

The distinction between conglomerate and breccia is made on the basis of the angularity of the coarse clastic materials. The term *breccia* is usually applied to rocks in which the coarse fragments are angular, whereas the term *conglomerate* is used when the fragments are subangular or more completely rounded. Inasmuch as angular fragments can be produced by processes other than sedimentation, the terms *sharpstone* and *sharpstone conglomerate* have been used for the angular fragments and breccias of distinctly sedimentary origin. The equivalent terms *roundstone* and *roundstone conglomerate* have been used for abraded coarse fragments and the conglomerates derived from them.

Descriptive Checklist

The following outline gives the features of conglomerates and breccias which should be looked for in making a complete lithic description of a rock. All the features outlined will not occur in any one rock so that only those portions applicable to a specific specimen need be considered.

Gross characteristics of the rock
 Color
 Evidence of bedding: sandstone lenses, cross bedding, cut-and-fill
  structures, imbricate pebble lenses, graded bedding, pebble
  orientation, etc.
 Framework: packing of pebbles

Orthoconglomerate: pebbles dominantly in contact throughout a sand matrix

Paraconglomerate: pebbles not in contact and scattered throughout a mudstone matrix

Sorting

Framework pebbles: uniform size or limited size range versus wide size range

Matrix: same

Physical characteristics due to grain packing and cementation

Friability

Porosity and permeability

Characteristics of the pebbles

Percentage of pebbles in the rock

Composition of pebbles: percentage of each rock type

Size: size range and average size

Note any correlation between size and rock type of the pebble

Shape

Roundness and sphericity

Orientation of pebbles due to shape, imbricate structure, flat lag pavement

Surface texture

Primary textures: mat, polish, faceting, impact scars, striations, etc.

Secondary textures: etching or secondary enlargement, crystal reflections, corrosion, replacement, etc.

Characteristics of the matrix

Sand-sized matrix

Percentage of the whole

Composition of sand-sized component: minerals, rock fragments, broken and abraded shell fragments, etc., and percentage of each

Size: size range and average size

Shape

Roundness and sphericity

Effects of secondary enlargement, corrosion, or recrystallization

Mudstone matrix

Percentage of the whole

Composition

Proportions of argillaceous, silt-sized, or sand-sized materials

Mineral composition of sand-sized component

Structure

Laminated mudstone matrix

Thickness and continuity of laminae

Relationship of pebbles and laminae

Nonlaminated, or massive, mudstone
Cementation and other postdepositional effects
Effectiveness of bonding
Friability
Breaking characteristics: through or around grains
Character of cement
Recrystallized detrital materials: argillaceous or calcareous muds
Introduced mineral cements: mineralogy, grain size, secondary overgrowth versus cavity filling
Hydrothermal or metasomatic effects

Discussion of Special Features

Framework. Unconsolidated gravels consist of a framework of coarse clastic particles (pebbles, cobbles, boulders) and interstitial voids (Fig. 3-29). The voids may be filled with finer-grained materials which were either deposited at the same time as the gravel or washed subsequently into the voids. Such rocks consist of a framework of pebbles which are in contact throughout the mass, and they indicate an abundance of coarse material accumulating in an environment of strong current action, such as a beach or stream. Such rocks have been termed *orthoconglomerates* by Pettijohn (Fig. 6-23).

In contrast, the rapid deposition of a mass of sediment with a wide size range, including gravel-sized material, results in a dominance of matrix material having randomly scattered coarse fragments, not necessarily in contact with each other. Such deposits are developed by glacial dumping without reworking by meltwater, mudflows, turbidity currents, etc. Similarly, scattered coarse material in a dominant fine matrix develops by ice or root rafting of erratics into an area of quiet-water deposition. Conglomerates of this type have been termed *paraconglomerates* by Pettijohn (Fig. 6-24).

Sorting. Studies of modern gravels indicate that sorting, both in the framework and in the matrix, is indicative of a depositional environment. Orthoconglomerates are moderately well to exceptionally well sorted, both in their framework materials and in their matrix. Beach gravels are exceptionally well sorted, often having as much as 90 percent of its coarse material falling in one ϕ class. Alluvial gravels are not so well sorted, having a wider range of ϕ sizes and the most abundant size rarely composing more than 25 percent of the rock. Paraconglomerates are very poorly sorted, the larger fragments ranging from many feet or tens of feet down to sand size. In fact some blocks are large enough to show up on standard field maps. The reason for the wide size range is that transportation is by high-viscosity (glacial ice) or high-velocity (mudflow and turbidity current) agents. Sorting in the matrix of paraconglomerates is also important. Ice- or root-rafted coarse fragments are usually dropped into well-sorted and even thinly laminated

Fig. 6-23. An orthoconglomerate. A framework of touching igneous-rock pebbles containing little interstitial clastic material. The rock has been cemented by coarse crystalline calcite (white). Lake Shore conglomerate of Keweenawan age, Eagle River, Michigan.

sediments. In contrast, the matrix of mud-flow or glacially deposited materials is as unsorted as the coarse fragments.

Composition of the pebbles. Any type of rock or mineral fragment may occur in the coarse clastic phase of conglomerates, as long as the material is physically durable enough or chemically stable enough to persist through the sedimentary processes. Pebbles are of considerable interpretative value, because they are the most indicative of the source rock of any clastic materials and are diagnostic of the rigor of the environment of their formation. Many conglomerates are composed of pebbles of only stable mineral and rock types, such as quartz, quartzite, quartz sandstone, and chert (Fig. 6-25). These are indicative of a source containing only stable materials or of the complete destruction of all unstable materials derived from the source. Other conglomerates contain a variety of rock types as pebbles, some of which are stable and others unstable (Fig. 6-26). Such a mixture indicates that unstable rocks were present in the source rocks and that the weathering and sedimentation environment either was not severe enough to destroy them

Fig. 6-24. A paraconglomerate. A wide range of pebble- to boulder-sized clasts in an abundant clay-silt matrix. The large clasts do not touch. Gowganda formation, Portlock Station, Ontario.

or was not of sufficient duration to destroy them. Conglomerates can thus be classified as mature or immature. Field naming of conglomerates should be based on size and rock type present or dominant in the conglomerate; thus: quartz pebble conglomerate, granite boulder conglomerate, limestone cobble conglomerate, etc.

Attention should be paid to the relative sizes and shapes of pebbles of different lithic types. Generally, if all the material has been derived from a common distant source, pebbles will decrease in size with decreasing stability. If chemically or mechanically unstable pebbles are larger than associated stable pebbles, it generally indicates a much closer source for the unstable material.

Composition of the matrix sand. Sand-sized material of almost any composition is the commonest void-filling material of conglomerates. Generally, there is a correlation between the lithic character of the pebbles and the sand, but this is not necessary, particularly where the matrix sand is carbonate organic debris. Sands often are not as mature as the associated pebbles. Thus quartz pebbles may have considerable feldspar in the associated

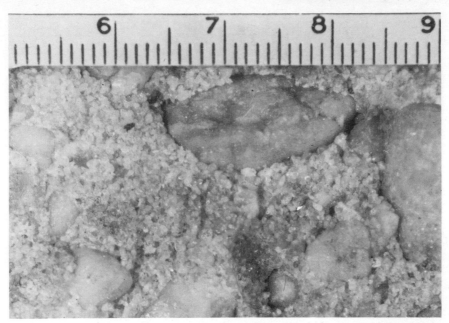

Fig. 6-25. A quartz-pebble conglomerate. All pebbles and sand grains are stable minerals — quartz and chert. The pebbles are rounded to subrounded and of a limited size range. The bright surfaces on the pebbles and sand are crystal faces developed by secondary enlargement. Olean conglomerate of Pennsylvanian age, Olean, New York. (*Ward's Universal Rock Collection.*)

sand. This is due to the greater mechanical durability of single-crystal sand grains as compared with polymineralic rock fragments in the abrasion and impact environment of a stream or beach.

Laminated mudstone matrix. Materials ranging from sand size up are often carried into a basin of quiet-water deposition by floating ice, animals, or wedging in tree roots. Floating ice may be of glacial origin, but it may also be river ice or shoreline ice, and glaciation need not be implied. When the fragments are released, they fall to the bottom and become embedded in the sediment. The impact of the fragment causes the sediment to be dished downward, and as a result, laminae under the pebbles are bent. Subsequent deposition of the fine-grained material will progressively cover the pebble. Frequently, the next few laminae will terminate abruptly against the pebbles, but ultimately they will arch over the fragment. The relationships of the laminae to the fragments clearly indicate a dropping in of the fragment from overhead.

Nonlaminated, or massive, mudstone matrix. Environments in which all sizes of materials can be deposited simultaneously give rise to conglom-

Fig. 6-26. A mixed granite-pebble conglomerate. The pebbles are mixed stable (vein quartz) and unstable (granite and greenschist) rock types. The matrix consists of arkosic sand and silt. Fern Creek formation of Archean(?) age, Sturgeon River Falls, east of Iron Mountain, Michigan.

erates having a nonbedded sandy-mudstone matrix. High-velocity currents, such as mudflows or turbidity currents, which end abruptly or high-viscosity transporting agents, such as glaciers or landslides, give rise to such unsorted materials (Fig. 6-24). In each case a wide size range of material is dumped. The shape of the deposit and the character of the associated sediments are the features which will identify the depositional environment. Turbidity-current paraconglomerates are usually sheetlike bodies associated with gray-wacke and shale. The conglomerates often show graded bedding, as do the associated graywacke beds. Mudflows are local fan-shaped or tongue-shaped deposits with high to moderate initial dips and are built up as distinct thick- to very-thick-bedded units. Cut-and-fill channeling is very common. The associated sediments are usually intermountain basin-fill sandstone and silt-stone, often with playa-lake evaporite materials. Clay-sized materials are not commonly associated, because mud flows develop characteristically in semiarid to arid regions, where clay is removed by deflation. Glacially dumped material (till) generally occurs as discontinuous sheets on an ero-sional unconformity of considerable relief. Till sheets are completely lacking in bedding, although several sheets may be superimposed with an erosional contact or an intervening weathered profile. A small proportion of the pebbles may be striated or faceted. There is usually very little correlation between durability and size of fragments as there is in the other paraconglomerates, where considerable impact between particles is common. Tills usually have associated well-sized bedded river and lake deposits and varved silt and clay. Landslide paraconglomerates are usually local arcuate bodies which are a jumbled mass of roughly broken rocks, largely of one or two lithic types.

Effectiveness of cementation. The degree and type of cementation can often be judged from the breaking character of the rock. Slight cementation of orthoconglomerates usually results in a friable rock, particularly where the cement is an introduced mineral cement such as silica or calcite. In the case of carbonate cements, leaching at the present surface may result in a friable rock, in spite of original complete cementation. Often in such rocks, the quartz-sand grains are highly frosted as a result of etching. The breaking character of completely cemented conglomerates depends on the relative strength of the cement and the clastic particles. If the cement is softer, or has a better cleavage, than the clastic material, the rock breaks around the pebbles and sand. If they are equally hard, the rock breaks through the clastic grains as easily as it does through the cement. Rocks in which lithification has occurred through the recrystallization of an argillaceous matrix are generally very tough because of penetration of the new clay-mineral flakes into the coarser clastic particles. Rupture is thus generally through the sand and pebbles due to their brittleness.

Types of cement. The common cements of conglomerates are of two basic types: introduced mineral cements and recrystallized detrital materials. They can be classified as follows:

Introduced mineral cements
 Silica
  Quartz: generally as secondary enlargement
  Chalcedony or opal: usually as interstitial coatings of grains
 Carbonates
  Calcite: coarse crystalline spar or secondary enlargement of detrital encrinal material
  Dolomite: crystalline cavity filling
 Hydrous iron or manganese oxides: most commonly as a minor staining of other cements but may form the primary cement, in which case banding and colloform structures are common in larger voids
 Minor mineral cements
  Anhydrite, gypsum, barite, pyrite, and other sulfides
 Asphaltic material
Recrystallized fine-grained detrital material

Carbonate muds: generally give an extremely fine-grained pulverlent
material filling interstices
Argillaceous muds

Lithologic Names for Conglomerates and Breccias

Rocks in this class should be named on the basis of size class and lithic
character of the dominant coarse clastic material. An adjective may be pre-
fixed to describe the character of the interstitial material or the cement if
that material composes more than 10 percent of the rock. Thus:

Quartz pebble conglomerate
Chert cobble conglomerate
Granite pebble conglomerate

each would consist of pebbles of dominantly one rock type and one size
grade and little interstitial material other than sand.

Argillaceous quartz pebble conglomerate
Argillaceous mixed granule conglomerate
Argillaceous limestone pebble conglomerate

each would have more than 10 percent interstitial clay. The mixed-granule
conglomerate would have several rock types of granule size.

Other modifying terms for the minor constituent are siliceous, calcareous, carbonaceous, bituminous, ferruginous, etc.

Special terms are:

Till and *tillite.* Glacially deposited unstratified material which is unindurated or indurated.
Fanglomerate. An alluvial fan deposit.
Agglomerate. A volcanic accumulation of material coarser than sand
size.
Peperite. Granule- to fine-pebble-sized volcanic material which has an
argillaceous or calcareous matrix. It is developed by submarine
eruption of volcanic material into unconsolidated sediments.

SANDSTONES

Definition

Those clastic sedimentary rocks in which fragments in the range $\frac{1}{16}$ to
2 mm compose more than 50 percent of the rock. Rocks in which this size
material is dominantly calcium carbonates or dolomite are excluded and are
included in the limestones, although texturally they may be identical to
sandstones.

Sandstones are classified on the basis of the mineralogic or lithic character of the sand-sized constituents. Three main types of materials occurring
as sand-sized grains are quartz and chert, feldspar, and rock fragments ex-

clusive of chert. A useful classification modified from Pettijohn is given below.

Descriptive Checklist

The following outline gives the features of sandstone which should be looked for in making a complete description of a specimen. Some of the contrasting features of the different types of sandstone are pointed out in the subsequent discussion. All the features outlined will not be applicable to any one specimen.

Classification of Sandstones*

Cement or matrix			Detrital matrix prominent (over 15%) to predominant; chemical cement absent	Detrital matrix absent or scanty (under 15%); voids empty or filled with chemical cement			
Sand or detrital fraction	Feldspar exceeds rock fragments	Graywackes	Feldspathic graywacke	Arkose	Arkosic sandstones Subarkose or feldspathic sandstone	Orthoquartzites	Quartz sandstone
	Rock fragments exceed feldspar		Lithic graywacke	Subgraywacke	Lithic sandstones Protoquartzite		
	Quartz content		Variable; generally < 75%	< 75%	75%–95%		> 95%

*Modified from F. J. Pettijohn, "Sedimentary Rocks," p. 291, Harper, 1957.

Gross characteristics of the rock
    Color
    Bedding characteristics and thickness
        Evidence of bedding
            Splitting or parting: thickness
            Internal lamination: color banding, compositional changes, grain orientation, grain-size change, etc; thickness of lamina.

Internal structure of strata
    Graded bedding: type and size range
    Cross bedding: scale
Features of bedding-plane surfaces
    Ripple mark: type and orientation
    Mud cracks, raindrop imprints, crystal impressions, etc.
    Current marking: sole markings, flame structures
Sorting
    Uniformity of size of sand grains
    Proportion of fine-grained detrital matrix
Physical properties due to grain packing and cementation
    Friability
    Porosity and permeability
Characteristics of the sand grains
    Percent of rock
    Composition of sand grains: percentage of each rock or mineral type
    Size
        Size range and average size
        Correlation between size and composition
    Shape
        Roundness and sphericity
        Effect of secondary enlargement
        Correlation between shape and mineral or rock type
    Surface textures
        Primary textures: mat, polish, frosted
        Secondary textures
          Euhedralism due to secondary enlargement
          Vagueness of grain boundaries due to replacement
Characteristics of detrital matrix
    Percent of whole
    Composition
    Texture
Cementation and other postdepositional effects
    Effectiveness of bonding
        Friability
    Breaking characteristics: through or around grains
Character of cement
    Recrystallized detrital cement
        Argillaceous mud
        Recrystallized carbonates
    Introduced mineral cement
        Mineralogy and size of grain
        Secondary overgrowth versus cavity filling

Hydrothermal or metasomatic effects
Organic Evidence

Discussion of Special Features

Bedding. The different types of sandstone are characterized by different types of bedding, because they represent different depositional environments. The graywackes commonly show a graded bedding which contains important amounts of fine-grained material throughout the stratum. The base of beds, where they rest on shale, commonly shows sole markings and flame structures. Chips and flakes of the underlying sediment are common near the base, especially if the underlying unit is shale. Graywackes very rarely show cross bedding, ripple mark, or evidence of shallow-water or continental deposition, such as mud cracks, raindrop prints, etc.

Quartz sandstone and lithic sandstones are commonly cross-bedded and ripple-marked (Fig. 6-27). Graded bedding is rare, except that scattered

Fig. 6-27. Quartz sandstone. Each quartz-sand grain has been coated with a film of minute iron oxide grains and, subsequently, secondarily enlarged by growth of quartz in structural continuity with the original grain. A minor amount of recrystallized clay is present in some interstices. Cambrian age, Potsdam, New York. Projection of a thin section. (*Ward's American Rock Collection.*)

pebbles or granules occur at the base of strata, and the type involving a grain-size reduction without appreciable clay-sized material occurs in scattered strata. Slump structures are particularly common in lithic sandstones of deltaic origin. Fossils are fairly common along bedding surfaces as tracks and trails and burrows.

Arkose generally shows indistinct bedding (Fig. 6-28). This is particularly true in basal arkoses, which grade downward into an underlying granite. The upper portions of such units may be partially reworked to give crude bedding and local cross bedding. Thick transported arkoses are usually poorly bedded, showing local coarse cross bedding and cut-and-fill structures. Evidence of continental or shallow-water origin is generally abundant in these arkoses and their associated fine-grained sediments. Mud cracks, raindrop impressions, footprints, etc., are very common, but fossils are rare.

Sorting. Sorting is characteristic of the different environments of sand accumulation and agents of deposition. Sand deposited under a continual

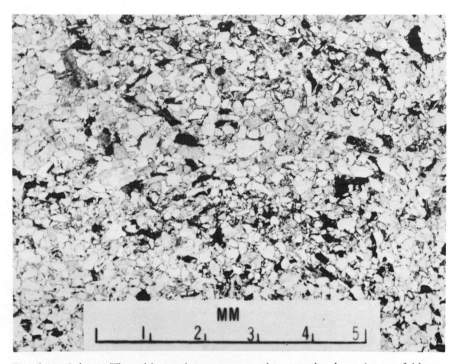

Fig. 6-28. Arkose. The white grains are quartz, the gray cloudy grains are feldspar, and the black grains are altered biotite and iron oxides. A few small accessory mineral grains and some clay also appear. Triassic age, Portland, Connecticut. Projection of a thin section.

current action, such as wave or wind currents, is well sorted both as to uniformity of sand size and absence or minor importance of interstitial detrital matrix. Thus beach and dune sands are well sorted without a matrix. Deeper-water sands, which are still shallow enough to be subjected at least to storm-wave action, are moderately well sorted as far as the sand size is concerned but have an infiltered quantitatively minor matrix. Intermontane basin-filling sand, as represented by many of the thick arkoses, is brought in by stream action of highly variable current velocity. Therefore the sand is often poorly sorted, especially in the coarser accumulations, but still has relatively minor infiltered matrix. A high percentage of fine-grained matrix material and a complete range of size from coarsest to finest are indicative of dumping in of sediment without current sorting. Graywacke characteristically is nonsorted in this fashion and is interpreted as the deposit from turbidity flows.

Composition of the sand grains. The classification of sandstone depends on the relative proportions of the different materials present as sand grains (see p. 367). The sand grains most readily recognizable megascopically are single-crystal grains of quartz, feldspar, micas, or other minerals. Rock fragments are common but are difficult to recognize; therefore the lithic sandstones are often misclassified. The important types of sand grains can usually be recognized with the aid of a binocular microscope.

1. *Quartz.* Usually glassy and colorless in sand-sized grains; therefore the color in the bulk rock depends on the background material or staining. Grains rubbed off the specimen onto a black surface are colorless, unless they have a surface stain or an abnormally high content of inclusions. Quartz grains, when pressed with a stiff steel point (such as a dental probe), are unaffected until a high pressure is applied, at which time a conchoidal fracture can be seen to develop within the grain if the grain is well embedded in the rock.
2. *Chert-sand grains.* Usually white to cream-colored and nontransparent. A granularity may be apparent in coarser material. Under pressure from a steel probe, the grains are unaffected or may break to give small irregularly shaped grains or conchoidal chips. The conchoidal fracture cannot be seen developing, as in quartz grains.
3. *Feldspar.* Characteristically white to pink, and usually in rectangular shapes. The shape is usually recognizable, even after fairly extensive rounding. Feldspars show all stages of alteration to clays so that cleavage surfaces vary from bright to unidentifiable. Under a steel probe the fresh feldspars yield flat chips of angular outline, chipping following the cleavage. Highly altered feldspar yields a white powder very readily under a steel probe. A full range of pressure behavior between these

extremes may be seen, although it takes relatively little alteration to give the powder effect and cleavage may still be recognizable.

4. *Sand-sized rock fragments.* Because of size limitation these fragments are derived from fine-grained rocks, particularly phyllite, slate, and shale. Color varies from light to dark, and the grains appear opaque. Because of their softness the rock fragments are generally smaller and better rounded than associated quartz or feldspar. Under pressure from a steel probe, most rock fragments yield easily as flakes or scales, often showing a greasy semiplastic effect distinct from the dry-powder effect of feldspar.

5. *Micas.* Both dark and light micas are common as sand grains and are readily recognized by cleavage. With a needle point the grains can be picked apart into flakes which are the diameter of the grain. This is in contrast to rock fragments, which give flakes much smaller than the sand grain.

6. *Accessory minerals.* May be recognizable as pink, green, blue, brown, or black grains. Usually they cannot be identified without microscopic examination, but the ones most commonly seen are:

  Pink: zircon, garnet
  Green: epidote, chlorite, glauconite
  Blue: tourmaline
  Brown: tourmaline, rutile
  Black: magnetite, ilmenite, spinel, tourmaline

Size and shape of sand grains. The size and size range for sand grains should be noted. Frequently, in sands composed of several minerals as sand grains, there is a noticeable correlation between size and composition. Thus in coarse arkose, feldspar is often coarser than quartz, probably because of relative sizes in the source rock and short distance of transport. In graywacke, quartz is frequently the coarsest-grained material, and rock fragments occur as smaller sizes because of their relative softness. Where secondary enlargement has not been important, sand-grain shapes are characteristic. In purer-quartz sandstone, quartz grains are usually rounded to very-well-rounded, indicating a long distance of transport or eolean transportation. In contrast, in lithic and arkosic sandstone and graywacke, quartz is usually angular, showing little effect of abrasion.

Grain-surface textures. Frosted, mat, and polished grain surfaces can be observed if individual grains can be freed and examined on a contrasting colored surface. A black surface is usually best. Cementation controls the ease with which grains can be separated from the rock. A mat or frosted surface is by far the commonest, but they are not necessarily related to transportation environment as they are in unconsolidated sands. Quartz commonly is pitted by chemical solution during cementation by a carbonate. Such surfaces give the appearance of frosting similar to that developed by desert sandblasting. Incipient secondary enlargement of quartz may also give a frosted appearance because of the development of many minute crystal facets on the grain. Such minute facets can usually be recognized, because they reflect light simultaneously from a large area of the grain. More extensive secondary enlargement can be recognized both in the rock and in individual grains by reflection from well-developed faces. This appears only if the interstices between the grains have not been completely filled by the secondary quartz. Larger quartz-sand grains in graywacke often appear to have indistinct or fuzzy boundaries in the rock due to partial intergrowth of the groundmass into the crystals during recrystallization of the groundmass. Originally polished quartz grains do not usually retain their polish during cementation, but grains of minerals not involved in the chemical processes of cementation or recrystallization may retain an original polish.

Detrital matrix of sandstones. The amount and character of detrital fine-grained interstitial material are most important in interpreting the environment of deposition and in classifying the sandstones (see the discussion on sorting of sandstones, p. 370). The common detrital-matrix materials are silt and clay or fine-grained carbonate material. Appreciable quantities, sometimes dominating the rock, are characteristic of graywacke and represent deposition from a muddy suspension in quiet water. Lesser quantities of silt and clay are characteristic of river-deposited sand. Detrital fine-grained carbonates are often produced by the grinding of shell material on beaches or in shallow wave-agitated water. It is usually accompanied by sand-sized shell fragments which are recognizable. Probably much of the interstitial carbonate is introduced rather than detrital, and the two are difficult to differentiate.

Cementation. The type and effectiveness of cementation markedly affect the physical character of the sandstone. Sands which have been incompletely cemented, or in which the cement has been removed by groundwater solution, are usually highly friable. Individual grains can be readily rubbed off the specimen by the thumb without breaking the grains. Thoroughly cemented sandstones will not allow grains to be rubbed off, but the character of breaking of the rock is indicative of the type of cement. If the cement is as hard and resistant as the sand grains, then the rock will rupture through the grains. In contrast, if the cement is softer than the sand grains, then the rock will break around the grains. Thus quartz-sand grains cemented by quartz will break through the sand grains, whereas quartz sand cemented by calcite will break through the cement and around the grains. This is not invariable; it also depends on the extent of filling of pore space by cement. Quartz sand with pore space only partially filled by quartz cement may break around the grains. Some idea as to the extent to which interstitial voids are filled by cement can be obtained by observing the rate at which the rock takes up fluids. A drop of water or acid placed on a porous sandstone will be taken up in the pore space and disappear rapidly, whereas a drop placed on a rock in which all voids are filled by cement will sit on the surface until it evaporates.

Types of cement. Cements may be recrystallized fine-grained detrital material deposited with the sand, recrystallized sand-sized particles, or introduced minerals from outside the sand. Detrital cements are most commonly clays or carbonates. Clay of graywacke or other argillaceous sandstone recrystallizes as a fine-grained aggregate of clay flakes, chlorite, quartz, and feldspar to make a very tough compact rock (Fig. 6-29). Frequently, silica is introduced in the process of recrystallization so that no voids are left open. Individual mineral grains in the matrix cannot be recognized megascopically, giving a dark-colored dense groundmass in which the sand grains are set. Recrystallization often results in intergrowth of the new flakes into the margins of sand grains, resulting in indistinct borders to the sand grains.

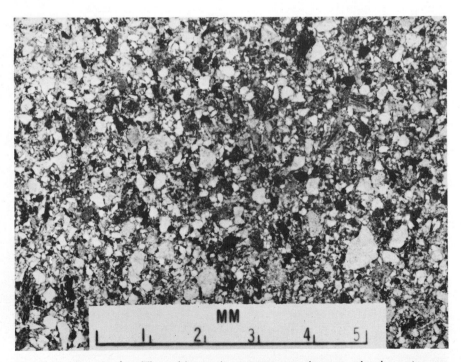

Fig. 6-29. Graywacke. The white grains are quartz, the gray cloudy grains are feldspar, and the dark grains are rock fragments in an abundant silt-clay matrix. Stanley formation of Mississippian age, Potter, Arkansas. Projection of a thin section.

Recrystallized detrital carbonate cements can be recognized when they are developed from sand-sized carbonate grains, such as shell fragments or crinoid-stem plates, or when carbonate mud composes an appreciable percentage of the original matrix. Under such conditions quartz-sand grains appear to be floating in the carbonate matrix and are not in contact with other sand grains (Fig. 3-43). As solution and recrystallization of the carbonate proceed, the quartz-sand grains are moved apart, moving into space originally occupied by the carbonate. In sediments which originally contained only a few scattered shell fragments, recrystallization results in small patches of calcite cement, each often consisting of only one crystal, surrounding several floating sand grains.

Introduced mineral cements consist of almost any mineral which is soluble in connate waters. The common cements are the various forms of silica, carbonates, and iron oxides; but barite, anhydrite, gypsum, pyrite, siderite, and halite are all moderately common. Replacement of one cementing material by another is quite common as the chemical environment in the sand changes during burial and lithification. The source of the cement is usually not clearly indicated; however, most is probably derived from adjacent sediments, particularly shales, during their compaction.

Silica cements take the form of opal, chalcedony, chert, or quartz. Most commonly opaline or chalcedonic cements cannot be recognized megascopically. They usually take the form of thin films, coating the quartz-sand grains and partially filling the voids. Both opal and chalcedony occur as botryoidal structures, with chalcedony showing fine fibers at nearly right angles to the grain surface. Chert appears as a white structureless matrix in which the quartz grains often appear to float. Replacement of the margins of the sand grains by chert often results in indistinct borders of the sand grains. Quartz most commonly occurs as secondary enlargements of the sand in crystallographic continuity with the original quartz (Figs. 6-25 and 6-27). Some silica for secondary enlargement may be derived from solution of quartz at points of contact between grains. In areas of solution of grains, the remnant grains are small and flattened, usually elongated parallel to the bedding. Highly irregular or sutured contacts between the grains can sometimes be seen. Secondary enlargement usually can be seen only where the voids are incompletely filled so that euhedral faces develop or where the original rounded grain was coated with a thin dust of clay or iron oxide before secondary enlargement. Coating of the detrital grain by dust often results in color contrasts between the original grain and the secondary coat of quartz. Secondary enlargement sometimes tends to concentrate any original interstitial clay into patches of white coarsely crystalline clay minerals in the centers of larger interstices.

Introduced carbonate cements can be identified by their having the sand grains in contact with each other and the cement occupying the interstices. Calcite assumes two common forms as a cement. The first is an interstitial microcrystalline cavity filling or coating of grains. When such rocks are broken, sand grains often fall out, leaving a fine-grained shell of carbonate behind. Second, calcite occurs commonly as macrocrystalline cement in the form of large single crystals including several sand grains. (Fig. 6-30). Cleavage reflections from the calcite can be seen over large areas, speckled by included sand grains. The large calcite crystals are not usually euhedral but are spherical or irregular equidimensional crystals, depending on mutual interference for their external form. The famous sand-calcite crystals of the Badlands of South Dakota are euhedral and of exceptionally large size (Fig. 3-44).

Iron oxides and iron sulfides are less common cements. Many iron-stained sandstones contain hematite or limonite in such small quantities that they cannot be considered a cement, even though they have dramatic

Fig. 6-30. Coarse crystalline carbonate cementation in a bituminous sandstone. In the light-colored areas fine sand is cemented by calcite crystals 5 to 8 mm in diameter. In the dark areas the cement is a bituminous material. The bright area in the left center is a reflection from a single calcite cleavage. Santa Cruz, California. (*Ward's Universal Rock Collection.*)

effects on the color of the rock. Fine-grained disseminated hematite or limonite colors rocks red, brown, or yellow and occurs as minute dust particles adhering to the surfaces of sand grains. Iron minerals do form the major cementing agent in some sandstones. The primary or secondary origin of these materials cannot usually be determined from megascopic examination. Iron oxides may be introduced as such or may develop as a result of surface oxidation with or without hydration from pyrite, marcasite, or siderite (Fig. 3-46).

Minor cements include barite as large euhedral crystals enclosing numerous sand grains (desert roses), anhydrite or halite as large crystals in some sandstones associated with evaporites, and asphaltic residues developed from oil seeps.

SILTSTONES, SHALES, AND ARGILLITES

Definitions

This group includes those rocks dominated by clastic material in the silt- and clay-size ranges. The maximum size is $\frac{1}{16}$ mm and extends downward to colloidal sizes. In practical terms these sizes are too small to be visible to the unaided eye; silt sizes can be detected by their gritty nature between the teeth. In contrast, clay sizes behave as a paste between the teeth, regardless of their mineral character and physical properties. The relationships of different rock types in the group are given below, after Twenhofel (1937):

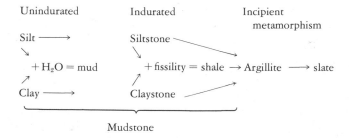

Mudstone is a collective term for the entire group of rocks, but it should be used only when there is doubt as to the precise character of the rock or when local variations between silt and clay are rapid or unsystematic. Mudstone is also used for an indurated clay rock without fissility.

The term *shale* should be applied only to those rocks which show bedding-plane fissility or laminations. *Argillite* is used variously for rocks in which recrystallization or incipient metamorphism has taken place without the development of secondary slaty cleavage and usually without fissility.

Descriptive Checklist

The following outline gives features of the fine-grained clastic rocks which should be checked in making a detailed examination. Inasmuch as individual grains are below the size limits for megascopic identification, the main features are in the gross characteristics of the rock.

Gross characteristics of the rock
    Color
    Bedding characteristics
        Evidence of bedding
            Laminations
                Compositional and textural contrasts
                Character of laminations, uniform or lensoid
            Nonlaminated claystones and siltstones
        Splitting characteristics
            Bedding-plane fissility
            Nonfissile rocks, blocky or conchoidal fracture
        Internal bedding structure
            Graded bedding
            Cross bedding
        Mineral segregations along bedding planes or laminations; pyrite, gypsum, opalized fish scales

Composition
    Silt versus clay fractions
    Siliceous rocks
    Calcareous shales
    Carbonaceous shales
        Oil shales
    Concretions in shales
    Organic evidence

Discussion of Special Features

Color. Color is more important in shales than in any other sedimentary rocks, owing in part to the fine grain size which prohibits detailed mineralogical work without special equipment. Color is due to a pigmentation of some type which usually composes only a minor fraction of the total bulk of the rock. The commonest pigments are ferric iron, as disseminated fine-grained hematite or limonite; ferrous iron, as siderite, pyrite, or silicates such as glauconite or chlorites; and carbon or carbonaceous material, as films of carbon or hydrocarbons. The more common colors and their source pigments are:

Red: Ferric iron as disseminated hematite; ranges from 3 to 6 percent total iron, two-thirds or more of it in the ferric state.

Purple: Dominantly ferric iron occurring as hematite; one-half to two-thirds of total iron is in the ferric state.

Green: Ferrous iron usually occurring as silicates and less commonly as siderite. Total iron is 3 to 6 percent, and more than two-thirds of it is in the ferrous state.

Gray to blue gray: Disseminated siderite occurring as very-fine-grained material; on exposure to the atmosphere the siderite oxidizes rapidly, and the color changes to buff or brown.

Black: This is usually interpreted as indicating a high carbon content; these are the carbonaceous shales. A black color may also be imparted by finely disseminated pyrite, often in an amorphous or colloidal form (melnikovite). Actually, many high-yield oil shales are brown rather than black, but in this case the carbon is combined as hydrocarbons rather than as amorphous carbon.

Light-gray or buff: These indicate a lack of or a removal of pigmenting agents. Calcareous shales are normally light-colored, as are many siliceous shales. However, some green shales, in which the color is the product of relatively minor ferrous compounds, become light yellow, tan, or brown on exposure to the atmosphere.

Laminations. Very thin bedding, or laminations, is characteristic of many clay and silt rocks. The rocks commonly part along the lamellae, but many times they do not. Thus fissility is not a necessary product of laminations. Most laminations are visible as a result of compositional or textural differences or both between adjacent thin units. The three most common differences causing laminations are: (1) alternations of coarse and fine materials, such as silt and clay; (2) alternations of light and dark layers of differing organic content; and (3) alternations of calcium carbonate content. The variations can be correlated with the differences of settling velocity of the introduced materials or with seasonal variations in the character of introduced materials. In the Pleistocene lakes the variations can be directly correlated with annual seasonal variations of introduced materials and can be properly termed *varves*. Unless the cyclical nature of sedimentation can be clearly related to annual cycles, the term varve should not be used. In some siltstones, laminations are seen to be due to variations in mineral composition of the silt particles. Heavy-mineral accumulations in certain lamellae, particularly where the minerals are iron-bearing and are subject to oxidation and rust-staining, produce noticeable color contrasts.

Persistence of laminations. Individual laminations may be of uniform thickness and persist over long distances, or they may be discontinuous and

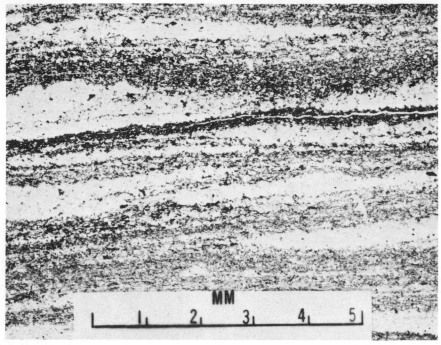

Fig. 6-31. Laminated silt and clay shale. Note the lenticular nature of the silty units. Hartshorne formation of Pennsylvanian age, Jenny Lind, Arkansas. Projection of a thin section.

lenticular. Persistent laminations are indicative of deposition in quiet-water environments, where wave action cannot cause bottom erosion. In contrast, lenticular laminations indicate some turbulence with minor scour and fill structures (Fig. 6-31). Such an environment is commonly developed on very gently sloping tidal flats. The presence of laminations indicates a paucity of burrowing bottom scavengers, since they tend to homogenize the sediment. Agitation which is too vigorous prevents fine laminations because, first, silt- and clay-sized particles are not likely to be deposited if they remain in suspension, and second, churning action homogenizes the sediment.

Nonlaminated claystone, mudstone, and siltstone. The absence of visible laminations in clay- or silt-sized materials may be attributed to several causes. First, the nonlaminated material may be a residual non-transported material. Such is the case for residual clays and laterites developed in place by chemical weathering of the underlying rock. These may be recognized by their downward gradation into bedrock, as in the Arkansas bauxite, where the texture of the parent syenite is often pre-served in the basal portions of the residual material. Another criterion for recognition is the presence of unabraded resistant materials from the parent rock, such as angular quartz grains and accessory mineral grains in residual clay derived from granite. A final criterion is an abnormal chemical composition as compared with normal transported shales. High iron or alumina hydroxide content is characteristic of the laterites. Similarly, a high concentration of trace elements, present in the underlying rock, is indicative of residual origin.

A second origin for nonlaminated sediments is the nature of the continuous-deposition environment of the uniform clastic material. Deposition either may be very slow or may be very rapid. Loess is an example of nonlaminated silt-clay-sized material in which deposition is a very slow and gradual accumulation of windblown dust in vegetation. Reworking by organic activity in the top layers of the dust may be a factor in the destruction of lamination. Some nonlaminated clays show evidence of rapid deposition in the nature of the preserved fossils. Straight nautiloids of large size have been found which have completely empty, but collapsed, living chambers. This indicates complete burial before the soft tissue could decay to prevent in-filling of the living chambers by clay. Such rapid deposition of uniform material would prohibit development of visible laminae.

A third cause involves the destruction of depositional laminations by contemporaneous reworking and homogenization. Reworking may be by mechanical agitation, such as waves or currents, or by burrowing scavengers. The sediments of many modern northern lakes show no visible laminations, in spite of a strong annual-depositional cycle which should produce varves. The activity of scavengers is a possible reason.

Finally, laminations may be destroyed by postdepositional chemical processes and diagenesis. Slowly accumulating fine volcanic ash, probably originally thin-bedded, may react chemically with contained and overlying water and be entirely reconstituted and recrystallized. Similarly, diagenetic reorganization of opaline tests and spicules or included organic material may completely or partially destroy primary laminations.

Splitting characteristics — fissility. Fissility is the ability of shales to split into thin sheets parallel to the bedding. This property is controlled either by parallelism of micaceous clay flakes or by alternations of texture or composition in the lamellae. Fissility frequently is not shown by completely fresh unweathered specimens but develops rapidly on surface exposure. Thus many shales encountered in well drilling yield solid cores of considerable length, but on standing in a core lab, they break down over a period of months to give individual thin disks of shale. Similarly, when some shales are freshly quarried, they yield large irregular or polygonal blocks without visible laminations. On standing exposed at the surface over one winter, these blocks will be reduced to a pile of chips a fraction of a millimeter to

2 mm thick and a few centimeters on a side. In both cases actual rupture into flakes is the product of the mechanical action of growing crystals: in the first case, salt from the desiccation of connate brine, and in the second case, ice crystals. However, in both examples the nature of the rupture and the shapes of the resultant fragments are controlled by some inherent property of the shale.

Fissility is dependent in part on the composition of the sediment. Highly argillaceous rocks tend to be fissile. As siliceous or calcareous impurities are added to the clay, fissility is lost and splitting becomes flaggy or slabby. Highly carbonaceous shales tend to be extremely fissile, splitting into very thin sheets, often of considerable area, which are thin enough to be flexible. Either organic carbon or hydrocarbons occurring as thin films or flakes parallel to the bedding enhance fissility.

Bedding-plane splitting—spurious fissility. Some claystones and many siltstones split parallel to the bedding, yielding layers $\frac{1}{2}$ to 12 in. thick. Many of these show internal laminations along which the rock will not split. Such rocks cannot be called fissile. These bedding-surface partings are controlled by some compositional difference which results in weakened bonding across the layers. In siltstones the zones of parting often follow thin clay seams or thin zones having an abnormal abundance of minute mica flakes. Discoloration of the parting surface is common and indicates that these were channels for groundwater movement. Thus they may represent zones in which rock material has been removed, such as thin carbonate beds or lamellae, pyritic zones, or seams of gypsum. In many mudstones which are not truly fissile, bedding-plane parting can be developed along zones of pyritized fossils, carbon films, or secondary gypsum seams.

Blocky and conchoidal-fracturing rocks. Many claystones and silts which have not been subjected to burial, with its associated compaction and diagenesis, rupture in directions independent of bedding. This applies to some finely laminated claystones of Cenozoic to Recent lake deposits as well as to massive nonlaminated residual clays. Apparently, extremely fine grain size has prevented orientation on settling, and lack of deep burial has prevented reorientation by compaction or diagenesis. Underclays of coal seams also commonly show no laminations or fissility but break with a conchoidal fracture. Highly siliceous shales approach chert in texture and characteristically break into polygonal blocks. The introduction or recrystallization of organic opaline silica or volcanic ash acts as a cement to tie clay flakes together, even where they show a strong parallelism.

Graded bedding. Silt and clay mixtures often show graded bedding where a mixed sediment is introduced into a sedimentary environment in batches, such as through floods or by storm waves stirring the bottom. The sediments accumulate on the floor according to settling velocity so that an individual bed will grade from silt on the bottom to clay on top. Where this is a batch process, an abrupt transition appears between the clay at the top of one bed and the silt at the base of the next. Frequently, evidence of scour of the top of a cycle is noticeable at the base of the younger cycle.

Varves are a type of graded bedding related to an annual winter-summer cycle. They may or may not show gradations from one pair of beds to the next. Glacial-lake varves consist of light-colored silty layers grading upward into darker-colored clay layers. The silts were presumably deposited in the summer and the clays in the winter. Pleistocene varves range from a few millimeters to a few centimeters in thickness. Varved lake deposits of Eocene age show an annual change from silt to clay to carbonate to organic-rich clay, representing climatic changes through one year. In the spring, rains bring in much clastic material, the coarse fraction of which settles out quickly. As the water gets warmer, plant life begins to develop, and carbonates begin to develop by the disassociation of bicarbonates. Settling of both dead plants and carbonates takes place, but in zoned lakes the plant material remains suspended at the contact of the cold- and warm-water layers, while the carbonate settles through. Thus on top of the earlier silty layers, successively finer clastics are deposited, grading up into a thin zone of carbonates. Finally late in the summer any organic material not completely decomposed settles out and forms the final layer of the year.

Cross bedding. Small-scale cross bedding is common in some siltstone and in the silt layers of lenticularly laminated rocks. They seem to be related to cut-and-fill structures developed on a small scale on tidal flats, alluvial flood-plain fills, etc.

Composition. Composition is indicated to a limited extent by color and splitting characteristics, as discussed previously. The proportions of silt- to clay-sized materials can be determined crudely by pulverizing the material between one's teeth. Silt-sized materials grit between the teeth and can be detected even when not visible with a hand lens. In contrast, clay-sized material does not grit between the teeth, regardless of mineral composition. Apparently, this is because the particle size is smaller than the pore size in the teeth. Clay-sized quartz and feldspar have been so tested by the author.

Siliceous shales are commonly hard and brittle, approaching chert in their physical properties. They therefore break with a conchoidal fracture, yielding blocky fragments even where fine laminations are visible. The silica is usually derived from the recrystallization of opaline sponge spicules or radiolarian tests or from admixed volcanic ash. Colors usually range from black, dark gray, to blue gray or are light tans, buff, or white. Color is usually dependent on the amount of organic material present.

Calcareous shales are transitional rocks between clay-shale and limestone. When carbonate and clay are in subequal amounts, the sediment is called a *marl* and the resulting rock a *marlstone*. Most shales contain relatively little calcite; however, 20 percent calcite materially modifies the character of the shale, destroying fissility and causing ready effervescence with acid. Calcite is usually present as very fine grains, either as chemically precipitated material or as microfossils. Colors generally are light tans and buff.

Carbonaceous shales are normally black to very dark gray, highly fissile, and highly pyritiferous. Concretionary structures are common, and fossils are often replaced by pyrite. The organic material is not normally visible megascopically, but carbon films may be seen as carbonized organic material. Thin laminations may be visible, but many highly fissile carbonaceous shales show no laminations.

Oil shales are brown to black and usually laminated (Fig. 6-32). They are often not fissile. Their characteristic feature is the presence of distillable hydrocarbons. This can usually be detected by burning with a match or cigarette lighter. The hydrocarbons will often burn with a small sputtering

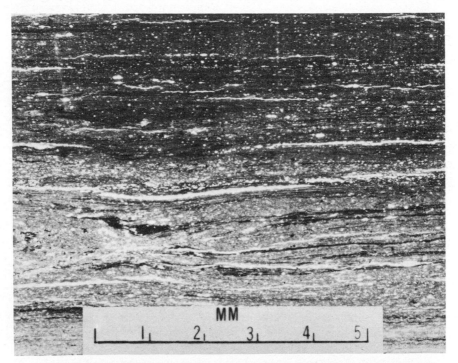

Fig. 6-32. Oil shale. Dark streaks of kerogen are present along bedding planes, and they discolor the darker portions of the rock. The light spots are silt grains. Green River formation, Garfield County, Colorado. Projection of a thin section. (*Ward's American Rock Collection.*)

yellow flame, even after the match is removed. A strong petroliferous odor is given off from the heated rock; this odor is often obtained on a cold freshly broken surface. Asphaltic residues are common in concretions, especially septaria, in oil shales.

Concretions. Concretionary structures are commonly found in argillaceous rocks. They may be primary, having laminations bending around them, or secondary, having laminations traceable through them. Concretionary bodies in shales take a variety of shapes, ranging from simple spherical bodies or disks to highly irregular bizarre shapes. The minerals segregated are commonly calcite, silica, iron sulfides, and siderite. Calcite-enriched concretions often form in shales, surrounding fragments of organic material, such as leaves or stem fragments. Here the concretionary body may merely represent a partial cementation due to modification of the chemical environment in the mud around the buried plant material. Pyrite and marcasite concretions and rosettes and clay-ironstone concretions and septaria are particularly common in black shales and shale associated with coals. Bizarre carbonate concretions are common in some varved glacial clays.

LIMESTONES AND DOLOMITES

Definitions

Limestone and dolomite are those rocks in which the carbonate fraction exceeds the noncarbonate fraction in total amount. The term *limestone* is used for those rocks in which the carbonate is calcite and the term *dolomite* for those in which the carbonate fraction is the mineral dolomite. The term limestone includes rocks of a wide variety of origins and textures, and its use tends to obscure any genetic interpretation. As a result a number of terms have been introduced to give genetic significance to the names of rocks.

A genetic classification of limestones and dolomites divides them into three major groups: autochthonous, allochthonous, and metasomatic limestones.

Autochthonous limestones are those which have formed in place by either organic or chemical processes. They include tufa and reef-core rocks.

Allochthonous limestones are those which have been derived from autochthonous material by mechanical fragmentation and which have undergone transportation and redeposition as solid particles. Allochthonous limestones can be subdivided on the basis of fragment size into calcirudites, calcarenites, and some calcilutite. Oolitic limestones belong in this group.

Metasomatic limestones include those which have undergone thorough recrystallization, particularly where introduction of magnesium is involved. The dolomites and dolomitic limestones are the main rocks in this group.

A classification based on the nature and origin of the constituent particles of limestone has been proposed by Folk (1959). This classification lends itself admirably to hand-lens work and has both descriptive and genetic significance. This system will be observed in the following discussion. The constituent particles are listed and described below.

A. Terrigenous constituents
    1. Sand
    2. Clay
B. Allochemical constituents
    1. Intraclasts
    2. Ooliths
    3. Fossils
    4. Pellets (not megascopically recognizable and therefore omitted here)
    5. Pseudoallochemical constituents
C. Orthochemical constituents
    1. Microcrystalline calcite ooze (micrite)
    2. Sparry calcite cement (spar)
    3. Replacement orthochems

Terrigenous constituents (Fig. 6-33). Those particles which are derived from outside the basin of deposition. The materials are derived from the land and brought into the depositional area as discrete solid particles. Quartz sand and silt, feldspar, and clay minerals are all common. If the terrigenous component exceeds 50 percent, then the rock is not a limestone but a calcareous sandstone or calcareous shale.

Allochemical constituents (allochems). Those materials formed by chemical or biochemical precipitation within the depositional basin. They make up the framework of most limestones. The particles have for the most part undergone transportation and redeposition.

1. *Intraclasts* (Fig. 6-34). Fragments of partly lithified limestone eroded from one part of a depositional site and redeposited in an adjacent area. They do not include pebbles of an older limestone eroded from the landmass or shoreline. The material of intraclasts is derived from the surface of an accumulating lime bed by wave action, mud cracking, or submarine slumping. The intraclasts range from soft lime lumps which become flattened out with vague boundaries on redeposition to fragments of sufficient resistance to show abrasion effects at their margins. Fossils or ooliths at the edge of the intraclast may be broken or abraded off, indicating an equal resistance in the groundmass of the fragment.

2. *Ooliths* (Figs. 3-42, 6-35, and 6-36). Spherical or ovoid bodies which are usually in the size range of sand and which show concentric or radial structure or both. Ooliths normally have a nucleus consisting of a quartz-sand or silt grain, a fossil fragment, or an intraclast. The calcite is gener-

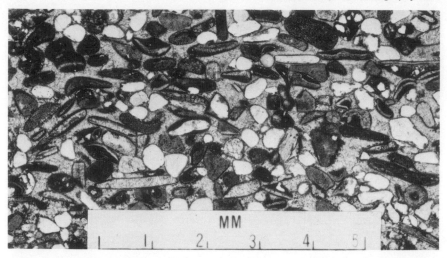

Fig. 6-33. Sandy biomicrite. Many of the abundant quartz-sand grains (white) have been partially replaced by calcite along their margins. The organic debris is broken, abraded, and rather well sorted. Note the parallelism of the platy fragments, probably indicating bedding. Hale formation of Pennsylvanian age, Gaither Mountain, Arkansas. Projection of a thin section.

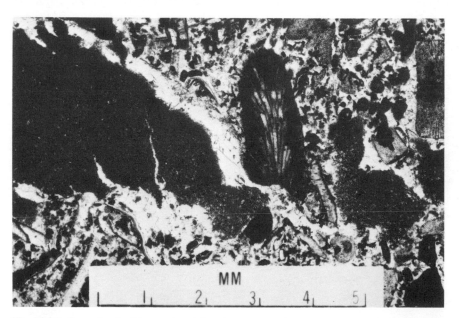

Fig. 6-34. Intraclastic biosparite. The large irregular micrite intraclast was soft when redeposited, as indicated by its fuzzy boundaries and fracturing. Crinoidal and bryozoan debris make up most of the allochems of the rock. Pitkin formation of Mississippian age, Wesley, Arkansas. Projection of a thin section.

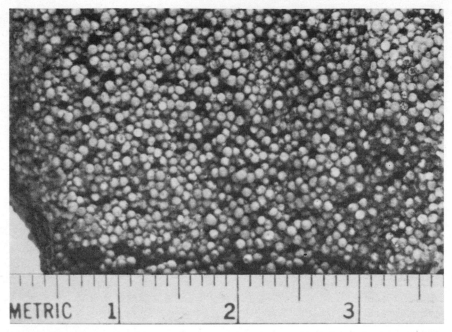

Fig. 6-35. Ooliths weathered out in relief. A spar cement has been removed by weathering. Boone formation of Mississippian age, Benton County, Arkansas.

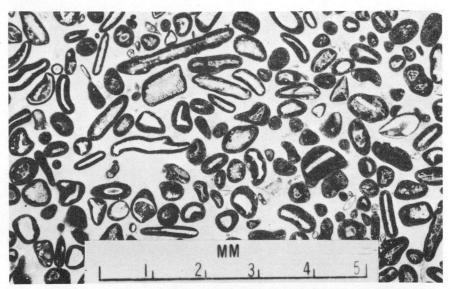

Fig. 6-36. Oosparite. Dark oolith coatings of fragmented organic debris set in a matrix of clear coarsely crystalline calcite. Brentwood limestone of Pennsylvanian age, Fayetteville, Arkansas. Projection of a thin section.

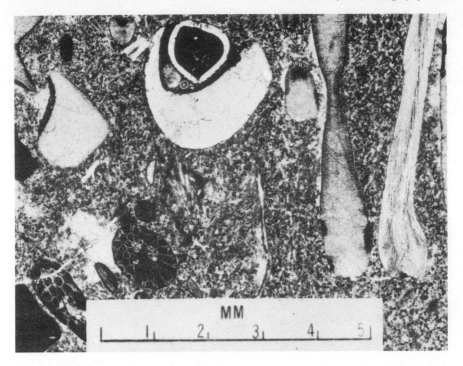

Fig. 6-37. Abundant pellets in a biomicrite. The large fragments are abraded bryozoan, crinoidal, and brachiopodous debris. The two black fragments at left are micrite intraclasts. Micrite with abundant dark pellets form the matrix. Pitkin formation of Mississippian age, Wesley, Arkansas. Projection of a thin section.

ally built up of one or more concentric shells around the nucleus. If only one shell is present, they are called *superficial* ooliths. Modern ooliths are apparently built of aragonite needles tangent to the surface of the nucleus. The radial structure, when visible, is probably the result of recrystallization of aragonite to calcite.

3. *Fossils*. All fossils are included as allochems, except coral and algal structures growing in place and forming a relatively immobile resistant mass.

4. *Pellets* (Fig. 6-37). Microstructures of spherical or elliptical shape composed of microcrystalline calcite and assumed to represent invertebrate fecal matter. They rarely exceed 0.1 mm in diameter and therefore cannot be considered in a megascopic classification.

5. *Pseudoallochems*. Rare textures and structures developed by recrystallization processes in place which mimic true allochems. Normally they could not be distinguished from the true features by megascopic means.

Orthochemical constituents (orthochems). Constituents of the limestone developed in place by precipitation or recrystallization. They show no evidence of transportation.

1. *Microcrystalline calcite ooze (micrite)* (Fig. 6-38). This consists of ultrafine-grained calcite precipitated chemically or biochemically and accumulated on the sea floor, either interstitially between allochems or as a separate deposit. It varies widely in color from white to black through grays, blue, and brown. After treatment with hydrochloric acid, it generally appears as a white powdery substance.
2. *Sparry calcite cement (spar)* (Fig. 6-36). Coarsely crystalline calcite which acts as a cement and fills interstices between allochems. After treatment with hydrochloric acid, spar is clear and glassy.
3. *Replacement and recrystallization orthochems.* These include the minerals developed after deposition and burial. Allochems and orthochems are subject to recrystallization, masking or destroying their original charac-

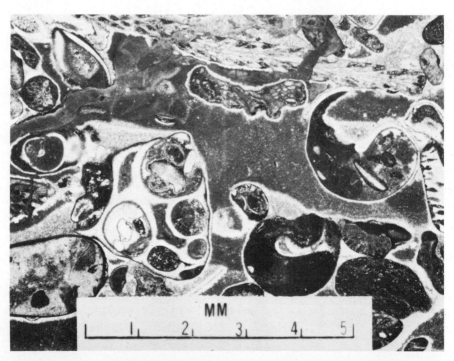

Fig. 6-38. Biomicrite. The dark area in the center of the field is microcrystalline calcite (micrite) which has been partially recrystallized at the top to a coarser light-colored matrix. Fossils include bryozoa, gastropod, crinoid, and ostracod remains. Brentwood limestone of Pennsylvanian age, Morrow, Arkansas. Projection of a thin section.

ter. Similarly, replacement, as in dolomitization, affects all the constituents and masks their identity. Both processes occur in place and are therefore orthochemical processes.

A suggested method of study of limestones to determine the constituents is to treat the rock surface with a few drops of dilute hydrochloric acid for a few seconds. After the treated surface has been washed and dried, the etch produced enhances contrasts in textures and grain sizes so that the constituents can be readily recognized. The acid etch has two effects: first, it removes pulverized material from the surface, and second, the different constituents dissolve at different rates, depending on grain size and orientation. An additional effective technique is the use of acetate peels. A sawed and ground rock surface is etched with dilute hydrochloric acid. A piece of acetate film slightly larger than the etched surface is placed on a hard flat surface and is wet with acetone. The rock is held firmly against the film until the acetone is evaporated. The film may then be trimmed, placed between lantern-slide glasses and projected on a screen with a slide projector. Textures in the limestone show up clearly, and large magnifications can be obtained so that small features, such as foraminifera, etc., can be recognized.

A classification of limestones and dolomites modified from Folk is given in Table 6-2 and is based on the proportions of the constituents shown at the top of the table. Pellets and pelletal rocks of Folk's original classification are omitted, because they cannot be recognized megascopically.

Descriptive Checklist

The following outline gives a suggested sequence of features to look for in making a detailed study of limestone specimens. Important features are discussed further in the text. All portions of the outline will not be applicable to any one specimen.

Gross characteristics
    Color
    Evidence of bedding
        Major partings
        Internal bedding features: alinement of fossils, fossil concentrations, change in constituent-particle character, size grading, change in proportion of terrigenous component
    Cross bedding
    Features of bedding surfaces
        Ripple mark
        Stylolites
    Sorting
        Uniformity of size of allochems

Table 6-2.   Classification of Limestones and Dolomites

Limestones, partially dolomitized limestones, and primary dolomites

Volumetric allochemical composition		> 10% allochems[1]		< 10% allochems[2]		Variable allochems	Replacement dolomites[3]	
		Spar > micrite (Sparry allochemical rocks)	Micrite > spar (Microcrystalline allochemical rocks)	1–10% allochems	< 1% allochems	Undisturbed bioherm rocks	Allochem ghosts	No allochem ghosts
> 25% intraclasts		Intrasparrudite / Intrasparite	Intramicrudite[5] / Intramicrite	Intraclast-bearing (micrite[5]) calcilutite	Calcilutite (micrite), if calcite; Dololutite (dolomicrite), if primary dolomite	Biolithite	Intraclastic dolomite	Crystalline dolomite
> 25% ooliths (< 25% intraclasts)		Oosparrudite / Oosparite	Oomicrudite[5] / Oomicrite	Oolith-bearing (micrite[5]) calcilutite			Oolitic dolomite	
Fossils[4] (< 25% ooliths)		Biosparrudite / Biosparite	Biomicrudite / Biomicrite	Fossiliferous (micrite) calcilutite			Biogenic dolomite	
		Intraclasts	Intraclasts	Intraclasts			Intraclasts	
		Ooliths	Ooliths	Ooliths			Ooliths	
		Fossils	Fossils	Fossils			Fossils	
		Most abundant allochem					Evident allochems	

[1] Upper name in each box refers to calcirudites and lower name to calcarenites.

[2] Inasmuch as the microcrystalline constituents are difficult to differentiate megascopically, the general field term calcilutite is preferable to micrite.

[3] Crystal size of dolomitized rocks should be designated as follows, based on grain size:

  $<\frac{1}{16}$ mm Microcrystalline $\frac{1}{4}$–1 mm Coarsely crystalline   > 4 mm Extremely coarsely crystalline

  $\frac{1}{16}$–$\frac{1}{4}$ mm Medium crystalline   1–4 mm Very coarsely crystalline

[4] If one fossil type is dominant, this may be indicated by a prefix modifier, such as crinoid-biosparrudite.

[5] Rare rocks.

Evidence of in situ burial of fossils
Porosity and permeability
Characteristics of terrigenous constituents
Composition and percentage
Size and shape
Distribution
Characteristics of allochemical constituents
Percent of total rock; packing
Composition of allochems, proportions of intraclasts, fossils, ooliths,
and characteristics of each
Intraclasts
Lithic type
Size and shape
Definiteness of borders
Fossils
Types of fossils and relative proportions
Evidence of extent of transportation, abrasion, and sorting
Orientation of fossil fragments
Character of filling of voids in articulated bivalves, gastropods,
cephalopods, crinoid calyxes, etc.
Ooliths
Size and size range
Character and size of nucleus
Characteristics of orthochemical constituents
Proportion of total rock
Ratio of spar to micrite
Recrystallization and metasomatic features
Recrystallization of allochems
Aragonite inversion to calcite
Secondary enlargement of echinodermata fragments
Recrystallization of microcrystalline calcite ooze
Dolomitization
Replacement by silica
Diagenetic features
Chert nodules
Pyrite segregations
Geodes and other crystal-lined cavities

Discussion of Special Features

Color. Commonly light; tan, buff, or gray, less commonly dark gray to black, bluish gray, or red. The color is a rough index of the amount and type of admixed impurity. Very-light-colored rocks are normally free of any appreciable amount of fine-grained terrigenous or organic components. Red and brown colors indicate disseminated iron oxide or hydroxides. These may be primary constituents or may have developed by oxidation of other primary iron minerals, such as siderite, glauconite, or pyrite. Dark-gray to black rocks generally indicate an abundance of organic hydrocarbons and pyrite and may be accompanied by appreciable clay. Highly argillaceous limestones tend to be intermediate in color.

Bedding characteristics. Limestones and dolomites range from thinly laminated to massively bedded rocks. However, they tend to split on only certain bedding surfaces and not on others. The feature which controls the splitting is often not identifiable, in that the limestone on either side of the parting may have identical characteristics. Frequently, the parting surfaces show evidence of having been avenues for groundwater migration and solution. In some limestones the parting surfaces are clearly clay seams.

Within the massive units, bedding may show up in a variety of ways. Thin laminations of slightly different color or texture are seen in some fine-grained rocks. Parallelism of large inequidimensional allochems often indicates bedding. Fragmented fossils, such as shells, fenestellate bryozoa, or single crinoid plates, flattened ooliths, and platy intraclasts line up parallel to the bedding. Concave shells tend to line up with their concave surfaces downward, the most stable position, and indicate the stratigraphic top in deformed strata. Changes in the character and proportions of allochems and terrigenous constituents are common in massively bedded units and may help define original bedding surfaces. Thus a single bed may become less sandy or less oolitic upward from the base of the unit. Size-grading through single units is also common, with the allochem becoming progressively finer-grained upward.

Cross bedding. This is often well developed in those limestones composed dominantly of sand- to granule-size allochems. Current transportation of fossil debris and ooliths is common along shallow lime banks and reefs in the form of giant "sand waves." Dune development on land of carbonate clastics is also common and would produce cross bedding. Commonly the cross bedding is not noticeable in unweathered outcrops of massively bedded material, but it becomes quite apparent by groundwater etching, particularly along joints.

Current and oscillation ripple mark. These develop in clastic calcareous sediments in the same manner as in sand. They are not as commonly preserved, however, due to recrystallization of the carbonate material in diagenesis. Large scale para-ripples are occasionally seen on limestone bedding planes, particularly in association with reef structures. These ripples have a wavelength of 15 in. to several feet and an amplitude of several inches. Strong tidal currents cause the development of these features.

Occasionally, there is a clear correlation between these large ripples and surge channels through reef mounds. Where a wide size range of clastic material is available, the larger sizes will be concentrated in the troughs and on the side of the ripple mounds facing the current.

Stylolites. A common bedding-plane feature in some limestones. They consist of irregular columns of the upper and lower beds penetrating each other. The stylolite surface is often marked by an accumulation of terrigenous constituents or other insoluble constituents of the limestone, such as manganese oxides or phosphatic material. Shell fragments in the columns are truncated abruptly by the columns. Such stylolites are the product of groundwater solution under a load of overlying rock.

Sorting of allochems. The average size and uniformity of size of transported constituents are as good an indicator of effectiveness of current action here as they are in noncarbonate clastic sediments. Uniform grain size and grain-to-grain contact between the larger fragments indicate an effective current action with winnowing of the fines, removing them for deposition elsewhere. Usually, such well-sorted material is cemented by spar. The presence of interstitial micrite indicates a variable current action, with an accumulation of clean coarse material under high current action and subsequent sifting in of fine material during intervals of weak current action. The presence of large transported allochems not touching one another in a matrix of micrite is indicative of intermittent strong currents sweeping a mixed batch of material into the depositional site and subsequent settling in a quiet environment. Graded bedding usually results.

Articulated bivalves and unabraded fossils of other types generally indicate little current action and occur in poorly sorted sediments. An exception is the fixed massive organisms which are reef builders. These are dominantly corals, encrusting bryozoa, and algae, which generally live in zones of strong wave action but form highly porous masses into which detrital materials of any size may settle. Recrystallization of reef materials, particularly algae, gives a fine-grained rock, masking the effects of current in sorting materials.

Terrigenous Constituents. Usually sand- or silt-sized quartz and clay minerals. These can best be studied as the insoluble residues from acid-treated limestone. In hand-lens studies these can best be studied by dissolving a chip of the specimen in dilute hydrochloric acid. On flat etched surfaces of the rock, the sand- and silt-sized grains stand out in relief and can be recognized by their shape, color, and transparency. Quartz-sand grains in limestones usually have a frosted surface due to attack by carbonates during recrystallization. Euhedral secondarily enlarged quartz grains are common as doubly terminated crystals. The original detrital-grain boundaries are often noticeable because of minute included carbonate grains.

Intraclasts. These are fragments of partially consolidated limestone materials which have been broken up by wave or current action or by slumping of semiconsolidated materials and have been redeposited in the same depositional area. Pebbles of older limestone eroded from bedrock are not included as intraclasts. These are often difficult to distinguish from intraclasts unless their lithic character is markedly different or their paleontological content is of a distinctly different age than that of the host limestone. Intraclasts can usually be recognized by their having a somewhat different texture or proportion of allochems than the host material. Abraded or broken fossil fragments or ooliths at the edge of the intraclast are common diagnostic features. The borders of intraclasts are sharp and distinct when the sediment was well consolidated before erosion. In contrast, borders may be fuzzy and the masses flattened when the intraclast was a semiconsolidated or pasty mass at the time of erosion. Platy fragments may be formed by desiccation cracking, and these may be swept together by currents to form an intraformational edgewise conglomerate.

Allochemical fossils. All fossils except massive reef-building organisms are included. Two general groups of fossil materials should be distinguished: broken, abraded, disarticulated materials and articulated fossils buried in place. The first indicates strong wave or current action. In these the uniformity of size, evidence of abrasion, and orientation with respect to bedding should be noted. The type of organism is, of course, often indicative of environment, and reference texts on paleoecology should be consulted where detailed interpretation is desired. Some basic features of commonly observed fossils are discussed below on the basis of their appearance on acid-etched surfaces.

1. *Echinodermata.* All calyx plates, spines, and stem plates consist of single calcite crystals. Crinoid stem plates are oriented with their *c* crystallographic axis parallel to the axis of the column. Stem plates are joined by an evenly sutured joint resembling the milled edge of a new coin. When a clastic fragment consists of parts of two or more plates, these sutures can be seen as an even zigzag pattern across the cleavage surfaces. Echinoid spines have a uniform radial and concentric porous structure which is brought out by etching. A similar porous structure is seen in crinoid stem plates, but it is not as regular and well defined as in the spines.
2. *Brachiopods.* These show a characteristic finely fibrous-appearing structure, with the fibers either parallel to the surface of the shell or cutting diagonally across the shell at low angles. In edgewise sections, brachiopod shells often show a zigzag appearance due to the radial plications. These shells most commonly appear to be calcite rather than aragonite so that recrystallization does not destroy their structure as readily as it does the structure of pelecypods.

3. *Pelecypods and Gastropods.* These are both important contributors to limestones. Their shells are most commonly either wholly or partially composed of aragonite, which recrystallizes readily and destroys the original texture. Both classes have shells of two distinct calcareous layers plus an outer organic layer which is seldom preserved. The outer calcareous layer consists of prisms set roughly perpendicular to the shell. The inner layer consists of fibers, or lamellae, which are very thin and are roughly parallel to the inner surface of the shell.

4. *Bryozoa.* These are very important limestone builders which occur in three distinct forms. The massive forms are quite distinct, and their structures show up well on the etched surface. The structure is characterized by fine tubes which start out parallel to the axis of the colony and then curve sharply outward to intersect the surface at nearly right angles. The lacy bryozoa (fenestellate) are commonly fragmented, making sand- to granule-sized chips lying parallel to the bedding and are easily recognized by their lacelike structure. Where only the edge of such a colony is visible on the broken surface, it appears as a gently curved line of dots on the etched surface. Tabulate bryozoa occur as large irregular growths which encrust surfaces, particularly in reef mounds. The structure consists of small tubes perpendicular to the growth surface and divided by transverse plates.

5. *Corals.* The larger tabulate corals, cup corals, and colonial masses are readily recognized when they are essentially whole. Fragmented corals are difficult to recognize, since the plates show no megascopically distinct structure. Flat to hemispherical masses of stromatoporoid corals are found in some limestones. They are often difficult to distinguish from similar bryozoan and algal structures with which they are often associated in reefs of Paleozoic age. The stromatoporoids are characterized by thin parallel platy or concentric layers supported by a network of plates which form innumerable polygonal cells.

6. *Foraminifers.* These are often preserved in limestone in recognizable form. The pinhead-sized forms can often be recognized on etched surfaces and show characteristic cellular structures consisting of more or less globular-shaped cells built up in a coiled superstructure. The variety of forms is immense, and photographs in a micropaleontology text should be consulted. The large foraminifers (numulites and fusilinids) are characteristic of certain limestones. Fusilinids show a characteristic "wheat-grain" shape and have a well-developed coiled internal cellular structure. Numulites are flattened discoidal shapes, also with a well-developed coiled cellular internal structure.

7. *Algae.* These are common limestone builders (Fig. 6-39). However, the calcium carbonate deposited is extremely fine-grained and therefore commonly undergoes extensive recrystallization. The algae most com-

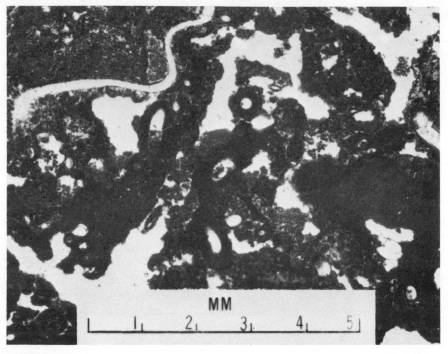

Fig. 6-39. Algal biolithite. A dark dense algal limestone with clear spar filling the original porosity of the deposit. Three brachiopod fragments are included. Algal reef core in Pitkin formation of Mississippian age, Wesley, Arkansas. Projection of a thin section.

monly recognizable megascopically are the encrusting forms. These occur as platy sheets to rounded globular forms, depending in part on water depth. Being plants, algae require sunlight for growth. Globular forms occur in shallow water, where the sun can penetrate during a long period of the day. In deeper water, where sunlight can penetrate only during the noon hours, the resulting algal colonies are flat sheetlike bodies. Sunlight cannot penetrate to depths of more than about 75 ft so that encrusting algae will not grow at greater depths. These structures are characterized by very thin lamellar plates separated by vertical plates which divide the layers into minute cells. One common form develops pellets of oolith to pisolith size. These can be distinguished from true ooliths and pisoliths by the fact that each layer coats only one side. Algal growth can take place on the other side only after the pellet has been rolled over by waves or current action. The pellet is built up of hemispherical layers which overlap on the edges. Such structures are called *oncolites* (Fig. 6-40).

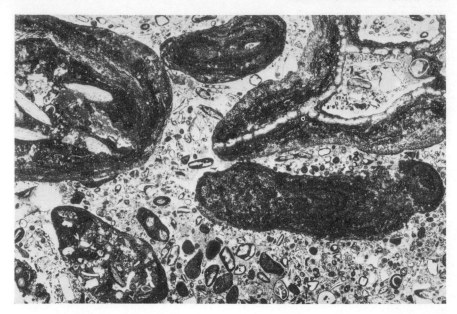

Fig. 6-40. Algal oncolites. Successive coatings of algal limestone built on a core of a fenestellate bryozoan fragment. The resultant oval bodies attain a diameter of 1 in. Pitkin formation of Mississippian age, West Fork, Arkansas. Projection of a thin section.

8. *Minor allochemical fossils.* These are constituents which are normally too scarce to make up any important volume or which are usually unidentifiable because of size or recrystallization; they include cephalopods, trilobites, ostracods, and sponges.

Filled or partially filled cavities in fossils are common and are often indicative of burial in place, the direction of horizontality, or the rate of sedimentation. Articulated bivalves, such as brachiopods or clams, are often buried in their living positions without disturbing the shells. The finer clastic sediments or lime muds sift into the body cavity, partially filling it. The upper portion, originally filled with trapped gases, commonly undergoes subsequent complete or partial filling with coarse crystalline calcite or other introduced mineral material. Tabulate coral cells are often filled with crystalline quartz in this way. The line of demarcation between detrital fill and spar indicates the original horizontal direction and may aid in determining bedding or initial dip of the sediment. Some articulated partially filled shells may be rolled over after the filling is partially lithified so that several fossils should be observed to determine uniformity of horizontality.

Snails are also incompletely filled, and gas bubbles are trapped in the upper portion of the coils. The cavities may subsequently be filled with spar.

The horizontal fill levels will usually be at different levels in the different whorls, but all will be parallel.

Open-ended shells, such as straight nautiloids, are occasionally indicative of the rate of burial. If the soft body tissue still occupies the living chamber at the time of burial, the living chamber may be exploded outward by developing gas pressure due to organic decay. In contrast, if gas could escape but the living chamber was buried before all organic tissue decayed, then the living chamber will often collapse. If decay was complete before burial, the living chamber will be filled with sediment, whereas the sealed chambers may be collapsed. All three relationships have been seen in shells deposited in rapidly accumulating muds.

Ooliths. These are spherical or ellipsoidal bodies of calcium carbonate showing radial or concentric structure or both. They range in size from $\frac{1}{4}$ to 2 mm in diameter but are most commonly between $\frac{1}{2}$ and 1 mm. Similar bodies of larger size are called *pisoliths*. They may occur as sparsely scattered ovoids in a calcarenite composed dominantly of fossil fragments, or they may be more abundant, in some rocks making up the entire clastic framework with a minimum of interstitial cement. The concentric layers vary in number from a single layer to many distinct layers. A nucleus is generally present, although it may be too small to recognize megascopically. The nuclei are generally sand- or silt-sized quartz grains, fossil fragments, or rounded pellets of microcrystalline calcite. Ooliths often break out of the rock, leaving a hemispherical hole consisting of one of the outer layers. The material lining the remaining hole is very fine-grained, white, and powdery. In shape, ooliths approach a sphere; however, if the nucleus is tabular or prismatic in shape, the oolith may be ellipsoidal. As more material is added to the oolith, growth is most rapid on the flattened sides so that a spherical shape is approached.

The origin of ooliths seems to be a process of combined aggregation and abrasion in a wave-agitated environment. Modern ooliths develop by the addition of minute aragonite needles, which have a random tangential orientation, to the surface of the oolith. Subsequent recrystallization of the aragonite yields radial calcite fibers which may or may not obliterate the concentric structure. The intensity of wave agitation controls the ultimate size of the bodies. Ooliths grow in size only while they are being actively moved about. When they become too large for wave action to move them freely, they become a portion of the bed upon which smaller ooliths are still being enlarged. In this manner the maximum size is controlled by the intensity of agitation. Superficial ooliths consist of a large nucleus with only one overgrown layer, which makes them the same size as associated ooliths. Apparently, they develop from larger nuclei. Pseudooliths are spherical bodies without concentric overgrowths. They are the same size or slightly larger than the ooliths with which they are associated. Generally, they are composed of microcrystalline lime mud and represent rolled and abraded fragments of a critical size. They were small enough to be moved but too large for material to be added for oolith construction.

Spar. A clear crystalline calcite which occurs interstitially between the allochems. Grain size ranges from barely megascopically recognizable individual crystals to those large enough to include several allochems in a single calcite crystal. Generally, spar represents introduced or recrystallized interstitial material similar in character to the mineral cement in sandstones. Generally, too, its presence indicates an original cleanly washed calcarenite, and it therefore occupies less than 30 percent of the rock. In rocks where it composes a greater volume, it indicates either a recrystallized micrite or a solution and recrystallization of allochems. Recrystallized micrite normally yields fine-grained spar, and the allochems show a sharp contact with the spar. The allochems form a continuous network of fragments in contact or, more commonly, not in contact with each other. Where spar has been derived from the allochems, the allochem boundaries are fuzzy and indistinct or show corrosion and embayment of a type which could not be the result of abrasion. Single-crystal allochems, such as echinodermata fragments, often show a secondary enlargement by calcite in crystallographic continuity with the plate. In such enlarged crystals, the original portion is usually cloudy in appearance, whereas the overgrowth is clear and colorless.

Micrite. A microcrystalline calcite in which the individual grains cannot be recognized megascopically. On etched surfaces it has the appearance of a white powder. The proportion of micrite in limestone varies from a minor interstitial filling between allochems to the entire rock. Lithographic limestones are composed entirely of micrite. In rocks in which it is interstitial between allochems, micrite represents clay-sized lime mud sifted down between the allochems in a manner analogous to clay in sandstones, and it is indicative of variable current action. Probably much of this micrite represents comminuted organic debris. Extensive lime-mud banks are currently being developed in many warm seas, probably dominantly by chemical precipitation. Such muds would give rise to fine-grained limestone free of detrital allochems. Fossils of organisms indigenous to the mud bank would be buried in place without abrasion, breakage, or disarticulation. Some chalks may represent micrite accumulations of chemical origin, but most give evidence of resulting from the accumulation of pulverized calcitic foraminiferal tests and algal material.

Dolomite and dolomitization. Dolomite is a carbonate rock in which at least one-half of the carbonate is in the form of the mineral dolomite. Most commonly the mineral dolomite composes more than 90 percent of the rock.

The bulk of dolomites is the result of replacement of limestones through the introduction of magnesia. Some show evidence of reorganization of an original magnesian limestone deposit.

Recrystallization and reorganization result in a uniform-grained crystalline rock in which the original textures and structures either are completely obliterated or are largely obscured. The resulting rock is fine-grained and crystalline, usually without recognizable clastic or organic structures. Where fossils are preserved, the fine structure of the hard parts is lost, and the borders of fossils are usually indistinct so that the fossil will not break out of the rock cleanly. Cavities in fossils which have been dolomitized often are lined with a druse of minute dolomite rhombs.

Dolomite usually accommodates some iron in the structure, whereas calcite does not. This results in the development of buff to brownish colored weathered surfaces on dolomite outcrops. The presence of iron makes several staining techniques effective in distinguishing between rock dolomite and limestone.

The distribution of dolomite in limestone is highly variable. A unit may be completely dolomitized, partially dolomitized with a sharp demarcation between dolomite and limestone, partially dolomitized with a transition zone between dolomite and limestone, or dolomitized in discontinuous patches or stringers which may or may not be related to obvious textures or structures of the host limestone. The patchy distribution can occasionally be seen to be related to the reorganization of magnesian calcareous material precipitated by algae. Some algae precipitate both magnesium and calcium carbonates which segregate during recrystallization into dolomite and calcite. Dolomitized reef rocks are frequently highly porous and give evidence that the dolomite is the result of either the selective resolution of calcite or the reaction of calcite with magnesium salts in sea water. Metasomatic dolomitization often accompanies hydrothermal mineralization, as in the Mississippi valley lead-zinc deposits. Here dolomitization completely destroys the original texture of the limestone, developing a dense fine-grained rock. Dolomitization in such situations is peripheral to the mineralization.

Chert nodules. During the diagenetic reorganization of limestones, silica is commonly reorganized into chert nodules, lenses, beds, and crosscutting veinlike bodies. The silica was probably originally disseminated through the carbonate deposit as amorphous or opaline organically precipitated material. The silica was dissolved or taken into a dispersed colloidal state owing to change of the chemical environment as organic material decayed. The silica was then concentrated into certain zones by migration of fluids and reprecipitated as gelatinous silica colloid. Subsequent dehydration and recrystallization resulted in the development of chalcedonic or cherty quartz.

The chert masses normally occur in ellipsoidal to highly irregular bodies and usually in definite planes in the limestone. Nodules either may be along bedding surfaces or may be in the middle of massive limestone beds. Stringers and veinlike bodies are seen cutting across bedding. Occasionally, chert becomes so abundant along a particular zone that the nodules coalesce to form a continuous bed.

Individual nodules are characterized by a white punky rind around the outside, usually equally developed on all sides of the mass. This evidence indicates that the chert does not develop as a clot of gelatinous colloidal silica on the sea floor which was subsequently covered with carbonate material. If this were the case, then the top and bottom of the chert nodules should have differing textures. Inside the rind, the material may be either light-colored chert or dark-colored flint. The cores of some large nodular masses may be hollow or filled with punky white material. Hollow interiors are characterized by curved surfaces which are concave outward and intersect in sharp ridges indicative of radial shrinkage outward, probably by desiccation. The light-colored cherts often show a mottled texture which is frequently related to unreplaced carbonate fossil remains. Weathering of such chert leaches out the carbonate material, leaving molds of the fossils. Some cherts contain abundant disseminated fine-grained carbonate which is removed on weathering, yielding a soft fine-grained granular silica called *tripoli*. Partially dolomitized limestone which has subsequently been replaced by chert often shows a selective replacement of calcite but not dolomite. Thus the chert contains small rhombohedrons of dolomite which on weathering leave rhombohedral holes — dolocasts.

Other mineral segregations in limestones. Pyrite and marcasite are common minor constituents of limestone. They usually occur in crystalline aggregates of roughly spherical or discoidal shape. The surface of pyrite masses usually shows crystal faces. Bladed crystals of marcasite are occasionally found lining cavities in limestone, but more commonly, marcasite occurs as druses of small crystals having an irridescent tarnish. Minute dendritic crystal clusters of pyrite are often seen in the insoluble residues. Other sulfides are much less common as cavity fillings. Sphalerite occurs in dark complexly twinned crystals, and hairlike brassy crystals of the nickel sulfide millerite occur as radiating tufts or random fibers in some cavities.

Geodes are crystal-lined cavities in limestone. The crystals are most commonly quartz, but rhombohedral carbonates and sulfides are present also. Geodes appear to have initially been unfilled body cavities in articulated bivalves and echinodermata. They appear to develop by radial expansion of chalcedonic replacement of the cavity wall. The crystal lining is probably developed after the crystallization of the chalcedony.

SILICEOUS SEDIMENTS

The siliceous rocks are those composed dominantly of chemically or bio-chemically precipitated silica. Other materials may be present but are quantitatively minor. These materials may include carbonates, clays, or sand grains as recognizable constituents. Most siliceous rocks were originally precipitated as amorphous opaline material, but recrystallization may convert them into chalcedony or microcrystalline quartz. Biogenic silica is precipitated as microscopic particles by diatoms, radiolaria, and siliceous sponges. Inorganic silica is precipitated by hot springs or may be derived by other processes, such as prolonged submarine leaching of volcanic ash or introduction by hydrothermal solutions. The direct accumulation of clots or beds of gelatinous silica is probably not important in the formation of these rocks. Most of the massive siliceous rocks are probably the result of recrystallization or diagenetic concentration of originally organically precipitated material.

On the basis of texture the relatively pure-silica rocks can be grouped into three categories: siliceous earths; chert, flint, or novaculite as nodules or bedded deposits; and siliceous oolites. Impure siliceous rocks include porcelanite and siliceous shales. Each of these is discussed below.

Siliceous Earths

Siliceous earths are friable porous rocks composed dominantly of micro-granular material, usually of organic origin. The material is usually radiolarian, diatomaceous, or spicular, or a mixture of the three with subordinate admixtures of clay or carbonates. These rocks are characterized by exceedingly high microporosity, resulting in very low density. Many of the dried earths will float in water until saturated. Recrystallization of the original opal and cementation are very incomplete so that the material is highly friable and mechanically weak. A white powder rubs off on the hands with the least bit of handling. Most of these earths are thinly laminated to thin-bedded. Laminations are the product of variable proportions of admixed clay or carbonate or different degrees of incipient recrystallization. Laminations are often difficult to see, except on a very fresh break, because of the friability. The color is generally white, cream, tan, or buff. Because of the chemical instability of opal, these siliceous earths can be expected only in rocks of young geologic age. Similar sediments of greater age have been recrystallized to chert. Similar-appearing siliceous earthy material, tripoli, may be present in the surface portions of outcrops of older chert. These are developed by groundwater leaching of chert, probably only where the chert contained considerable disseminated carbonates.

Chert, Flint, and Novaculite

These are microcrystalline massive rocks having a conchoidal to splintery fracture. They occur either as nodules and discontinuous beds in limestones or as bedded layers with intercalated shales or siderite. Cherts associated with limestone are secondary diagenetic products of the reorganization of disseminated organic silica in the original calcareous sediment. These have been discussed under the limestones. Bedded cherts are primary in origin, but the nature of the original material is a subject of debate. They may be recrystallized organic silica sediments equivalent to radiolarites or diatomites. Some may be the product of inorganic precipitation of colloidal silica. Still others appear to be the products of prolonged submarine leaching of pyroclastic material from which all but the silica of the original ash has been removed.

The bedded primary cherts usually occur as thin units, which are themselves thin-bedded, with alternating strata of chert and shale. The chert-shale alternations are usually 1 to 2 in. thick and appear to be quite consistent. Detailed examination, however, usually shows pinching and swelling along a layer, with units lensing out or bifurcating. The alternations therefore do not indicate a simple oscillating cyclical sedimentation; they may indicate a diagenetic differentiation of a relatively homogeneous siliceous-clay deposit into clay-rich and chert-rich zones. The fact that most of the bedded cherts have the same appearance on both the upper and lower bedding surfaces is evidence for diagenetic separation rather than rhythmic sedimentation, where there should be some grading into underlying shale and abrupt transitions into overlying shale. The bedded cherts are usually dark in color, showing shades of red, green, black, gray, or brown. The color depends on the nature of the impurities.

The novaculite of Arkansas and Oklahoma is an abnormally thick and abnormally pure chert. It is extremely fine-grained, dense, and translucent on thin edges. It ranges in color from blue-white to light grays and white. Some portions are discolored by admixed material, becoming darker gray, black, reddish, or cream-colored. The bedding is normally massive, having single beds several feet to several tens of feet thick. Normally, no evidence of bedding is seen within a single unit, but thin laminations are visible occasionally. The novaculite breaks with a well-defined conchoidal fracture, but in outcrop it is usually extensively jointed and occurs in polygonal blocks.

Oolitic Cherts

These rocks are clearly the product of chertification of an initial oolitic limestone. They are relatively rare but are important, in that they give positive evidence of replacement. The oolitic texture of the original limestone is usually perfectly preserved, with both radial and concentric structure of the ooliths recognizable (Fig. 3-47). The central nuclei of the ooliths are often quartz-sand grains which have been secondarily enlarged during silicification. Replacement of the ooliths usually precedes replacement of the matrix so that the oolith may be embedded in carbonate or in chert. Weathering of the carbonate may leave the ooliths separated by interstitial voids. The surfaces of such ooliths generally have a crystalline-quartz druse.

Impure Cherts

Porcelanite is an impure chert containing disseminated clay or carbonates. These inclusions result in a microporous porcelaneous-appearing surface. These rocks are generally light-colored and show a subconchoidal fracture. Where carbonate was initially abundant and has been removed by leaching, the resulting rock is soft, white, and often powdery. Some tripoli is clearly of this origin.

Siliceous shales are rocks in which clays and microcrystalline to colloidal-sized silica occur in subequal amounts. The silica is not clastic in origin but is probably derived from volcanic ash or organically precipitated opal. The rocks are generally very hard, dense, and brittle. They often show thin-lamellar bedding but are nonfissile, breaking along clay-rich partings and across the bedding with a subconchoidal or splintery fracture. The resulting shape is generally polygonal blocks. The color is generally dark: blue-black, gray, or brownish black. Oval white spots are often noticeable, having the appearance of skeletal snowflakes. These may be relics of primary radiolaria which have been enlarged by addition of silica.

PHOSPHATIC SEDIMENTS

Phosphorite, or rock phosphate, is a sediment composed dominantly of phosphate minerals. The dominant materials approach apatite in composition, but many of the minerals have essential carbonate as well as phosphate. The material is collectively called *collophane*. There are a number of possible origins, including inorganic precipitation, replacement, accumulation of organic phosphates, or residual concentrations by removal of limestone concentrating disseminated phosphate.

Typical *bedded phosphorite* consists of distinct layers 1 mm to 2 m thick of dark blue-gray to black pelletal material. The pellets tend to be flattened ovoids of uniform size in any one horizon. They may show concentric banding and are therefore ooliths, but most commonly they appear devoid of internal structure. The pellets are usually in contact with one another, and there is minor interstitial amorphous phosphate, calcite, iron oxides, clay, and carbonaceous material. Nuclei of the pellets and ooliths may be recognizable as detrital clastic grains of a variety of types. Fossils are rare in such deposits and are usually crushed, abraded, or replaced to such an extent that they are unidentifiable. Glauconite is a common accessory in the phosphates and is usually replaced by collophane to a greater or lesser degree.

Phosphatic "bone beds" are a second characteristic type. Here slow accumulation of fish bones, teeth, and scales are related to diastems and unconformities. The organic material has had most of the nonphosphatic material removed by leaching or replacement by amorphous collophane. Bone material has usually been fragmented into a breccia or microbreccia, and the bone structure may or may not be preserved. These rocks are generally dark-colored due to hydrocarbons.

Phosphatic nodules are forming currently on the sea floor under environments of very slow sedimentation. Similar nodules are encountered in many marine sedimentary rocks. They range from sand-size grains to nodules many centimeters in diameter, and they are characteristically black, extremely fine-grained, have a subconchoidal fracture, and have a pitchy luster. The nodules are generally discoidal in shape but may have a complex surface texture of knobs and small cavities. The surface has a smooth glazed hard appearance, and the glaze extends into the cavities as well as onto the protuberances. Detrital inorganic grains and fossil fragments are common in some of the nodules.

IRON-BEARING SEDIMENTS

Iron in some form is a normal constituent of almost all sediments so that a definition of "iron-bearing sediments" is difficult. Included here are only those rocks in which iron-rich minerals, sulfides, carbonates, silicates, or oxides make up more than one-third of the volume of the rock. Generally, this implies more than 20 percent iron oxides. The chemical state of the iron-rich minerals is an important indicator of the oxidizing versus reducing conditions in the sedimentary environment. The iron sulfides pyrite and marcasite indicate strongly reducing conditions and are usually associated with sediments rich in organic material. The iron carbonate siderite and the sedimentary iron silicates glauconite and chamosite and their relatives indicate mildly reducing conditions. The oxides and hydroxides are indicative of oxidizing conditions and are rarely associated with organic materials.

Iron Sulfide-rich Sediments

Pyrite and marcasite are very common accessory minerals in many sediments, particularly black shale and coal and less abundantly limestone. Only rarely are they sufficiently concentrated to give a pyrite rock; rather, they are generally disseminated as single crystals, small crystal clusters, or concretionary bodies. Pyritized fossils are common in shale rich in disseminated pyrite.

Sideritic Sediments

Siderite occurs as three distinct sedimentary types. The most common is the clay-ironstone concretions of shale. These concretions consist of an extremely fine-grained dense aggregate of siderite with appreciable admixed clay. They are diagenetic in origin and often show septarian structures. Their fine grain texture, hardness, high specific gravity, and dark color are characteristic. Weathering usually converts the siderite to a yellow, ocherous limonite which often causes the spalling of concentric sheets.

The second occurrence of siderite is as small pellets and rosettes in certain sediments. These are also diagenetic in origin but are apparently an early product, in that they are common in semiconsolidated sediments as well as in thoroughly lithified rocks. They range up to a few millimeters in diameter but differ from the clay-ironstone concretions in the coarser radial crystalline texture of spherulites. These bodies are most common in clay and silt.

Bedded siderite, often with interbedded chert, is the third and most important class. Much of the iron formation of the Precambrian age of the Lake Superior region was originally of this character. The unmetamorphosed and unoxidized rock is characteristically thin-bedded, having alternating beds of dense fine-grained siderite and chert. The siderite varies from pale greenish gray to dark gray, but on weathered exposures it is extensively oxidized and discolored by hematite or limonite. The discoloration usually extends into even the densest rock on minute fractures so that completely fresh material is difficult to obtain from surface exposures. Chert varies from light greenish gray to nearly black and is also usually iron-stained by oxidation. The beds vary from a fraction of an inch to several inches in thickness and often show minute laminations superimposed on the bedding. Individual chert and siderite beds may have sharp contacts between them or may appear to grade into each other. However, grading does not appear to be of a "graded-bedding" type with sharply defined pairs of lamellae. The siderite layers are often more or less cherty so that after etching with hot HCl there is an interstitial network of chert having etched-out rounded masses of fine-grained siderite. Similarly, the chert beds may contain more or less siderite, either as rounded masses or as euhedral rhombohedra up to 1 mm in dimensions. Stylolites are common in some sideritic beds. They are of small vertical dimensions, usually limited to one thicker siderite unit. The stylolite surface is often coated with minute flakes of hematite and graphite. Minor minerals present include iron silicates, hematite, magnetite, and graphite. These give the cherty layers their dark color.

Iron Silicate Sediments

The iron silicate minerals of sediments may be divided into two characteristic types. First, those in which the silicates are in pellets or ooliths and, second, the finely laminated nonpelletal materials. The mineralogy is usually complex and impossible to distinguish megascopically. The granular silicates include greenalite, glauconite, and chamosite. All three occur as fine-grained green granules. Greenalite and chamosite are characteristic of some of the Precambrian iron formations and are admixed with chert and iron oxides. Glauconite is a common constituent of many sediments, particularly sandstone and limestone. It is the dominant constituent of greensands. The laminated iron silicates are a variety of members of the chlorite minerals, including stilpnomelane and thuringite. They usually occur as clay-sized to fine-sand-sized flakes with abundant admixed chert, iron oxides, siderite, etc. A yellowish or greenish color is characteristic of the rock. Iron-staining is common on weathered surfaces. Stilpnomelane is often associated with magnetite so that the rock is magnetic.

Sedimentary Iron Oxides

Hematite and limonite are the dominant iron oxides, although magnetite may be sedimentary in origin. As dominant minerals the iron oxides are the product of several processes. Evidence indicates that some of the banded chert-hematite rocks are primary sedimentary products. Others are the product of oxidation or oxidation followed by metamorphism of chert-siderite rocks. Oolitic hematite may be either a primary depositional product or a replacement product of primary calcareous ooliths. Hematite replacement of fossils is clearly a secondary product, although replacement probably followed deposition almost immediately. Iron-ore bodies of hematite, magnetite, or limonite produced by secondary modification of iron-rich sediments will not be considered here.

The bedded chert-iron oxide units will be considered here, regardless of their primary, secondary, or metamorphic origin. All consist of alternating units of chert and iron minerals in beds a fraction of an inch to a few inches in thickness. Brecciation of the chert beds is quite common in all, with filling between the fragments by iron oxides. Iron silicates may be present as fine greenish or brownish fibers of the amphibole grunerite or stilpnomelane. Chert is either a fine-grained dense light-colored chert or red jasper or a more coarsely crystalline granular white chert. The coarser material is usually associated with magnetite, whereas the finer is associated with hematite. Hematite may be soft, fine-grained, and red; hard, dense, and blue; or coarsely crystalline and micaceous. Magnetite is usually granular, having individual anhedral grains around 1 mm in diameter.

Oolitic hematites of either primary or replacement origin are similar megascopically. They consist of rounded or flattened bodies of microcrystalline hematite $\frac{1}{4}$ to 2 mm in diameter. Limonite and iron silicates may be present in varying amounts. The interstices between the ooliths are filled with microcrystalline hematite plus varying amounts of carbonates. Clastic grains are fairly abundant and often occur as nuclei of ooliths.

The fossiliferous hematite consists of a replaced coquina of crushed and abraded shell fragments plus more or less ooliths and flattened ooliths. Fragments of brachiopods, bryozoa, and crinoids are often recognizable. The fragments are generally flattened parallel to the bedding. Hematite is extremely fine-grained, and much carbonate may remain in the rock.

EVAPORITES

Evaporites are those minerals precipitated from saline solutions by evaporation. They consist of the most soluble salts and are often of very complex mineralogy. Inasmuch as evaporites are highly soluble, they rarely occur in outcrop, and therefore only the major types will be dealt with here and then only briefly. The three major types are gypsum, anhydrite, and halite rocks.

Rock gypsum varies from fine- to coarse-grained. Primary deposits are normally fine-grained and may be massive or lamellar. Lamellae often show a coarsening of grain size upward through the lamellae, giving the reverse of graded bedding. Gypsum rocks are generally light-colored, but they vary widely, depending on the type and amount of impurity. Secondary gypsum is either extremely fine-grained, alabaster (Fig. 6-6), or coarse crystalline, selenite or satin spar. The hydration of anhydrite to give gypsum results in a large volume increase, which is evidenced by intense crumpling of the beds.

Primary anhydrite rock is normally fine-grained and laminated. The layers of anhydrite are separated by thin partings of impurities, such as clays or carbonates. In coarser-grained rocks the individual crystals are in the form of rectangular blocks stacked like a pile of bricks. Recrystallized anhydrite tends to be darker in color than primary anhydrite. Dark gray and blue-gray are common in the recrystallized material. Thin parallel plates and rosettes of plates are common in the recrystallized rocks (Fig. 6-5).

Rock salt normally occurs as coarsely crystalline material which usually shows evidence of bedding (Fig. 6-4). Different beds vary in color owing to varying amounts and types of impurities, both in the salt crystals and between them. The salt crystals are usually anhedral, but viscinal faces which are indicative of an original hopper-shaped skeletal crystal can be seen developed between grains. Liquid-filled cavities are common in the crystals and may also include gas bubbles. The cavities may be regularly distributed in the crystal on growth faces, again often indicating hopper-crystal development. Fragmentation of the skeletal crystals as they accumulate results in a wide range of grain size and shape. Rock salt usually has associated anhydrite, gypsum, and dolomite so that the proportions of these four rock types vary greatly.

7 METAMORPHIC-ROCK PETROGRAPHY

TEXTURES AND STRUCTURES

The distinction between texture and structure as applied to metamorphic rocks is so difficult that many authors have ceased to make a distinction and now substitute the term *fabric*. The difficulty arises because many important features may be either textural or structural in origin. An example is the term *foliation*, which may refer to parallelism of individual mineral flakes and thus be textural, or it may refer to the parallelism of mineral aggregate layers, which would be structural. In this chapter *texture* is taken to mean those features of the rock dependent on the size, shape, and orientation of individual mineral grains, whereas *structure* is taken to mean those features dependent upon the size, shape, or orientation of polygranular units of the rock. Thus a lenticular body of polygranular quartz in a schist would be described in structural terms, whereas the shape and relationship between individual grains in the lenticular body would be described in textural terms.

Many terms applied to metamorphic rocks are derived from similar structures or textures of igneous rocks. The prefix "blasto-" or the suffix "-blastic" are added to the root term to indicate metamorphism. The prefix "blasto-" is generally applied to a feature inherited from the parent rock and still recognizable in the metamorphite. The suffix "-blastic" is applied to a feature developed in the metamorphic environment. Thus blastoporphyritic implies a metaigneous rock in which an original porphyritic texture can be recognized, whereas porphyroblastic implies a metamorphic rock containing large crystals of a mineral set in a finer-grained groundmass.

SHAPES OF INDIVIDUAL MINERAL GRAINS

The terms *euhedral*, *subhedral*, and *anhedral* are often applied to metamorphic-mineral grain shapes. They have the same connotation as in igneous rocks, that is:

1. *Euhedral.* A crystal bounded by its own normal crystal faces.
2. *Subhedral.* A crystal bounded in part by its own crystal faces but with

some boundaries controlled by mutual interference of adjacent growing crystals.

3. *Anhedral.* A crystal whose boundaries are entirely controlled by the boundaries of adjacent grains. This may be the result of mutual interference of adjacent crystals of identical growth characteristics, or it may be the result of adjacent crystals with higher crystalloblastic strength controlling the boundaries of the weaker growing crystal. It may also be the result of mechanical crushing.

The euhedralism of crystals in metamorphic rocks does not have a time connotation in a crystallization sequence as it does in igneous rocks. During metamorphic recrystallization all crystals are involved in the process to some degree, and therefore all are essentially of contemporaneous formation. The terms *idioblastic* and *xenoblastic* may be preferred, because they imply a metamorphic origin for the crystal.

1. *Idioblastic.* A crystal of metamorphic origin bounded by its own crystal faces.
2. *Xenoblastic.* A crystal of metamorphic origin not bounded by its own faces. It is equivalent to anhedral as described above.

The shapes of individual crystals, whether idioblastic or xenoblastic, are described in standard terms relative to dimensional relationships as follows:

1. *Prismatic* (Fig. 4-11). One dimension is markedly greater than the other two, forming a prism with or without terminations. Dimensions are on the order of $a:b:3-5c$.
2. *Acicular* (Fig. 4-13). Slender needlelike crystals with or without visible crystals faces. Dimensions are usually $a:b:10-20c$, where a and b are a fraction of a millimeter.
3. *Bladed.* Three distinctly different dimensions, one of which is usually much larger than the other two. Dimensions are usually $a:2-3b:5-10c$.
4. *Tabular.* Two dimensions are markedly greater than the third; the crystal is usually bounded by two flat parallel faces and therefore has a uniform thickness. Dimensions are usually $3-10a:3-10b:c$, with a and b either equal or nearly equal.
5. *Lenticular.* A lens-shaped crystal, thickest in the middle and tapering to the edges, often to a very thin edge. Dimensions are usually $2-5a:2-5b:c$.
6. *Equant, or equidimensional.* All three dimensions are equal or subequal, but the crystal has an irregular shape.

7. *Blocks, or blocky crystals.* All three dimensions are equal or subequal, but the crystal has roughly planar sides.
8. *Spherical.* Equidimensional grains which have rounded boundaries.

Euhedral or idioblastic crystals should be described in terms of their crystallographic forms, and any recognizable forms of subhedral crystals should be identified.

TEXTURES OF ROCKS NOT SHOWING STRONG PREFERRED ORIENTATION

Equigranular rocks are those in which all grains are of approximately the same size. These textures are particularly common in monomineralic rocks such as quartzite and marble. Preferred orientation of grains is very often present but not recognizable megascopically.

1. *Mosaic texture* (Fig. 4-10). Crystals are equigranular and equidimensional and are generally polygonal in shape, with simple straight-line or gently curved intergranular boundaries.
2. *Sutured texture.* Crystals are generally equigranular and equidimensional, or they are lenticular and have highly irregular boundaries and much interpenetration of each grain into its neighbors.
3. *Mylonitic texture.* A very-fine-grained product of mechanical crushing without recrystallization of the primary minerals. Such rocks usually show directional features as thin lamellae. The crushed material is lithified by an introduced or hydrothermal cement.

Inequigranular rocks are those in which there are two distinct grain sizes or noticeably different grain habits. These textures are of two fundamental origins: (1) recrystallization in a polymineralic rock as a result of metamorphism without directional stresses, and (2) incomplete mechanical crushing not accompanied by development of a strong preferred orientation.

1. *Crystalloblastic texture* (Figs. 4-11 and 4-12). A crystalline texture due to metamorphic recrystallization. This term is equally applicable to those rocks with strong preferred directions. It implies the essentially simultaneous growth of all mineral grains in a solid environment. The sizes and shapes of grains then depend on the relative ability of the different minerals to make room for themselves, that is, on their crystalloblastic strength.
2. *Porphyroblastic texture.* A crystalloblastic texture with two or more distinct grain sizes, equivalent to porphyritic texture in igneous rocks. The individual large crystals are called *porphyroblasts*, equivalent to phenocrysts in igneous rocks.

3. *Poikiloblastic texture = sieve texture* (Fig. 4-5). A texture in which large porphyroblasts include numerous small crystal grains. During growth the large crystal apparently grew around mineral grains which could not be accommodated in the composition of the growing large crystal.

4. *Decussate texture* (Fig. 4-13). The crystalloblastic texture of polymineralic rocks in which there is no preferred orientation of grains.

5. *Cataclastic or autoclastic textures.* Textures produced by mechanical crushing without essential recrystallization. These rocks may or may not show directional features, and there is considerable overlap in terminology. Only those that commonly show no directional features are considered here.

6. *Mortar texture.* A texture consisting of larger mineral fragments set in a groundmass of crushed material derived from the same crystals. The individual mineral grains often show effects of mechanical warping, such as curved cleavages, bent twin lamellae, ragged or shattered borders, etc. The rock produced may be called a *crush breccia.* Complete crushing produces a mylonite having mylonitic texture.

STRUCTURES AND TEXTURES OF ROCKS SHOWING STRONG PREFERRED ORIENTATION

Preferred orientation may be either planar or linear, and either may be expressed by individual mineral-grain shapes and orientation or by the shapes and orientation of polygranular assemblages of minerals. A single rock may show both linear and planar features at both single-grain and polygranular aggregate levels. Therefore an accurate description of features is more important than terminology. Current usage is to describe "planar elements" and "linear elements" in which the element is a unit of the rock, whether it be a single crystal, a polycrystalline body, or a direction defined by fractures.

Foliation

Foliation is the general term used to include all planar textures and structures of metamorphic rocks which were developed during metamorphism. The foliation may be defined by layering of contrasting mineralogies (gneissosity), by planar preferred orientations of individual grains (schistosity), by planar fracture surfaces (cleavage), or by any combination of these three. In British terminology foliation is restricted to layering of contrasting mineral assemblages.

Cleavage. Structures which show preferred direction in metamorphic rocks and are defined by more or less closely spaced fractures. Several types of cleavage are recognized. The orientation of cleavage to regional structure varies with the type of cleavage, and reference should be made to structural geology texts for interpretation.

1. *Fracture cleavage* (Fig. 7-1). Physically distinct fractures developed by mechanical rupture. In this type of cleavage the fractures are distinct breaks separated by unfractured rock which will break readily in all directions. Apparently, fracture cleavage does not involve parallel orientation of mineral grains or aggregates.
2. *Slaty cleavage.* Closely spaced distinct fractures developed by mechanical deformation. In this type of cleavage all portions of the rock tend to split more readily in one direction than in any other. Rocks with slaty cleavage are microcrystalline but show a characteristic sheen on the fracture surface. Reorientation of flaky grains or development of lenticular grains parallel to the cleavage is characteristic.
3. *Flow cleavage = schistosity.* The ability of metamorphic rock to split preferentially in one direction as a result of parallelism of megascopically visible micaceous mineral flakes.
4. *Slip cleavage.* Fracture cleavage involving visible displacement along the fracture. Micaceous minerals may be dragged by slippage into positions parallel to the fractures.

Fig. 7-1. Fracture cleavage cutting metaargillite. The bedding parallel to the hammer handle shows as light-colored silty zones. Siamo formation, Negaunee, Michigan.

5. *Linear cleavage.* The ability to split into long pencil-like fragments, usually due to intersection of two cleavages or one cleavage and bedding.

6. *Bedding cleavage.* Slaty, or flow, cleavage parallel to the bedding. This may be due to load or to isoclinal folding and should not be confused with bedding-plane fissility of unmetamorphosed rocks.

Schistosity (Fig. 7-2). A planar texture defined by parallelism of megascopically visible platy, prismatic, or lenticular crystals. A *schistose texture* is one which shows schistosity. Parallelism of flakes of micas, chlorites, talc, graphite, hematite, etc., are characteristic. The flakes may or may not show a linear parallelism superimposed on the planar. There may be associated mineral

Fig. 7-2. Chlorite schist showing well-developed parallelism of flakes to give schistosity and crinkle folds across the schistosity to give lineation. Chester, Vermont. (*Ward's Universal Rock Collection.*)

segregation into bands or streaks, but only those portions dominated by parallel platy crystals can be said to be schistose.

Planar structures (Fig. 4-3). Structures in which the planar element is represented by mineral segregation into bands, lenses, and irregular planar masses. The segregation of minerals may be of several origins. Metamorphic differentiation results in segregation by concentrating the more mobile mineral phases in shear zones, leaving the unsheared portion of the rock deficient in the mobile constituents and enriched in the relatively immobile phases. The growth of porphyroblasts is often accompanied by secretionary growth of more mobile constituents. Intimate magmatic injection gives rise to planar structures of contrasting mineral assemblages which are difficult to differentiate from metamorphic products. Primary structures such as pebbles, concretions, bedding, pillow structures, etc., result in mineral banding which may be difficult to distinguish from metamorphic products at some scales. Finally, mechanical deformation without extensive recrystallization may result in similar-appearing structures. The determination of which process is dominant usually involves field relationships or detailed petrographic study or both.

1. *Gneissic, or gneissose, structure* (Figs. 7-3, 7-4, and 7-5). A banded structure with alternating bands of an equigranular mosaic texture, usually of

Fig. 7-3. Amphibolitic gneiss. Thin hornblende-rich bands alternating with thin quartzo-feldspathic-rich bands. Note the concentration of garnet in the hornblende-rich bands. Windsor, Vermont. (*Ward's Universal Rock Collection.*)

Fig. 7-4. Well-defined banding in gneiss. The dark bands are rich in hornblende and represent more calcareous beds of the original sediment. Note the pattern of passive flow folding. North of Parry Sound, Georgian Bay area, Ontario.

quartzo-feldspathic material, and a schistose texture, usually dominated by micas or hornblende. The bands are not usually continuous over long distances but lens out. They are, however, thin compared with their lateral extent.

2. *Augen structure* (Fig. 4-4). A lenticular structure composed of an aggregate of grains which in cross section has the shape of an eye. Most commonly, augen structures are developed by the growth of a large porphyroblast forcing aside the surrounding rock and allowing the secretionary growth of mobile constituents in the corners of the eyes.

3. *Flaser structure.* A structure developed by dynamic metamorphism in which lenticular masses of coarse-grained or uncrushed material are separated by wavy zones of finer-grained foliated material.

4. *Boudinage structure* (Fig. 4-17). A lenticular structure developed when competent units embedded in a yielding matrix are pulled apart by deformation. Boudinage structures are often developed from small dikes or veins in schistose rocks or from quartzite or limestone units in deformed argillaceous rocks. The individual pillow-shaped unit is called a *boudin.*

Fig. 7-5. Diffusely banded gneiss. This outcrop shows much less distinct banding than that of Fig. 7-4 but shows the same pattern of passive flow folding. South of Parry Sound, Georgian Bay area, Ontario.

5. *Ptygmatic structures* (Fig. 7-6). Contorted veins or dikelets which show irregular patterns. The veins are usually quartz or pegmatite embedded in foliated rocks. They may be concordant with the host-rock structure but are most commonly discordant. The folds are usually thickened on their axes and thinned on their limbs. The individual deformed body is called a *ptygma.*

Lineation

Lineation is a general term for the parallel orientation of textural or structural features that are linear. Prismatic, bladed, or rod-shaped elements, or lines of intersection of planes, have their longest dimension parallel. Linear textures are linear elements which show as a result of parallelism of elongated single crystals. Prismatic crystals, such as hornblende, and bladed crystals, such as kyanite and sometimes biotite, are the more common examples. The linear texture is usually superimposed on a planar element. Linear structures are polygranular units which show parallelism of an elon-

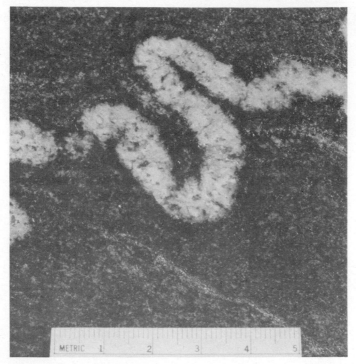

Fig. 7-6. Ptygmatic folding of a pegmatite dike in gneiss. From a boulder near Denver, Colorado.

gated dimension. The linear element may be a deformed relict structure or the result of new crystal growth and deformation.

1. *Lineation of relict structures.* Pebbles, fossils, ooliths, and amygdules are often sufficiently well-preserved so that they can be readily recognized. Their deformed shapes can be recognized by comparison with normal shapes, and a direction of elongation can be determined.
2. *Lineation of metamorphic structures.* During metamorphism under an environment involving a strong one-dimensional deformation, mineral segregations may be pulled apart into pencil- or cigar-shaped parallel masses.
3. *Rodding structures.* Rods of quartz, usually with a mosaic texture, which have formed along the axes of folds. Apparently, during close folding, quartz is segregated along the fold axes, or quartz which formed in bands prior to folding is concentrated on the axis and pinched out on the limbs of the folds.
4. *Mullion structures.* Like rodding structure, but it involves the country rock rather than quartz segregations and occurs on a larger scale.

5. *Linear boudinage structures.* Boudinage in which one dimension of the boudins is much greater than the other two. It is not confined to the axial portions of folds as is mullion structure.

6. *Linear structures evidenced by intersections.* The intersection of any two planar elements gives a linear element. The intersection of bedding with cleavage is a common and useful example. The intersection of foliation with the axial plane of small folds or crenulations is common in schistose rocks (Figs. 7-7, 7-8, and 7-9).

DESCRIPTIONS OF ROCKS

The common stable assemblage of minerals in metamorphic rocks can be determined for any basic rock type from the charts of Chap. 4. The problem, however, usually is to determine the parental rock type and the type and intensity of metamorphism from an identifiable assemblage of minerals. Therefore, in the discussion below, rocks are grouped on the basis of texture or structure or both and are further grouped on the basis of general bulk composition. The character of each mineral is discussed and the associated mineral-species minerals are listed. Chemical relationships of each mineral to lower- and higher-grade minerals are pointed out. Rocks are considered according to the following outline.

Fig. 7-7. Lineation of biotite crystals on the schistosity plane. The large randomly oriented crystals are staurolite. Biotite appears as elongated flakes oriented from upper left to lower right. Byron, Maine.

Fig. 7-8. Lineation in phyllite due to crinkle folds. Note the lustrous sheen of the phyllite surface. Mariposa County, California. (*Ward's Universal Rock Collection.*)

I. Aphanites (with or without visible porphyroblasts)
   A. With marked foliation
      1. Slates
      2. Phyllites
   B. With little or no foliation
      1. Hornfelses
      2. Spotted slates
II. Phanerites
   A. With marked foliation
      1. Schistose rocks
         a. Derived from noncalcareous sediments
         b. Derived from calcareous sediments
         c. Derived from igneous rocks
      2. Gneissose rocks
         a. Clearly derived from sediments
         b. Not clearly derived from sediments
   B. With little or no foliation
      1. Quartzites

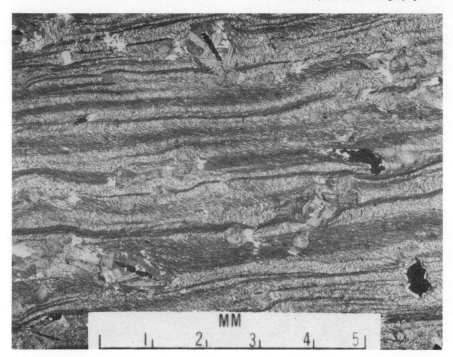

Fig. 7-9. Crinkle folds in muscovite-chlorite schist. A few scattered porphyroblasts of biotite and iron ores are apparent. Waites River formation, Walden Heights, Plainfield quadrangle, Vermont. Projection of a thin section. (*Courtesy Dr. R. H. Konig.*)

    2. Calcareous source rocks
       a. Marbles
       b. Calc-silicate rocks
    3. Granulites
    4. Serpentine

SLATES

Slates are a low-grade metamorphic product of argillaceous rocks characterized by strongly developed cleavage throughout the rock. The cleavage can be shown microscopically to be due to parallelism of clay- to silt-sized micaceous flakes and lenticular grains. Thus, as the preferred orientation extends throughout the rock, a cleavage separation may be developed anywhere in the rock. This feature distinguishes slaty cleavage from fracture cleavage and is a requisite for true slate. Rocks with fracture cleavage but without slaty cleavage may be termed *semislates*, or *argillites*.

    Because of the parallelism of micaceous flakes, the cleavage surface of slate normally has a sheen which is lacking on fracture-cleavage surfaces and on argillites. Commonly, a freshly cleaved surface will show a plumose structure, which consists of very shallow ridges and grooves, branching and curving outward from a central zone like an ostrich feather. Each plume structure is usually bounded by joints which intersect the slaty cleavage. The origin of the plume structure is unknown; however, it may be initiated by the presence of lenticular silt-sized grains of quartz. The ridges of one cleavage face will fit precisely into the grooves of the opposite face, and a minute granule can often be seen at the beginning of each ridge.

The relationship of slaty cleavage to bedding is of great importance in deciphering the structure of a region. Slaty cleavage is not normally parallel to bedding, and therefore the line of intersection of cleavage and bedding gives a lineation which is usually subparallel to the axis of the fold. The recognition of bedding is therefore important. Regular color banding is usually a good indication of bedding (Fig. 7-10). Color banding is most commonly a reflection of varying contents of iron or organic carbon, since these two are the most common pigments in argillaceous rocks. Textural variations are more positive evidence of bedding. Thin silt layers are commonly present and show up as distinct lamellae or as a thin streak of slightly more

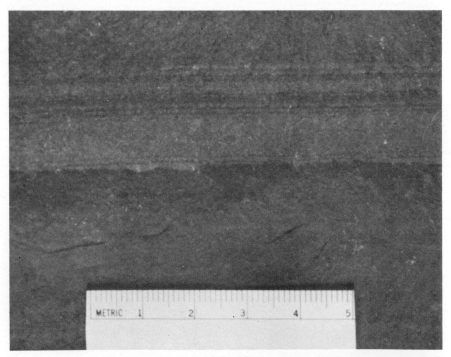

Fig. 7-10. Slate showing well-defined bedding cutting across the cleavage. A poorly developed plumose structure is visible in the upper portion. Bangor, Pennsylvania. (*Ward's Universal Rock Collection*.)

granular material. Fossils, particularly graptolites in black slate, may be present and will define bedding. A thorough search for bedding should always be made in slates, and usually some evidence will be found. Care must be taken in hand specimens not to confuse the ridges and grooves of the plumose structure for bedding.

The color of slates is their most descriptive feature. Colors generally are green, red, purple, gray, and black. Green slates are generally indicative of iron in the reduced ferrous state. Red or purple colors indicate iron in the oxidized ferric state. Many green or red slates will be mottled, spotted, or streaked with the other color. Thus green slates will often be oxidized to red along joints, and red slates will become spotted with green on exposure because of reduction of iron, possibly by organic material. Gray and black are generally interpreted to indicate a rather high content of organic carbon or hydrocarbons. This is generally, but not necessarily, the case. Disseminated iron sulfides will also produce a black color. Light-colored slates are relatively rare. Apparently, clays low in iron or organic material recrystallize more readily and yield phyllite or schist rather than slate.

Very little material in slates can be identified megascopically. Occasionally, minute flakes of mica, silt-sized grains of quartz, or small pyrite euhedra can be seen. Pyrite also forms small recognizable dendritic or skeletal crystal growths or films on the slaty cleavage. Siltstone lenses and lamellae are usually composed of quartz with or without mica flakes. Microscopically, slates can be seen to be composed of quartz, sericitic and chloritic flakes, variable amounts of iron-ore minerals, minute carbonate granules, feldspar granules, and carbon or hydrocarbon films. The mineral assemblage is thus in the chlorite zone or the muscovite-chlorite subfacies of the greenschist facies.

The distinction between slate and fissile shale is important. In shale the fissility is parallel to bedding, whereas slaty cleavage is normally at an angle to bedding. This feature, as well as the field association of metamorphic and sedimentary rocks, is the main criterion. In a hand specimen under pressure from a sharp needle point on the cleavage, slate appears to flow somewhat like paraffin, causing an outward thrusting of thin plastic scales. In contrast, shale under a needle point usually breaks up as a powdery substance rather than flow. A final refined field distinction is that when struck with a hammer, slates "klink," whereas shales "klunk."

PHYLLITES

Phyllites are strongly foliated rocks intermediate in grain size between slate and schist. The bulk of the grains is not megascopically recognizable as individuals, but they are coarse enough so that the flaky or scaly nature of the mineral grains is recognizable. The coarser size and abundance of micaceous minerals give phyllite a characteristic luster. Phyllites dominated by muscovite, sericite, paragonite, and the hydromicas are light-colored with a silvery luster (Fig. 7-8). Those dominated by chlorite are some shade of green and have a satiny or velvety luster. Those dominated by talc (soapstone) have a greasy luster. Graphitic phyllites have a lead-gray metallic luster.

Phyllite, except for soapstone, has a highly developed foliation as a result of both parallelism of flaky minerals and banding due to mineral segregation or variation in mineral composition. Commonly superimposed on foliation is a second planar structural element in the form of minute folds, slip bands, and crenulations. The folds and crinkles are often of such small amplitude that they are barely identifiable megascopically, but they may have amplitudes of several centimeters. The intersection of the microfolds and crenulations with the foliation gives a lineation to the rock which should be noted. Associated with the larger-scale folding may be the segregation of more mobile constituents such as quartz, resulting in a small-scale rodding structure.

Porphyroblasts are fairly common in some phyllites as megascopically identifiable crystals of garnet, biotite, chloritoid, magnetite, pyrite, or carbonates. These porphyroblasts may transect the foliation of the micaceous layers, indicating a postmetamorphic origin, or they may apparently force aside the micaceous materials, indicating growth of the porphyroblast during the development of the micaceous matrix. In the latter case there may be segregation of quartz in the pressure shadows of the porphyroblast, developing small-scale augen structures. The porphyroblasts often give a lineation in the rock as a result of parallelism of elongated crystals. Tiny biotite or chlorite flakes often show this when visible, but care must be exercised not to confuse such true lineation with apparent lineation when the rock has fractured on a surface which is not quite parallel to foliation. In such a case thin flakes which approach a circular or square plate will appear as thin lines. Mechanically strong porphyroblasts, such as garnet, pyrite, and magnetite, produce a third type of lineation on the foliation surfaces in the form of ridges and grooves. The effect is as though the porphyroblast had stood still while the phyllitic groundmass had moved past it, flowing around the resistant mass. The ridges and grooves often extend on both sides of the porphyroblast for a distance of several millimeters to a very few centimeters.

Mineralogically, the phyllites belong in the chlorite or biotite zones. These zones are equivalent to the muscovite-chlorite and muscovite-biotite subfacies of the greenschist facies. When garnet appears, it is probably a manganiferous variety in the spessartite-almandite series and not the typical almandite of the higher facies. Andalusite and cordierite may appear as porphyroblasts in phyllitic-appearing rocks, but they belong to the albite-epidote-hornfels facies of thermal metamorphism and are described below with the spotted slates. Phyllitic rocks of cataclastic origin are known as *phyllonites* and are difficult to differentiate megascopically from phyllites in the hand specimen. In outcrops they could be recognized by their gradation into larger cataclastic structure.

Mineralogy of Phyllites

White micas. Sericite, paragonite, muscovite, and hydromicas all occur in phyllites and are megascopically indistinguishable from each other. No clay minerals are normally present unless they are the product of postmetamorphic weathering or hydrothermal alteration. Individual mica flakes are wispy and microscopic in size, usually in the silt range of sizes. Occasionally, a flake will be large enough to be recognizable. However, the micas are commonly segregated into lenses or distinct bands alternating with quartz-rich mica-poor bands. Then micaceous flakes which have the appearance of muscovite, in spite of being composed of several mica flakes, can be cleaved off with a knife blade. The visible segregation of micas varies from the vaguest suggestion of banding to distinct rather sharply bounded lamellae.

Chlorites. Greenstones of phyllitic texture are common products of low-grade metamorphism of intermediate to mafic flows, dikes, and ash beds. Normally, no minerals are of recognizable development; however, chlorite is indicated as the dominant mineral by its foliation, green color, and hardness. Commonly associated minerals in silt-sized particles are albite, epidote, quartz, magnetite, and carbonates. Some talc may be present. The chlorite phyllites are commonly color-banded in various shades of green, probably due to variable proportions of the essential minerals.

Relict igneous-rock textures and structures may be preserved, particularly in greenstones developed from flows. Phenocrysts in the flow can be seen as local spots of contrasting color. Plagioclase phenocrysts appear as more or less rectangular lighter-colored patches of fine-grained material. Some are soft and claylike, whereas others are hard, dense, and light green and are presumably composed of saussurite, a microgranular mixture of albite and epidote. Amygdaloidal structures are often preserved but show strong deformation. Most commonly, they appear as strongly triaxial rounded patches of chlorite, which is darker green than the matrix. Deformed carbonate, zeolite, and quartz amygdules can be expected.

Quartz. Quartz is a dominant constituent even in those phyllites derived from basalts, but it is recognizable only in those rocks in which it is very abundant. In the greenstones, quartz appears because both chlorite and epidote are lower in silica than the primary igneous-rock-forming minerals, but it is usually too fine-grained to be recognized. In the white-mica phyllites, quartz is usually in silt-sized grains, and unless it is segregated into distinct bands, it may be overlooked. In the segregated bands it generally occurs as equidimensional fine grains having a mosaic texture. When the phyllite has been crumpled across the foliation, coarser-grained segregations of quartz may be seen along the axes of the folds—a rodding structure.

Graphite. Graphitic phyllite is a common product of regional metamorphism of carbonaceous black shale. The presence of abundant graphite apparently inhibits metamorphism so that graphitic phyllite may be found in association with mica schists bearing garnet or even staurolite. The graphite does not occur as recognizable flakes, but the rocks are lead-gray, have a submetallic luster, and will stain the fingers readily.

Graphite appears to form thin films, coating the other minerals of the rock. Pyrite or pyrrhotite or both, derived from the organic sulfur in the black shale, commonly occur as recognizable grains, lenses, and films in the phyllite. Pyrrhotite is a higher-temperature mineral than pyrite, since pyrrhotitic-graphitic phyllite is associated with garnet and staurolite schist.

Talc. Talc is a common minor mineral in phyllitic greenstones but is usually not recognizable in this occurrence. Soapstone is fine-grained talc or talc-carbonate rock derived from ultramafic igneous rocks. It commonly shows very little foliation, although the talc flakes may have good parallelism. The rocks have a characteristic soapy feel, are white to gray or gray-green, and are often veined by carbonates. Carbonates as porphyroblasts of recognizable size are fairly common as euhedral rhombohedrons. Magnesite, dolomite, and calcite are most common; ankerite and siderite are less common.

Biotite. Biotite is usually a minor constituent of phyllite, but it does occur as megascopically recognizable irregular brown flakes or as bladed straight-sided crystals. In the latter habit the crystals have dimensions around $3a:6b:c$ and are alined with their cleavages in the foliation and their long dimension parallel, giving a lineation. Biotite is more common in chlorite-poor phyllite. Where there is a mineral segregation, biotite occurs with the white micas rather than with the quartz-rich zone.

Garnet. Garnet is not common in phyllite, but when present, it occurs as sharply defined euhedral porphyroblasts up to a few millimeters in diameter. It is usually dark red to black and is presumably high in the manganese molecule—the variety spessartite. The crystals are typical dodecahedrons or trapezohedrons or both. A thin zone in which the groundmass is bleached can often be seen surrounding the garnets. These zones are usually less than half the garnet diameter in width and appear to be portions of the matrix which have been impoverished in ferromagnesian materials as the garnets grew.

Chloritoid. Chloritoid appears as small porphyroblasts in phyllite derived from sediments high in iron and alumina and low in potash. Its associated minerals are therefore sericite, quartz, iron ores, ±chlorite. Chloritoid porphyroblasts are small euhedral tablets or blades, usually having a random orientation with respect to the foliation. The crystals are dark green to black, have a greasy vitreous luster, and have one good cleavage. Greater hardness and lack of micaceous cleavage distinguish the small chloritoid crystals from similar-appearing biotite.

Tourmaline. The only accessory mineral which is commonly identifiable is tourmaline. It generally occurs as slender elongated prismatic black crystals. Most commonly, the prisms lie in the plane of foliation, but they are usually randomly oriented within the plane. The rounded triangular cross section of the prisms is diagnostic. The appearance of tourmaline may involve boron metasomatism, but many shales contain sufficient boron so that no introduced material is required for their formation.

HORNFELSES

Hornfels is a fine-grained dense rock produced by thermal metamorphism of shale, calcareous shale, graywacke, slate, phyllite, and to a less important degree, igneous rocks. The rocks are typically hard and break into blocks, often with a subconchoidal fracture. They vary widely in color from light green, brown, or purple to dark gray or black. Many have the appearance of a fine grained dense basalt. The lighter-colored varieties often show color banding, probably as a relict of bedding or foliation and representing variations in mineral proportions.

Hornfels represents a very wide temperature range, even involving fusion of a portion of the rock along some intrusive contacts. As a result, many minerals are characteristic of a hornfels. However, they are typically so fine-grained and dense that no minerals can be recognized megascopically. Any of the typical thermal-metamorphic minerals outlined in the tables of Chap. 4 may be present, along with relicts of the parent-rock minerals which have been incompletely modified by the metamorphism.

Mylonite is megascopically very similar in appearance to hornfels. It is the product of extreme crushing by dynamic metamorphism, particularly along fault zones. The crushed-rock material is thoroughly cemented, most commonly by the introduction of silica. This does not normally involve recrystallization of the primary mineral fragments. Uncemented material of the same origin is the silt- to clay-sized material of a fault gouge. Mylonite often shows color banding due to shearing, variations in cementation, or incomplete homogenization of the fault gouge. Microscopically, cataclastic textures can be recognized. Megascopically, they can be recognized only when the mylonite grades into less completely granulated material such as crush breccias or rocks with flaser structure.

In the field, hornfels and mylonites are easily distinguished by their occurrence. Hornfels occurs in immediate contact with intrusive igneous bodies and grades away from the contact into unmodified wall rock. There is commonly an intermediate zone between hornfels and country rock which is usually a spotted slate, described below. Mylonite occurs as sheetlike bodies which pinch and swell along the strike. They often grade laterally into crush breccias or rocks with flaser structure and vary in character along the strike as the fault passes from one stratigraphic unit into another. Mylonite does not require an associated intrusive igneous rock, as does hornfels.

SPOTTED SLATES

The spotted slates are a characteristic feature of contact-metamorphic aureoles around large intrusive bodies. They are typically developed from argillaceous rocks and grade toward the intrusive into hornfels. They consist of an aphanitic groundmass of slaty, phyllitic, or massive character through which are scattered oval spots of contrasting color or porphyroblasts of characteristic thermal-metamorphic minerals, such as andalusite or cordierite. Foliation is usually not as perfectly developed as it is in slate or phyllite, and it is usually a texture inherited from the parent rock.

The groundmass of spotted slate is generally dark and aphanitic. Textural, structural, and mineralogical features are similar to those described for slate and phyllite, except that lineation is rare and foliation much less well developed. The planar features, such as bedding laminations, fissility, and cleavage, are parallel to the structures of the parent rock. Parallel orientations of micaceous flakes in the parent rock control the orientation of the new metamorphic minerals as they grow. Foliation in the matrix is deformed by the growth of porphyroblasts, particularly andalusite, so that the foliation flows around the porphyroblasts.

The spots in spotted slates are of two basic types: (1) diffuse oval spots of somewhat different colors but lacking any recognizable porphyroblast, and (2) spots caused by distinct identifiable porphyroblasts (Fig. 7-11). The first type is due to localized incipient recrystallization of the rock, causing segregation of some constituent as microcrystalline aggregates. Segregations and recrystallization of graphite, iron oxides, or chlorite are the common causes but are recognizable only microscopically. Segregation of graphite develops in black shale and yields dark spots in a lighter groundmass. Recrystallization of limonite or hematite in shale to magnetite in the spotted slate results in bleaching of the recrystallized zones. Minor amounts of minute flakes of hematite or limonite have a very strong pigmenting effect, whereas the same amount of iron as minute magnetite granules has little pigmenting effect. The result is that at the first stage of recrystallization, the areas of magnetite granules are light-colored in a red or brown groundmass. At a later stage, after all iron has been converted to magnetite, the effect is slightly darker spots containing magnetite in a light-colored matrix. Incipient chlorite recrystallization yields green spots.

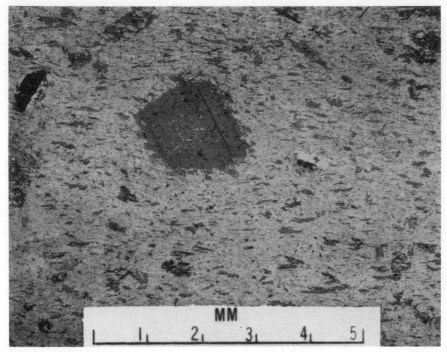

Fig. 7-11. Spotted slate with ragged porphyroblasts of biotite. Waites River formation, East Calais, Vermont. Projection of a thin section. (*Courtesy Dr. R. H. Konig.*)

Mineralogy of Porphyroblasts

Andalusite. The commonest mineral to occur as porphyroblasts in spotted slate is andalusite. At an early stage of its development, andalusite appears as ovoid spots clouded with inclusions. These can be recognized megascopically by their poorly developed cleavage across the entire spot. With further development, andalusite grows into well-defined prismatic crystals having the characteristic dark maltese cross of chiastolite. This results from the development of a strong growth anisotropy and a high crystalloblastic strength so that the bulk of the crystal is able to clear itself of inclusions which cannot be incorporated in the structure of andalusite. The cleared portions of the crystal become white, light gray, or yellowish. The dark cross represents those portions of the crystal having low crystalloblastic strength in which the foreign inclusions accumulate. Andalusite prisms are normally completely random in their orientation in the slate or phyllite matrix (Fig. 7-12).

Andalusite often alters to muscovite or sericite during retrograde metamorphism. The result is a prismatic or cigar-shaped aggregate of micaceous flakes oriented perpendicular to the margins of the mass. Relict

Fig. 7-12. Spotted phyllite containing andalusite porphyroblasts. New Hampshire.

granules of yellow or pink andalusite may be present in the middle of the mica aggregate.

Cordierite. Cordierite also develops porphyroblasts initially as ovoid spots clouded with inclusions. These can be distinguished from the incipient crystallization spots by their weathering characteristics. The cordierite ovoids weather less readily than the matrix, resulting in knots on the weathered surface. Unlike andalusite, cordierite does not develop euhedral crystals nor does it clear itself of inclusions as the crystals grow. Rather, as the crystals grow in size they remain ovoid and strongly poikiloblastic. The inclusions may become large enough to be recognizable granules of quartz and flakes of micas and graphite. Large cordierite porphyroblasts may show their characteristic blue color, but it is usually masked by the inclusions and the crystals are at best a bluish gray.

SCHISTS

Schists are rocks composed of dominantly megascopically identifiable mineral grains developed as a result of metamorphism and showing well-developed foliation dominantly as a result of the subparallel orientation of tabular, flaky, or prismatic crystals. Micas, chlorite, talc, and graphite are the dominant flaky minerals, and the amphiboles are the dominant prismatic minerals of schists. Quartz is commonly present, often making up the bulk of the rock. Important accessory minerals, which are indicative of the bulk composition of the parent rock and are critical for the determination of intensity of metamorphism, are chloritoid, garnet, staurolite, kyanite, sillimanite, epidote, albite, other plagioclases, and potash feldspars. Many other minerals may be present in identifiable grains, including iron oxides, iron sulfides, carbonates, tourmaline, and corundum.

Parallelism of the micaceous or prismatic crystals is the dominant textural feature. Superimposed on this is very commonly a compositional banding, giving a foliated structure. Banding usually consists of alternations of quartz-rich and mica-rich zones. The accessory minerals may be concentrated in either. Bands may be continuous and uniform over distances of many feet, but they are more commonly lenticular to indistinct. Most commonly, they do not represent relict bedding but are the products of metamorphic differentiation. Lineation may also be superimposed on the schistosity. The linear element may be the result of parallelism of elongated crystals, parallelism of small fold axes, or parallel elongation of segregated bodies. Linear texture due to elongated crystals is not as common in schist as it is in phyllite, particularly as the schist becomes coarser-grained. As prismatic minerals, such as tourmaline, staurolite, and hornblende, begin to appear as small crystals, they may show good lineation, but as they grow larger, they commonly lose linear orientation. This may be related to increase of plastic behavior of the rock mass due to recrystallization. The mechanical folding, which is very common in phyllite, continues to be important in schist, particularly in the biotite and garnet zones of metamorphism. At higher intensity of metamorphism the folds are lost, beginning with the smallest-scale crinkling, because of the increased mobility of the minerals through recrystallization. Segregation of the more mobile phases into the axes of folds and thinning of these segregations on the limbs of folds result in the development of distinct rodding structures and simultaneous loss of banding and, therefore, recognizable folds. In schist of high grades of metamorphism, the only folds recognizable are many centimeters in amplitude and are therefore lost on a hand specimen.

Porphyroblasts are characteristic of most schists. The minerals which form the porphyroblasts may be either the normal metamorphic minerals characteristic of a specific pressure-temperature environment for a particular bulk composition or the product of metasomatic introduction of material into the recrystallizing rock. The porphyroblasts are large relative to their matrix as a result of their high crystalloblastic strength in the particular environment, but they may or may not form euhedral crystals. Euhedralism depends on the little-known factor of form energy plus the mechanical history of the particular crystals, as discussed with respect to garnet (see p. 440). Augen structures are commonly developed with the porphyroblasts.

Mineralogically, the schists may represent the entire range of the metamorphic environment. However, as the intensity of metamorphism increases, the proportion of micaceous minerals generally decreases as they yield more equidimensional minerals. Therefore schistosity decreases with increasing metamorphism and is generally replaced by gneissosity. Schistose textures may persist, however, in rocks of exceptional composition in which the equidimensional minerals cannot develop. An example is a quartz-sillimanite schist developed from a silica- and alumina-rich rock deficient in potash so that feldspar cannot form. Mineralogically, the schists can be divided into three groups: those derived from argillaceous rocks, those derived from calcareous rocks, and those derived from intermediate to mafic igneous rocks. The mineralogy of these three groups is discussed below.

Schists Derived from Argillaceous Rocks

Muscovite. Recognizable muscovite ranges from the dominant to a minor mineral in these schists. The porportion of muscovite depends on the amount of available potash, the ratio of iron plus magnesia to alumina, and the grade of metamorphism. Typical assemblages in which muscovite is important to dominant are:

> Muscovite-quartz $\pm$ feldspar
> Muscovite-chlorite-quartz
> Muscovite-chloritoid-quartz $\pm$ chlorite
> Muscovite-biotite-quartz $\pm$ chlorite
> Muscovite-biotite-almandite-quartz $\pm$ feldspar
> Muscovite-biotite-almandite-staurolite-quartz $\pm$ kyanite $\pm$ feldspar
> Muscovite-biotite-sillimanite-feldspar-quartz-almandite

Muscovite generally occurs as highly irregularly shaped thin flakes which may be curved, bent, or even distinctly folded. The flakes are irregular in outline, rarely showing any recognizable development of faces in the prism zone. Frequently, there is sufficient interleaving or overlapping of the flakes so that the limits of one crystal cannot be clearly distinguished from its neighbors. Very perfect parallel orientation of the flakes is characteristic through the low and middle grades of metamorphism. At higher grades, where muscovite becomes less important owing to its breakdown to form feldspar and other minerals, the flakes tend to become less perfectly alined, but individual crystals become more distinct. Muscovite is often segregated along with other micaceous minerals into distinct bands which alternate with quartz-rich bands. Structural relationships indicate that the micaceous material, including muscovite, behaved as a residual chemically immobile phase of the rock.

Muscovite develops originally from either the recrystallization of the potash-bearing clay minerals or the breakdown of primary or detrital feldspars. It is involved in many of the subsequent reactions in the rock. Therefore its proportion in the rock may change markedly. Some of the important relationships are:

Muscovite + chlorite → biotite + chlorite
Muscovite + biotite → almandite + potash feldspar
Muscovite + chloritoid → staurolite + biotite
Muscovite + quartz → kyanite + potash feldspar
Muscovite + quartz → sillimanite + potash feldspar
Muscovite + garnet → biotite + sillimanite + quartz

Muscovite, or its fine-grained equivalent sericite, is a common product of retrograde metamorphism, and as a consequence, muscovite replaces or at least coats many of the high-grade alumina-rich minerals, such as andalusite, kyanite, and corundum. Sericitization of feldspars is also a characteristic retrograde product.

Chlorite. Chlorite is not normally recognizable as large crystals nor does it occur in great abundance in these rocks. The presence of abundant chlorite is generally indicative of mafic igneous-rock or abnormal sediment compositions. In typical metamorphosed clay shale, chlorite is fine-grained when abundant and minor in amount when present as coarse crystals. Typical assemblages derived from argillaceous sediments are:

Chlorite-sericite-quartz phyllite
Chlorite-muscovite-quartz
Chlorite-chloritoid-quartz ± muscovite
Chlorite-muscovite-biotite-quartz ± garnet

The assemblages in which chlorite is dominant are typically developed from mafic igneous rocks. These include:

Chlorite-albite-epidote-quartz
Chlorite-albite-actinolite-quartz ± biotite

and often chlorite is the only megascopically recognizable mineral.

In typical low-grade metamorphites derived from sediments, chlorite occurs as minute scales and flakes which are barely recognizable. As metamorphic environment increases in intensity, the chlorite develops as recognizable flakes, usually in one of two habits. The major proportion of chlorite will occur as small highly irregular flakes alined parallel to the foliation and interleaved with muscovite. A less abundant habit of chlorite is as a larger lenticular body of coarsely crystalline chlorite in which the flakes are approximately perpendicular to the foliation. The lenses of chlorite lie in the foliation plane. As metamorphism increases further, the amount of chlorite decreases rapidly through involvement in the production of new minerals. It is often still present with biotite and occasionally still present as a primary mineral with almandite.

In some sediments of abnormal composition, chlorite is a dominant mineral through a wide range of metamorphic environments. In sediments rich in iron and alumina, poor in lime, and very poor in potash, the characteristic assemblage is chlorite with chloritoid or almandite or both. In such rocks chlorite can develop as large crystals, often having rather random orientation. The chlorite persists in these rocks into the staurolite zone but diminishes in amount.

Chlorite is involved in two important reactions, each of which diminishes the amount of chlorite left in the rock:

Chlorite + sericite → biotite + a different chlorite
Chlorite → almandite

Chlorite is also an important product of retrograde metamorphism in these rocks. Chlorite partially or completely replaces biotite, garnet, and hornblende. Chloritization of biotite is generally recognizable as the loss of luster and the development of green color in the biotite. Chloritization of almandite usually involves merely a peripheral skin, but occasionally, chlorite pseudomorphs after garnet develop. In the Michigan examples of chloritized garnets, the chlorite forms a network roughly following the dodecahedral parting in the garnet, leaving large numbers of small residual masses of garnet.

Biotite. Recognizable biotite is an important constituent of most schists derived from argillaceous rocks. Once it appears, it persists to high grades and is important as long as the rock can be classified as a schist. It decreases in amount at high grades of metamorphism because of reactions converting it to other minerals. Typical assemblages involving biotite are:

Biotite-muscovite-quartz ± feldspar
Biotite-muscovite-almandite-quartz ± feldspar
Biotite-staurolite-quartz
Biotite-muscovite-garnet-staurolite-quartz
Biotite-garnet-kyanite-feldspar-quartz ± staurolite
Biotite-feldspar-sillimanite-quartz-garnet
Biotite-andalusite-muscovite-quartz

Biotite generally occurs in one of three typical habits. The most common form is as highly irregular flakes interleaved with a lesser amount of muscovite and having very well-developed parallelism. The margins of individual flakes are often difficult to define because of the interleaving of adjacent crystals. The flakes are commonly bent or curved either because of mechanical folding or because of the growth of adjacent porphyroblasts. The flakes commonly grow to a diameter of several millimeters but are characteristically thin in comparison with their diameter. A second, much less common, biotite habit associated with the first is as lenticular pods oriented parallel to the schistosity but having the cleavage oriented across the lenses (Fig. 7-13). The biotite occurs as relatively coarse, well-defined "books" which are often as thick as the pod is wide. Thus the pod appears to be composed of several roughly equidimensional blocky books of biotite which are oriented randomly within the pod. The lenses are usually a few millimeters thick and have a diameter 5 to 10 times their thickness. The third typical biotite habit occurs most commonly in schists which are relatively poor in biotite and is as euhedral plates, often alined to give a linear texture. Typical crystals have the proportions $5a:10b:1-3c$, although much less distinct elongation is common (Fig. 7-7). The flakes may show a well-defined elongated hexagonal outline, but more commonly they are somewhat frayed on the edges. When small, these crystals can be easily confused with other minerals, such as staurolite, tourmaline, or hornblende, and should be checked with a point for cleavage and hardness.

In general as metamorphism increases, biotite crystals become more distinctly defined and larger sized. The crystals are still most commonly

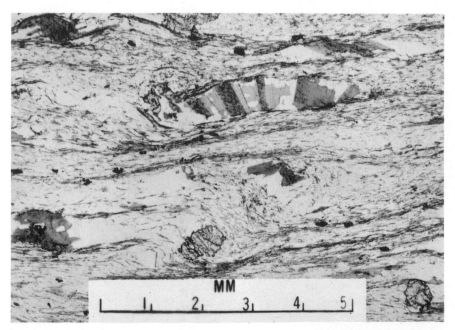

Fig. 7-13. Lens-shaped pod of biotite in a garnet-mica schist. Charlemont, Massachusetts. Projection of a thin section. (*Ward's American Rock Collection.*)

anhedral flakes with ragged edges, but there is less interleafing of flakes. The individual flakes may be curved or bent by either mechanical folding or the thrust of growing adjacent porphyroblasts. At high grades of metamorphism, biotite becomes less important quantitatively as it is converted into other minerals. However, biotite remains the dominant micaceous mineral in argillaceous rocks, and as it decreases in amount, the schistosity of the rock decreases and banded gneissose structures become dominant.

Biotite is involved in two reactions at the high grades of metamorphism:

Biotite + muscovite + quartz → potash feldspar + garnet
Biotite → potash feldspar + hypersthene

It is also probably involved in other reactions, yielding sillimanite and potash feldspar, because there is a common inverse proportion of biotite and sillimanite at high grades. During retrograde metamorphism, biotite is commonly altered to chlorite. The alteration is recognizable as the loss of luster and the development of a distinctly green color. Biotite itself is not commonly produced by retrograde effects.

Chloritoid. Megascopically recognizable chloritoid crystals are commonly developed in the early stages of metamorphism of argillaceous rocks which are abnormally rich in iron and alumina and poor in potash. It is thus indicative of sediments which have received an important contribution of material of lateritic origin. Chloritoid persists into the middle grades of metamorphism, where it is replaced by staurolite. The common mineral assemblages including chloritoid are:

Chloritoid-chlorite ± sericite
Chloritoid-sericite ± pyrophyllite
Chloritoid-chlorite-biotite
Chloritoid-muscovite-almandite ± chlorite

Each may have more or less quartz, and kyanite may occur in a few assemblages.

Chloritoid appears in well-developed euhedral crystals which range in size from a fraction of an inch to several centimeters. Most commonly, they are small dark-green to nearly black tabular crystals with a crude hexagonal outline and a good (but not micaceous) cleavage parallel to the flakes. The orientation of the chloritoids may be quite random with respect to schistosity, often resulting in knots in the chlorite and muscovite as a result of porphyroblastic growth of the chloritoid. Augen structures on a small scale may result. Large chloritoid crystals are usually distinctly poikiloblastic with minute inclusions of quartz or micas, resulting in a luster duller than normal and a cleavage poorer than normal. Rosettes of chloritoid plates are common in some chloritoid sericite schists. The individual plates are a few millimeters in diameter and are arranged in an overlapping petal pattern. This habit is characteristic of the manganiferous variety ottrelite.

In the staurolite zone, chloritoid becomes unstable and yields staurolite. Often the two minerals coexist through a narrow zone. Chloritoid in this zone decreases rapidly in amount and becomes rounded to irregular in shape rather than assuming its typical euhedral tablet shape. The important reactions are:

Chloritoid + muscovite → staurolite + biotite
Chloritoid + chlorite → staurolite + garnet

Neither of these reactions involves direct replacement of chloritoid by staurolite but, instead, involves the direct growth of new crystals. Chloritoid accommodates more iron than does staurolite, but neither accommodates much magnesium; therefore iron liberated in the breakdown of chloritoid is accommodated in compositional adjustments of other ferromagnesian minerals or in the formation of biotite from muscovite.

Garnet. The most widespread indicator of the middle grades of metamorphism is the appearance of an iron-rich garnet, almandite. The manganese garnet spessartite appears in the lower grades, however. As metamorphism increases in grade, garnet can accommodate more of the magnesium molecule, resulting in pyrope. Thus the sequence of garnets of argillaceous rocks ranges from spessartite-almandite solid solutions to almandite to almandite-pyrope solid solutions. The determination of the particular garnet composition requires accurate measurements of index of refraction, density, and unit-cell dimensions and is therefore outside the capabilities of megascopic study. Typical assemblages involving garnet are:

Manganiferous garnet-sericite-chlorite-quartz
Almandite-biotite-muscovite-quartz ± chlorite
Almandite-chloritoid-muscovite-quartz ± chlorite
Almandite-staurolite-muscovite-biotite-quartz-feldspar
Almandite-kyanite-biotite-quartz-feldspar ± muscovite
Almandite-sillimanite-quartz-feldspar ± biotite

The garnets appear in a wide range of size, abundance, and habit. Most generally, however, the crystals approach an equidimensional shape approximating a sphere (Fig. 7-13). Euhedral crystals are common and have the dominant habits of a dodecahedron, a trapezohedron, or a combination of the two (Fig. 7-14). Most commonly, they are either very numerous and small or fairly widely spaced and large. The abundance and size seem to depend on the environment at the time of nucleation of the garnet. The crystals have a high crystalloblastic strength relative to quartz, micas, and feldspar so that micaceous minerals are commonly physically shoved aside by the growing garnet, and augen structures often result.

Fig. 7-14. Large garnet prophyroblast in mica shist. Minas Gerais, Brazil.

The garnets of fine-grained schists are usually small, euhedral, abundant, and often surrounded by a narrow bleached zone. This is particularly common where the quartz-mica groundmass is almost phyllitic. Here a 2-mm garnet may show a surrounding zone $\frac{1}{2}$ mm wide in which there is much less dark mica or chlorite than is present further from the garnet. It would appear that these bleached borders represent the zone of rock which supplied the ferromagnesian material required for the garnet's growth.

In coarser garnet-mica-quartz schists there is a common tendency for garnet to occur more abundantly in zones which are enriched in quartz than in zones which are enriched in the micas. It may be that the growth of several garnet crystals in a zone of favorable composition causes the segregation of quartz, the most mobile phase, into coalescing augen structures. At this stage the garnets often show no crystal faces but are irregular equidimensional to somewhat flattened grains. It is not often clear what the cause of the anhedral form is, but in some specimens it can be seen to be the result of mechanical crushing of the crystal by shearing rotation. Trains of included mineral grains can occasionally be seen megascopically in garnet and are commonly visible microscopically (Fig. 7-15). The inclusion trains often make an S or Ƨ pattern and indicate physical rotation of the crystal during its growth. Less commonly, a garnet can be seen to have had its edges granulated during rotation so that the fragments extend out from the crystal core in a similar S or Ƨ pattern.

Fig. 7-15. Garnet porphyroblasts with curved trains of opaque inclusions. The groundmass is quartz and muscovite.

In the metamorphic rocks the spessartite-rich garnets tend to be dark red to black. On a weathered surface they are almost always black due to surficial oxidation of manganese. In contrast, the almandite-rich and almandite-pyrope solid-solution garnets are consistently red to red brown, and small crystals or chips are generally a clear pink. Pyrope-rich garnets are typically purplish. Poikiloblastic inclusions of quartz, mica, and graphite are common and tend to darken the color. The garnets low in spessartite are mechanically and chemically resistant so that they stand out in relief on a weathered surface and are rarely discolored by oxidation.

Garnet is rarely developed in retrograde metamorphism but is often replaced more or less completely by chlorite. In the Michigan examples the chlorite has developed along the margins of the garnet crystals and along the dodecahedral parting directions so that remnants of garnet are embedded in a network of chlorite.

Staurolite. Staurolite is a readily recognized mineral developed in the upper middle grades of metamorphism of iron- and alumina-rich, potash-poor argillaceous rocks. It has a high crystalloblastic strength and therefore usually develops as large readily recognized crystals. Two basic associations are common:

Staurolite-biotite-quartz ± garnet
Staurolite-muscovite-biotite-garnet-quartz ± kyanite ± feldspar

The first assemblage is the most familiar and consists of a fine-grained schistose groundmass composed dominantly of quartz and biotite, as megascopically recognizable crystals, in which there are large porphyroblasts of staurolite as euhedral simple or twinned crystals (Fig. 7-16). The staurolite crystals are often randomly oriented with respect to schistosity, and the micaceous groundmass flows around the staurolite, forming knots. The staurolite itself is generally crowded with small inclusions of quartz and biotite so that the crystals are rather dull and lusterless, and the cleavage is difficult to observe. As a consequence, on a freshly broken surface the staurolite crystals can be easily overlooked, except for the knotlike mass of the groundmass. On the weathered surface of these rocks, however, the staurolite crystals stand out in strong relief, since the groundmass has been weathered away. The well-developed euhedral form, with its characteristic twinning, is then very apparent.

In the second assemblage, including rather abundant muscovite, the staurolite shows a very different appearance, in that it is free of inclusions (Fig. 7-17). Staurolite then has its true appearance as highly vitreous-lustered, translucent, yellow-brown, hard, bladed crystals. In this form it is easily confused with biotite, hornblende, or tourmaline, unless the hardness, cleavage, and crystal outlines are checked.

Fig. 7-16. Staurolite with quartz and biotite in schist. Little Falls, Minnesota.

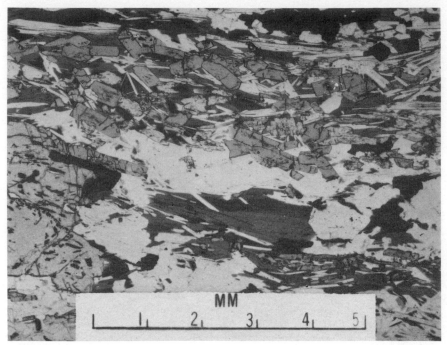

Fig. 7-17. Staurolite-garnet schist. Staurolite occurs as small well-formed crystals in a zone through the middle of the figure. The large white crystals are plagioclase. The opaque plates are graphite. Manhattan Island, New York. Projection of a thin section.

Staurolite is developed in those rocks which at a lower metamorphic grade included chloritoid. Staurolite accommodates less iron than does chloritoid, and therefore, with the transition of chloritoid to staurolite, other iron-bearing minerals are produced (Fig. 7-18). Staurolite is stable over a relatively limited range of metamorphic environments, and at high grades it breaks down to form almandite and kyanite or sillimanite. Kyanite associated with staurolite is not normally of this origin but is the product of the breakdown of muscovite to form kyanite and orthoclase. As micaceous minerals decrease in this manner, staurolite-bearing rocks may become gneissose rather than schistose.

Retrograde metamorphism with its associated introduction of potash converts staurolite to a fine-grained aggregate of sericite. Under surface weathering, however, staurolite is stable and therefore stands out in relief on the weathered surface.

Kyanite. Kyanite is the index mineral for recognizing the beginning of high grades of metamorphism in argillaceous rock. It may appear at much lower grades in rocks of unusual composition, but these are relatively rare.

Fig. 7-18. Staurolite developed in the argillaceous beds of a graded sedimentary sequence. The original sediment was deposited in beds that graded from silt to clay. The silt portion has been metamorphosed to a fine-grained micaceous quartzite. The clay portion has been metamorphosed to a coarse staurolite-mica schist. Thus the original size grading of the sediment has been reversed by metamorphism. Byron, Maine.

Kyanite normally develops from the breakdown

> Muscovite → kyanite + potash feldspar

Therefore, with the appearance of much kyanite the texture of the rock is dominated by equidimensional rather than micaceous minerals. Kyanite, then, is relatively minor in true schists and is more characteristic of gneisses. Typical mineral assemblages are:

> Kyanite-muscovite ± quartz ± feldspar
> Kyanite-biotite-garnet-feldspar-quartz ± staurolite

In the kyanite-muscovite assemblage, kyanite occurs as slender prismatic euhedral crystals which may appear almost fibrous. The crystals, even at this size, have a distinct bluish cast. They are commonly very perfectly alined in the foliation of the mica so that the rock has both strong planar and strong linear elements.

In the more common assemblage with biotite, garnet, and often staurolite, the kyanite appears as well-defined euhedral to subhedral prismatic crystals. The terminations of the prisms may be very irregular. The prisms are characterized by ridges and grooves parallel to their length and by visible cross fractures. The distinctive difference of hardness parallel to the length and across the prisms is diagnostic. The crystals are generally light-colored, very nearly white, and most commonly show a bluish cast. Inclusions of other minerals affect the color, making the crystals gray. Kyanite, unlike staurolite, tends to be oriented in the schists with the (100) face in the plane of schistosity. A linear orientation may or may not be noticeable.

In normal metamorphic environments, kyanite is replaced by sillimanite as temperature increases. However, under high pressures kyanite is the stable form even at very high temperatures. Thus kyanite is most commonly indicative of the early stages of high-grade metamorphism and disappears at higher temperatures through inversion to sillimanite. Occasionally, cores of kyanite surrounded by rims of sillimanite fibers are found. Similarly, sillimanite can be replaced by kyanite through either an increase of pressure or a decrease of temperature.

Kyanite is occasionally altered to a fine-grained aggregate of micaceous materials as a retrograde effect, but generally it is unaffected. In the weathering environment it is stable, and kyanite crystals commonly stand out in relief on weathered surfaces.

Sillimanite. Sillimanite is the index mineral of the highest grade of metamorphism of argillaceous rocks. Generally, by the time sillimanite is abundant, the micaceous minerals have all been converted to equidimensional minerals, and the rock is no longer a schist but, instead, is a gneiss. The reactions involved are:

Muscovite $\rightarrow$ sillimanite + potash feldspar
Biotite $\rightarrow$ almandite + potash feldspar

However, in those rocks which contained kyanite, sillimanite may form by inversion prior to its formation by the breakdown of mica. Therefore we may have schistose assemblages such as:

Sillimanite-muscovite-quartz
Sillimanite-muscovite-biotite-almandite-quartz $\pm$ feldspar

Sillimanite characteristically appears in the form of fibrous crystals. They are usually white or colorless, unless stained, and show a brilliant vitreous luster on the cleavage surfaces. Sillimanite developed from preexisting kyanite generally appears as clumps of fibers more or less perfectly preserving the outline of the kyanite blade, and these fibers are generally parallel to the length of the kyanite blade. Where sillimanite has developed through the breakdown of muscovite, it generally appears as swarms of minute fibers scattered irregularly through the rock. These fibers are often particularly concentrated in quartz segregations, and on weathering, they give such quartz a bone-white fibrous appearance. This habit of sillimanite is much more characteristic of the gneisses than of the sillimanite-bearing schists.

Quartz. Quartz varies widely in proportions in schists: from the dominant mineral to a minor constituent. Some rocks of abnormal composition contain no quartz; metamorphosed bauxite or other residual laterites are the main examples. In a rock of fixed bulk composition, the proportion of quartz in the specimen varies as metamorphism proceeds. This is because silica either is taken up or is liberated in many reactions; as examples:

Muscovite + chlorite A → biotite + chlorite B + *quartz*

Muscovite + biotite + *quartz* → potash feldspar + almandite

Muscovite + *quartz* → potash feldspar + andalusite (sillimanite or kyanite)

Muscovite + hornblende → biotite + anorthite + *quartz*

Chlorite + *quartz* → garnet

Chloritoid → staurolite + *quartz* + FeO

Staurolite + *quartz* → kyanite + almandite

Of course, not all silica required for a given reaction need come from quartz nor all silica yielded in a reaction go to make quartz. Generally, two or more reactions are taking place simultaneously in metamorphism so that the silica required for one reaction may be equal to the silica yielded from a different reaction. Precise balance should be rare, and therefore the effects of migration and recrystallization of quartz should be evident at every stage of metamorphism.

Quartz in metamorphic rocks is most commonly colorless, although it may be green due to disseminated included chlorite flakes, red due to finely divided hematite, milky due to minute fluid inclusions, bone white and silky due to sillimanite fibers, or even black due to included graphite. A true smoky quartz is rare, however. In a hand specimen the apparent color varies according to the minerals which surround it. Thus colorless transparent quartz appears dark when embedded in biotite, and so on. Grain size of quartz varies from silt-sized up to masses several millimeters or even a few centimeters in dimensions. It is often difficult to determine whether the larger masses are composed of a single crystal or are a composite aggregate. The continuity of conchoidal fracture across the mass is commonly the only indication. Quartz is one of the weakest minerals in the crystalloblastic series, and therefore it rarely develops any crystal faces. The crystals tend to be lenticular in shape, with their long axes in the foliation plane. Microscopic analysis of crystal orientation generally shows a preferred alinement of the *c* crystallographic axes. This is, of course, not determinable megascopically. The boundaries between adjacent quartz grains are generally somewhat irregular, but not highly sutured, on a megascopically visible level. Contacts between quartz and the micaceous minerals are often very irregular to hazy because of growth of the micas into the quartz or the filling of quartz into zones between mica lamellae. Minerals of high crystalloblastic strength generally have sharp contacts with quartz and are euhedral against it. An exception is seen in garnets which have been fractured by rotation and have had quartz deposited in the minute fractures.

The segregation of quartz in schistose rocks is common at all grades of metamorphism. Distinct uniform lamellae of quartz-rich material are often seen in phyllite and low-grade schist. This may be a relict of primary bedding. At higher grades, thin quartz-rich lamellae generally disappear, and instead, quartz-rich segregations tend to be distinctly lenticular or rodded. Distinct uniform bands are less common than lenses. Grain size in one lens often varies widely from irregular anhedra several millimeters in diameter to silt-sized grains. Segregation of quartz into augen structures is particularly common, with porphyroblasts of garnet or feldspar. In these structures quartz again is normally anhedral but more nearly equidimensional than lenticular. Grain size varies widely.

Feldspar. The proportion of feldspars in schistose rocks varies widely. Most schists in the low and middle grades of metamorphism contain no megascopically recognizable feldspar, even though feldspar may be present and fairly abundant. At the higher grades of metamorphism, feldspar becomes coarser-grained and more readily recognizable (Fig. 7-17). As it becomes more abundant, the rocks lose their schistose character and become gneissose. Feldspar is often developed as large porphyroblasts as a result of metasomatic introduction of alkalis, and very large feldspars may form augen structures scattered through schistose rocks (Fig. 4-4).

At low grades of metamorphism, albite is the only common stable soda mineral. Paragonite, indistinguishable from sericite, may proxy for albite in phyllite. All plagioclase is in the albite range of composition with very little anorthite in solid solution. Anorthite at this stage is proxied for by epidote. Albite is usually very difficult to recognize megascopically, since crystals are generally very small, anhedral, and untwinned so that the small unstriated cleavage reflections are difficult to find in the micaceous groundmass. The potash feldspar microcline is also stable at low grades but is very rare in metaargillites, because it requires considerably more potash in the rock than is normal. Most potash feldspar visible in rocks of low to middle grades is of metasomatic origin.

In the middle grades of metamorphsim, plagioclase in the range of oligoclase is stable, but any more-sodic plagioclase may occur if there is a paucity of lime in the rock. The reactions

Epidote + kyanite + quartz + albite → plagioclase + water
Epidote + muscovite + quartz + albite → plagioclase + microcline
+ water

become possible, and they proceed as far as the allowable plagioclase composition for the particular P and T will allow. Thus the association of plagioclase and epidote is common. Again potash feldspar is stable but is relatively rare, except where there has been metasomatic introduction.

At high grades of metamorphism, any composition of plagioclase is stable, and the exact composition will depend on the ratio of soda to available lime. Epidote can no longer coexist with plagioclase. Potash feldspar now becomes the stable potash mineral as the micas become unstable:

> Muscovite + quartz → potash feldspar + kyanite (or sillimanite)
> Biotite + muscovite + quartz → garnet + potash feldspar
> Biotite + sillimanite + quartz → potash feldspar + garnet

These are progressive reactions so that as the potash feldspar, usually orthoclase, becomes more abundant and coarser-grained, micas become less abundant. The rock through these reactions becomes gneissose rather than schistose.

The plagioclases are generally very light-colored or white. Since they have low crystalloblastic strength, the crystals are rounded to irregular, usually roughly equidimensional, and commonly show no preferred orientation. Polysynthetic twinning is the exception rather than the rule, as it is in igneous rocks. This is probably due to continual recrystallization and modification of composition as metamorphism advances. In the higher grades twinning becomes more common, probably as a result of continued minor shearing during retrograde processes. Large plagioclase porphyroblasts are almost always indicative of soda metasomatism. Large white ovoid albite porphyroblasts having more or less well-developed augen structures develop in this environment. Available soda in migrating solutions increases the crystalloblastic strength of sodic plagioclase, but the form energy remains low. Therefore the crystals can grow by concretion, shoving aside wall rock but remaining anhedral. In the absence of twinning, it is best to call such porphyroblasts alkali feldspars, thus making no interpretation as to the sodic or potassic nature of the mineral. Optical checks will, of course, allow more definite determination.

Potash feldspars are usually microcline through the low and middle grades, when present, and orthoclase at the high grades. Potash feldspars are rare or minor in the low and medium grades, except where potash has been introduced metasomatically. Then the feldspars appear as large oval porphyroblasts, very commonly with augen structure. The crystals are normally white or pink and in random orientation. Twinning of any type, including the Carlsbad type, is rare, apparently because continual recrystallization heals even growth types of twins. Perthitic intergrowths are common and are often megascopically visible. As with plagioclase, potash feldspar under closed-system metamorphism has low crystalloblastic strength and forms small anhedra. In a metasomatic environment having introduced potash, the potash feldspars have a high crystalloblastic strength and grow by concretion, shoving aside the matrix. Feldspar augen may be either randomly scattered through the rock or concentrated along selected bands. The growth along specific bands ultimately develops a banded gneiss.

Amphiboles. Amphiboles are relatively minor minerals in schists derived from argillaceous rocks. However, hornblende may appear at the medium grades if the original shale was somewhat calcareous and rich in ferromagnesian constituents. In the absence of the ferromagnesian materials, epidote is the lime-bearing mineral of low grades and plagioclase of the upper grades. Tremolite and actinolite are very uncommon in such highly aluminous rocks. Hornblende appears characteristically as elongated prismatic crystals, usually with a well-developed set of prism-zone faces but irregular terminations. The crystals are almost always well oriented in schists, with the axis of the prism zone lying in the foliation. A linear alinement may be superimposed on the planar alinement. The hornblende generally appears black under the hand lens, showing excellent cleavage surfaces which usually have a fibrous appearance due to the intersection of the cleavages. In small grains the hornblende can easily be confused with biotite, inclusion-free staurolite, or tourmaline. The greater hardness and the fact that it yields splinters rather than flakes under a probe distinguishes it from biotite. Inclusion-free staurolite is a translucent brown under strong illumination and lacks prismatic cleavage, whereas hornblende lacks translucency and appears black, greenish black, or brownish black. Crystal habit and lack of cleavage distinguish tourmaline from hornblende.

Epidote group. Epidote and zoisite are the common lime-aluminum silicates of the low and medium grades of metamorphism. At higher grades they break down to yield the anorthite molecule for plagioclase. Epidote or zoisite form early in metamorphism by the reaction of calcite and clay minerals or by the breakdown of detrital plagioclase. Epidote requires some iron, whereas zoisite is iron-free. In most schists derived from argillaceous rocks, these minerals are minor and occur as small granules which are easily overlooked. They have relatively high crystalloblastic strength, however, so that they shove micas, quartz, and feldspars aside. They do not have high form energy, and therefore they form granules rather than euhedral prismatic crystals. Epidote is commonly light yellow-green to moderately dark green; zoisite is commonly white, gray, or yellowish. Zoisite is actually more common in calcareous argillaceous rocks than is epidote, but because of its lack of a distinctive color and its small size, it is usually overlooked. Minor calcite may be associated with the epidote group in the mica-garnet schists of low to low-middle grades and with minor hornblende in the middle grades of metamorphism.

Tourmaline. Normally a minor accessory mineral, tourmaline may become quite abundant when metamorphism has involved metasomatic introduction of boron. Tourmaline occurs as small black prismatic crystals of high crystalloblastic strength and high form energy. This results in their being euhedral, usually with their terminal faces well developed. The trigonal prism dominates their habit and is modified by the hexagonal prism, giving a very characteristic rounded triangular cross section. The crystals are usually oriented to lie in the foliation plane and may also be linearly oriented in that plane. Boron metasomatism results in the development of large abundant crystals which lie in radial clusters in the foliation plane (Figs. 7-19 and 4-5). These are generally associated with quartz-muscovite-rich rocks in which all the ferromagnesian material has been taken up by the tourmaline. These crystals are often markedly flattened in the foliation plane, destroying the characteristic rounded triangular cross section, but their distinctive habit, association, and lack of cleavage are diagnostic.

Graphite. Graphite is not abundant in rocks which are true coarsely crystalline schists. When the rock contains much carbon, it tends to carry a phyllitic texture into the high grades of metamorphism when it becomes a graphitic gneiss. Few graphitic rocks contain more than 10 percent graphite, but as little as 2 to 3 percent will markedly affect the rock. Most commonly, graphite occurs as films coating other micaceous minerals or as discrete minute flakes. The coarser the graphite, the less marked is its effect on the rock.

Fig. 7-19. Tourmaline-mica schist. The tourmaline crystals lie in one plane but are random in that plane. Near Custer, South Dakota. (*Ward's Universal Rock Collection.*)

Other accessory minerals. Many other accessory minerals occur in schist but usually in minor amounts and rarely in megascopically recognizable crystals. Zircon and apatite are common detrital minerals which are stable and recrystallize during metamorphism. They are rarely of megascopic size. The titanium minerals are common, but only sphene is ordinarily recognizable. Rutile is abundant in some shale as minute hairlike crystals derived from the breakdown of titaniferous ferromagnesian minerals such as biotite and augite. Much of this titanium is taken back up in silicates during metamorphism in the early stages, but sphene may appear at high grades. The crystals are euhedral wedge shapes and are usually brown, but yellow and green are not uncommon.

Iron oxides and sulfides are common in many schists. One of the first noticeable changes in metamorphism is the recrystallization of all iron oxides, with a consequent drastic change of color. Magnetite may be present at any grade of metamorphism, but generally it is minor in the metaargillites, since most of the iron is taken up in ferromagnesian silicates. Magnetite often appears as very perfect octahedra, especially in chlorite-rich schist. The crystals are often oxidized to hematite, giving the variety martite. Pyrite is also common in the low-grade rocks, usually forming euhedral cubes which cut across the foliation indiscriminately. Pyrite is also common in the graphitic rocks, usually occurring as lenses, films, and finely crystalline aggregates rather than as large euhedra. At higher grades of metamorphism, pyrite is replaced by pyrrhotite, usually as lenses and films.

Schists Derived from Calcareous Rocks (Calc-schists)

Highly schistose rocks in which calcite is a dominant to important mineral are relatively uncommon. They represent the metamorphic equivalent of highly calcareous shale and argillaceous limestone and dolomite. The character of the clay-mineral impurity is dominant in determining whether micaceous minerals, and therefore a schistose texture, can develop. Kaolinitic or montmorillonitic clays are very low in potash and therefore do not result in the development of micas. Illitic clays, in contrast, are high in potash, which enhances the development of micas. In the eugeosynclinal environment the montmorillonitic clays tend to be dominant because of the abundance of mafic and intermediate volcanics. It is these rocks which tend to be most intensively deformed and metamorphosed. Therefore the calc-schists are uncommon rocks, and in their place, more massive rocks containing calcite, pyroxenes, amphiboles, plagioclase, and epidote appear.

Calcite. True calc-schists contain abundant calcite associated with muscovite or sericite, biotite, graphite, and minor amounts of epidote, tremolite or actinolite, diopside, lime garnets, and vesuvianite. The calcite usually varies in grain size from silt particles to grains several millimeters in maximum dimensions. The coarser grains are usually lenticular in shape and flattened parallel to the schistosity. In these grains twinning is either very abundant or rare. The twinning is apparently the product of deformational gliding. If recrystallization continues after deformation is complete, then the twinning is removed and only a few twin lamellae are present. In contrast, if deformation persists after recrystallization could no longer take place, then curved or bent lamellae become abundant. Twin lamellae in calcite are parallel to the long diagonal of the cleavage rhomb; that is, they bisect the acute angle between the cleavage cracks. This distinguishes calcite from dolomite.

The borders of calcite grains are generally highly irregular. Most of the other minerals present have a higher crystalloblastic strength and are euhedral against calcite. Fine-grained calcite often appears to be the product of granulation of coarser calcite crystals. Often this results in alternating bands of fine-grained granulated material and coarse crystalline material. Larger grains embedded in the fine-grained matrix show granulation of their borders.

Dolomite. Dolomite is less common in calc-schists than in calcite, because dolomite reacts readily at an early stage with any available silica to form tremolite. However, when all silica is tied up in stable silicates, such as the micas, dolomite appears as relatively euhedral crystals, generally of rounded rhombohedral form (Fig. 7-20). On a weathered surface these rhombs stand out in relief. The dolomite crystals tend to be much more uniform in size than the associated calcite and have simple borders, whereas calcite grains tend to have irregular to highly sutured borders. Twinning in dolomite is less common than in calcite and is parallel to a different direction. The result is that polysynthetic-twin lamellae in dolomite are parallel to both the long and short diagonals of the rhomb rather than to just the long diagonal. Thus on a cleavage surface of dolomite, the striations bisect both the acute and obtuse angles formed by the visible cleavage cracks.

Micas. Muscovite, or its fine-grained equivalent sericite, is the common micaceous mineral of calc-schist. Chlorite and biotite are less common. Muscovite forms silvery shreds and flakes which have rather indefinite borders when abundant. They may be either preferentially concentrated in thin layers or films or uniformly distributed through the rock. Biotite or phlogopite generally form more distinct flakes and may show relatively euhedral outlines against the carbonates. Generally, the flakes are minute and uniformly distributed through the rock. A golden-brown or bronzy color is more common than in the biotites of common schists. Chlorite generally is minor and anhedral and appears as greenish to greenish-brown flakes.

Amphiboles. Tremolite is a very common constituent of calc-schist and is often a dominant mineral in displaying the schistosity. The presence of

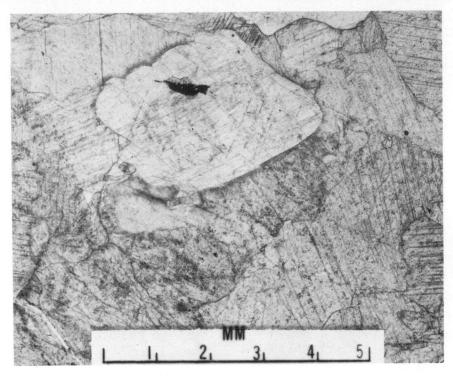

Fig. 7-20. Rounded dolomite rhomb in calcite marble. Gouverneur, New York. Projection of a thin section.

tremolite is indicative of an initial moderate- to high-magnesia content. Thus it develops in calc-schist derived from siliceous-dolomitic limestone or dolomite or from argillaceous limestone in which the initial clay was montmorillonitic. When tremolite is abundant, the associated mica is usually muscovite rather than biotite. Tremolite occurs as elongated crystals, ranging from fine fibers to distinctly prismatic crystals. Coarser crystals usually show a rounded lensoid cross section due to vicinal alternation between the prism and the (100) pinacoid so that they are flattened on the *a* axis and elongated on the *c* axis. The color is normally white to pale gray or pale green. If the rock contained appreciable iron, it is usually taken up in the amphibole to make green tremolite or darker-green actinolite. Preferred orientation in the amphiboles is usually good with either planar or planar plus linear elements.

Pyroxene. Minute granules of pale-colored diopside may be abundant in calc-schist which was initially magnesian and has been metamorphosed to a medium grade. Commonly, when diopside is abundant the rock shows very poorly developed schistosity and is massive. However, diopside can coexist with abundant tremolite. Diopside granules are not normally notice-able megascopically if the rock is schistose, but they may weather out in relief or be present in insoluble residues from such rocks.

Epidote group. Epidote and zoisite are common in calc-schist derived from rocks which initially contained more alumina than can be accom-modated in the micas. The group's appearance thus depends on the potash to alumina ratio of the initial clay; epidote is more common in initially dolomitic rocks, whereas zoisite is more common in calcite rock. This is because of the required iron in the epidote composition and its absence in zoisite. Epidote is readily recognized as yellow to yellowish-green granules which are usually roughly equidimensional and anhedral. Zoisite is difficult to recognize megascopically, because it lacks the characteristic color of epidote. It occurs as small elongated gray prismatic crystals which can be easily confused with tremolite. Zoisite has only one cleavage, whereas tremolite has two at 60°.

Quartz. Quartz varies from a dominant to a very minor constituent of these rocks. The quartz may represent recrystallized detrital grains, chert, or silica released by reactions. Usually, it occurs as colorless glassy grains which may be either equidimensional with highly irregular boundaries or lenticular.

Schists Derived from Igneous Rocks

Both silicic and mafic igneous rocks give rise to schistose rocks at low-grade metamorphism. Those derived from silicic rocks are dominated by musco-vite, quartz, potash feldspars, and minor chlorite and biotite. Mafic igneous rocks yield schists dominated either by chlorite, talc, and actinolite, with abundant but inconspicuous albite, epidote, and quartz or by hornblende, plagioclase, quartz, and epidote. Intermediate igneous rocks approximate the composition of average shale very closely, except that they have a higher alkali content. Therefore their mineralogy should be similar to schist derived from shale, except that soda and potash feldspars will appear, and those minerals dependent on a paucity of soda and potash for their appear-ance will be absent.

Chlorite. Chlorite is normally the dominant mineral of schist derived from basaltic flows, pyroclastics, and small intrusives in the low grades of meta-morphism. Chlorite is derived from the original ferromagnesian consti-tuents, and it so dominates the rock that the rocks are called *greenschists* (Fig. 7-2). Typical assemblages in which chlorite is dominant include:

    Chlorite
    Chlorite-talc
    Chlorite-albite-epidote-quartz ± sericite
    Chlorite-actinolite-plagioclase-quartz ± biotite ± talc ± carbonates

The characteristic assemblage is chlorite-albite-epidote-quartz, but commonly only the chlorite is recognizable megascopically.

Chlorite occurs as innumerable minute anhedral flakes (chlorite phyllite) to flakes several millimeters in diameter. The chlorite is commonly dark green, and even thin flakes have a green color in transmitted light. Two different chlorite minerals may be present, both differentiatable on the basis of color. This is particularly common where the original rock was an amygdaloidal basalt and the amygdules were partially filled with chlorite. The metamorphosed product will then show ovoid spots of chlorite having a color contrast to the groundmass. Individual flakes of chlorite are generally thin irregular plates without any noticeable elongation in one direction. They are, however, very well alined to develop a strong planar element. Small-scale crinkling and folding of this schistosity are common and produce a linear element. The deformation of amygdules also produces a linear element, in that the long axes of the deformed bodies are generally parallel. In the coarser chlorite schists, the orientation of chlorite flakes may be more random than in the finer-grained rocks.

Chlorite is involved in the production of new minerals as metamorphism increases. Some important reactions are:

Chlorite + sericite $\rightarrow$ biotite
Chlorite + calcite $\rightarrow$ actinolite + epidote
Chlorite + epidote + quartz $\rightarrow$ actinolite + anorthite
Chlorite + actinolite + epidote $\rightarrow$ hornblende

Thus as amphiboles appear and become more abundant, chlorite disappears.

Chlorite is a common retrograde product from ferromagnesian minerals. Chlorite-rich schists can develop from amphibole- or pyroxene-rich rocks and from such minerals as garnet. The retrograde origin can be recognized only where a characteristic mineral is pseudomorphed by chlorite. Thus chlorite pseudomorphs after garnet, which retain their isometric habit, in a chlorite schist is indicative of retrograde metamorphism, or polymetamorphism.

Sericite and muscovite. In metamorphic equivalents of mafic igneous rocks, sericite is the potash-bearing mineral. However, it is usually so interleaved with chlorite that it is not recognizable megascopically. Occasionally, with the appearance of actinolite in coarse crystals, recognizable muscovite books may be seen. Usually, however, at this stage biotite is the stable potash-bearing mineral.

In silicic igneous rocks, sericite and muscovite are abundant metamorphic products. On metamorphism the potash feldspars break down in part to form muscovite and quartz with an associated loss of potash. The amount of muscovite present measures the degree to which potash was lost.

Typical assemblages are thus:

Sericite-quartz-potash feldspar ± epidote ± albite ± chlorite
Muscovite-quartz-potash feldspars-biotite-albite-epidote
Muscovite-quartz-potash feldspar-biotite-plagioclase

Sericite first appears as thin silvery shreds. If individual flakes are visible, they tend to be anhedral and highly irregular in outline. The sericite, and later the coarser equivalent muscovite, is usually concentrated in thin zones which separate small lensoid masses of quartz and feldspar. The more or less continuous mica films bifurcate around the lenses and rejoin, giving the appearance of an augen structure. At this stage of metamorphism these are not true augen, because they are not the products of crystalloblastic growth of the feldspars but are, rather, flaser structures and the result of dynamic action.

At higher grades, sericite and muscovite are involved in reactions which result in their becoming less abundant. In the mafic igneous rocks, sericite reacts with chlorite to give biotite in reactions similar to those of argillaceous schists. In the silicic igneous rocks, muscovite breaks down to form potash feldspars and an aluminum silicate: kyanite or sillimanite. The extent to which the aluminum silicates form measures the extent to which potash was lost in the initial breakdown of feldspar to sericite.

Talc. Talc is a minor constituent of many of the chlorite-rich schists. It represents the excess magnesia which chlorite cannot accommodate. Generally, the talc appears as fine scales interleaved with chlorite and is not noticeable. However, it may occur in distinct patches as though it had developed from a particular phenocryst of the parent rock. Here it is generally white and finely scaled. As metamorphism increases in intensity so that the amphiboles become important to dominant, talc becomes more readily recognizable as coarsely crystalline books, often interstitial between coarser actinolite prisms, with random orientation of the flakes. It may also occur with well-defined planar alinement contributing to the schistosity. Less commonly, lenticular pods of coarse talc occur with individual talc flakes oriented across the pod.

Talc is restricted to the lower temperatures of metamorphism and, with increasing grade, gives rise to pyroxene and olivine. When this occurs the rock generally loses its schistosity and becomes gneissose or massive.

Amphiboles. Actinolite and hornblende are common products of the upper low grades to middle grade of metamorphism of igneous rocks. Tremolite is less common, because there is usually sufficient iron present to form actinolite or hornblende. Typical assemblages involving amphiboles are:

Actinolite-albite-epidote-chlorite ± quartz ± biotite ± talc
Hornblende-epidote-albite ± quartz ± biotite ± garnet
Hornblende-plagioclase ± quartz ± biotite ± potash feldspar

The amphiboles generally occur as elongated prismatic crystals having poor terminations. They may be completely anhedral, but their good prismatic cleavage generally gives them a subhedral appearance. The terminations are generally highly irregular and adjacent crystals penetrate each other on the ends. The amphibole prisms are generally well alined in a plane, giving a strong planar fabric element (Fig. 7-21). They may be alined with their long dimensions parallel within that plane, giving a superimposed linear fabric element. Occasionally, in rocks which are high in chlorite and talc, actinolite is present as lenticular knots of coarse crystals having random or radiating orientation (Fig. 7-22). In such a situation it appears that the shear stresses on the whole rock were taken up in the chlorite-talc-rich zones so that the actinolite was not subjected to much directional pressure.

Actinolite is generally present in bright-green blades which usually show a rounded lenticular cross section. Good prismatic cleavage at 60° is usually obvious and gives the crystals a fibrous appearance in the finer-grained rocks. A probe yields translucent green needles. The common hornblende is usually very dark green to black. Brown to brownish-black hornblende is more characteristic of higher-grade gneisses or igneous amphiboles. The prisms are usually small and markedly elongated. A fibrous appearance is characteristic as a consequence of this prismatic cleavage. Splinters chipped off with a probe may show a dark-green translucence on the thin edges, but more commonly they are almost black.

At higher grades of metamorphism, hornblende persists as well-alined crystals, but the development of abundant feldspar as segregated bands renders the rock gneissose rather than schistose. Actinolite, however, does

Fig. 7-21. Actinolite blades in chlorite schist. The actinolite blades lie in the plane of schistosity but are random in that plane. Chester, Vermont. (*Ward's Universal Rock Collection.*)

Fig. 7-22. A pod of coarse-bladed actinolite with talc in chlorite schist.

not persist, because with advancing metamorphic grade, it yields hornblende of much greater chemical complexity. At the highest grades, hornblende becomes unstable and yields pyroxene. The pyroxene formed is most commonly an orthorhombic one near hypersthene. The lime and alumina from hornblende combine with quartz to yield calcic plagioclase. Thus at this step the amount of ferromagnesian minerals decreases and plagioclase increases. The reaction may be written

Hornblende + quartz → hypersthene + calcic plagioclase

Epidote group. Epidote and its iron-poor relative zoisite are very common minerals in all schists derived from igneous rocks, because these minerals proxy for the anorthite molecule in plagioclase. Therefore the amount of epidote appearing will vary with the amount and composition of the initial plagioclase. In granitic rocks relatively little epidote will develop, whereas in basaltic or gabbroic parent rocks a major proportion of the epidote minerals will form. The reaction can be written

Anorthite + water → 2zoisite + 2silica + alumina

The alumina released is usually taken up by chlorite or sericite. Typical assemblages including epidote (or zoisite) are:

Epidote-chlorite-albite-quartz
Epidote-chlorite-albite-actinolite
Epidote-hornblende-sodic plagioclase-quartz

Epidote appears in a number of characteristic forms in these assemblages but is often megascopically unrecognizable. In the early stage of metamorphism, before shear has destroyed the original igneous texture, a very fine-grained light-greenish mass of epidote and albite replaces plagioclase. The original crystal outline of the plagioclases may be preserved, but the mass shows no cleavage and often has an earthy luster. Such material is called *saussurite* and was once thought to be a homogeneous single mineral. With increasing shear, the albite-epidote mixture becomes streaked out through the specimen and loses its distinctive characteristics.

In highly schistose rocks, particularly those with an abundance of chlorite, epidote becomes unrecognizable. The grains are minute and so coated with chlorite that individual epidote granules are hidden. Even when chlorite is moderately coarse, the associated epidote is usually difficult to find. Recognizable grains usually first appear with the appearance of hornblende. Then small glassy anhedral granules of yellow-green to greenish-yellow epidote can be seen scattered through the hornblende-rich zones. The grains are usually irregular to rounded and roughly equidimensional. Less commonly, they form somewhat elongated prisms which usually show no preferred orientation.

In the metamorphic equivalents of silicic igneous rocks, epidote is usually less abundant but is more readily recognized, because it is less masked by chlorite. Saussuritization of feldspars is common, and in some rocks there appears to be almost a complete epidotization. This usually indicates metasomatic modification.

With advancing metamorphism, epidote is progressively converted into anorthite molecules which are taken up in solid solution with albite to form plagioclase. The composition of the plagioclase stable is dependent on temperature. Therefore with slow temperature rise, the amount of epidote gradually decreases, and the anorthite content of the plagioclase increases. Epidote generally begins to disappear with the appearance of hornblende. At lower temperatures the stable assemblage is albite-epidote. The range of temperature through which the epidote-plagioclase assemblage is stable then depends on the initial plagioclase composition of the parent rock. Granitic plagioclase, oligoclase, yields a large amount of albite and a relatively smaller amount of epidote. In metamorphism the epidote breaks down as rapidly as it can be taken up in plagioclase. In contrast, plagioclase from mafic igneous rocks, labradorite, yields subequal albite and epidote, and therefore epidote persists longer during metamorphism.

Feldspars. Albite, potash feldspars, and later sodic plagioclase are all common minerals of these schists. More calcic plagioclase may occasionally persist as relicts of original crystals of mafic igneous rocks. The relationships of the feldspars have been detailed above and will be summarized briefly here. Potash feldspar is common in schist derived from silicic igneous rocks, both as relict grains from the parent and as recrystallized material. At higher grades, potash feldspar is produced by the breakdown of muscovite. Albite is the plagioclase of the low grades of metamorphism in association with epidote. More-calcic plagioclase appears progressively as metamorphism increases. The characteristic assemblages are:

Albite-chlorite-epidote
Albite-chlorite-actinolite-epidote
Albite-hornblende-epidote-quartz
Sodic plagioclase-hornblende-quartz
Potash feldspar-albite-epidote-muscovite-quartz-chlorite
Potash feldspar-albite-epidote-muscovite-biotite-quartz
Potash feldspar-sodic plagioclase-muscovite-biotite-quartz
Potash feldspar-sodic plagioclase-sillimanite-biotite-quartz

The first four are characteristic of mafic igneous rocks, whereas the last four are characteristic of silicic igneous rocks. The final assemblage is more likely to be gneissose than schistose.

Albite, even in relatively coarse grains, is difficult to recognize in these rocks owing to a lack of characteristic polysynthetic twinning. Normally, polysynthetic twinning appears to be the result of shearing of the crystals, and therefore it would be expected to be abundant in schistose highly sheared rocks. However, in the metamorphic environment all minerals are undergoing continuous recrystallization. Twinned crystals are less stable than untwinned crystals, and therefore they recrystallize readily to give untwinned grains. Twinning is common only where shear deformation continued while temperature fell. Twinning is thus not a criterion for recognizing plagioclase. In the highly chloritic rocks, albite is usually unidentifiable megascopically. The grains are usually very small and coated with chlorite. The same is generally true for the actinolite-rich rocks. Occasionally, however, larger anhedral porphyroblasts of pale bluish-white albite are abundant. These crystals are equidimensional, rounded, and have irregular boundaries. Augen structures may develop around them as the schistose layers are wedged apart by the growing crystal. The eye corners are usually filled with quartz.

The potash feldspar in schists derived from silicic igneous rocks is usually abundant and readily recognizable. In the early stage of metamorphism it persists as lenses of granulated quartz and feldspar developed by crushing of the parent rock without recrystallization. Cataclastic textures are common in the lenses, with highly irregular remnant fragments of crystals embedded in a fine-grained crushed product from the crystal. Generally, the color of the remnants is the same as the parent, but often the crushed material is much paler or white.

With increasing metamorphism, the potash feldspar begins to recrystallize into anhedra of microcline. Porphyroblasts of considerable size may develop and produce augen structures. This feldspar is commonly pink, but it may be a yellowish-white to white. A blue-white color is usually seen in albite. Porphyroblasts are generally rounded and anhedral, and often several porphyroblasts combine from the nucleus of an augen.

The sodic plagioclases associated with hornblende, and resulting from the instability of epidote, are usually untwinned, white, and anhedral. The grains are small and equidimensional and are usually concentrated in certain bands along with quartz and minor hornblende. Other bands are dominated by parallel-oriented hornblende prisms with minor plagioclase and quartz. The rocks are becoming gneissose rather than schistose. In this situation it is usually impossible to distinguish plagioclase from potash feldspar, but usually rocks which give rise to such an assemblage are low in potash and therefore most of the feldspar is probably plagioclase.

Feldspars persist relatively unchanged into higher grades of metamorphism. More feldspar generally forms at higher grades as a result of the breakdown of muscovite to yield potash feldspar and an aluminum silicate, biotite to give potash feldspar and pyroxene or garnet, or hornblende to give calcic plagioclase and hypersthene. The metaigneous schists thus grade into the metaigneous gneisses through the increasing importance of feldspar and the decreasing importance of micaceous or elongated prismatic minerals.

Quartz. Quartz may be present in a wide range of proportions in these schists. Generally, there is more quartz in the schist than in the parent rock, because many of the minerals formed at low grades of metamorphism are silica-deficient relative to their parent minerals. Chlorite, hornblende, muscovite, and epidote are all relatively silica-poor minerals.

In rocks rich in chlorite or muscovite, quartz may be hidden. Generally, in muscovite-rich rocks the quartz can be found by observing a surface perpendicular to the schistosity. There quartz will appear as abundant thin layers and lenses of silt-sized material. In some muscovite-quartz schists, quartz has recrystallized to a coarser-grained colorless to milky matrix in which the coarse muscovite crystals are embedded. Quartz in the chlorite-rich rocks is rarely recognizable, since it is masked by the chlorite.

In rocks with abundant recognizable feldspar, quartz is usually recognizable. Feldspar porphyroblasts generally have an aggregate of quartz associated with them. The quartz is usually colorless to slightly smoky, anhedral, and has either highly intergrown and sutured borders or smooth curved borders, giving polygonal crystals. Usually, a range of sizes of quartz grains will be present even in a local area, such as in an augen. Quartz in the quartz-feldspar lenses of a flaser structure is generally granulated, showing cataclastic texture. Quartz in the hornblende-plagioclase-quartz assemblage occurs as small equigranular and equidimensional anhedral grains. In these fine sizes it is often difficult to distinguish from feldspar, but the lack of clouding due to alteration and the presence of a conchoidal fracture rather than cleavage distinguish quartz from feldspar.

The proportion of quartz in the rock varies with metamorphic grade for one initial bulk composition. This is because silica released in some of the early reactions is taken back up in some of the higher-grade reactions.

Minor minerals. Many minor minerals occur as recognizable crystals in these schists. Very common minerals are magnetite, pyrite, and rhombohedral carbonates. Others less common are not discussed here.

Magnetite as small lustrous black octahedra is very common in chlorite schists. Generally, the octahedra range from a fraction to a few millimeters in dimensions, are perfectly euhedral, and cut across the schistosity. Oxidation and replacement by hematite are common, giving the material called martite. The octahedral form is perfectly preserved by the hematite. Other spinels may occur in these schists, so a check of hardness and streak is necessary if the crystals are nonmagnetic.

Euhedral cubes of pyrite are also common, cutting across the schistosity. The crystals usually average several millimeters in dimensions and consist of the cube striated by the pyritohedron and occasionally show small octahedron faces on the corners. There is usually no displacement of the chlorite by the pyrite, but the pyrite crystals cut across the schistosity. Limonite pseudomorphs after pyrite are occasionally seen.

Rhombohedral carbonates form euhedral porphyroblasts, usually as a simple rhombohedron, or make up an important portion of the groundmass. Veins of carbonates in greenstones are quite common. Calcite, dolomite, magnesite, and ankerite are all relatively common. Groundmass calcite or dolomite may be identified by its effervescence in cold dilute HCl. Ankerite, an iron-magnesium-calcium carbonate solid solution, may be recognized by its tendency to rust on a weathering surface. Rhombohedral ankerite porphyroblasts up to a few centimeters in dimensions are present in some greenstones.

GNEISSES

Gneisses are characterized by a planar structure which may be more or less distinct. The planar structure varies from distinct bands of contrasting mineral assemblages to essentially massive rocks through which some minor mineral shows a well-defined planar orientation (Fig. 7-23). Intermediate between these extremes are rocks with lenticular (Fig. 7-24) or rodded mineral segregations (Fig. 7-25). Mineralogically, the gneisses are dominated by minerals which tend to be equidimensional, such as quartz, feldspars, garnets, and pyroxenes. Hornblende and biotite may be quite abundant, and layers rich in these minerals may be quite schistose. However, in such rocks the segregation of minerals into distinct layers is the dominant structural feature.

The determination of the original character of the rock, whether igneous or sedimentary, is primarily on the basis of bulk composition. This is often rendered very difficult because of metasomatic introduction of materials, particularly the alkali metals. The best criteria for the distinction between metaigneous and metasedimentary origin of gneisses are the rare preservation of primary structures and the tracing of the gneiss in the field to its source rock. The presence of either recognizable pebbles, cross bedding, concretions, channelfills and the like or relict beds or lenses of distinctly sedimentary rock types such as chert, calcareous zones, or pure quartzites is the best evidence of sedimentary origin. In the absence of such features a sedimentary versus igneous origin is often difficult to prove. Some sedimentary

Fig. 7-23. A block of banded gneiss cut on three sides to accentuate the banding. The edges have been marked with ink. Hornblende-biotite granite gneiss near Pikes Peak, Colorado.

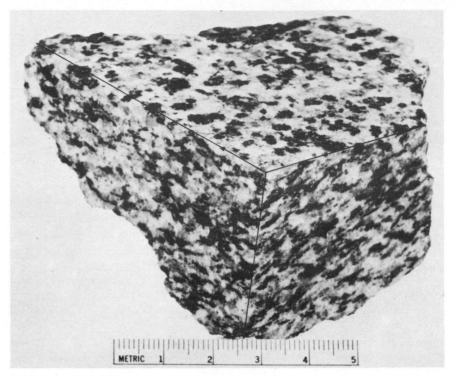

Fig. 7-24. A block of lenticular biotite gneiss cut on three sides to accentuate the lenticular shape of the biotite pods.

rocks have a sufficiently distinctive bulk composition so that their gneissic equivalents are recognizable.

Gneisses Clearly Derived from Sediments

Average shale differs chemically from igneous rocks in having a deficiency of alkalis. During the advanced grades of metamorphism, this shows up in the presence of minerals which are alkali-deficient, such as kyanite, sillimanite, garnet, cordierite, and staurolite. The dominant minerals are quartz or feldspars or both, and micas are minor. Segregation of the minerals into distinct bands may be very crude, and often the gneissosity cannot be recognized in hand specimens.

Sillimanite. Sillimanite is the index mineral for the highest grades of metamorphism of argillaceous sediments. Normally, it is a product of the instability of the micas, muscovite breaking down to yield potash feldspar and sillimanite. It generally occurs as minute fibers to slender prisms in one of two habits. Disseminated fibers of sillimanite in quartz is one characteristic form. The fibers are usually randomly oriented and often difficult to detect

Fig. 7-25. A block of rodded gneiss cut on three sides to accentuate the rodding of the quartzo-feldspathic material. Uxbridge, Massachusetts.

megascopically. On a weathered surface, however, such quartz stands out in relief and has a dull bleached-bone-white appearance. Generally, the fibers can then be found on a freshly broken surface with a hand lens. The second typical occurrence is as slender prisms, all lying in a common plane in selected planes of the gneiss. These zones apparently represent earlier muscovite-rich zones. The prisms are usually colorless and very lustrous and have one perfect cleavage. Discoloration to a yellow or brown by iron-staining is common. The prisms may be parallel, radiating, or random in their orientation.

A high sillimanite content is indicative of an original sediment which was very poor in potash. The clay minerals were probably of the kaolinite or gibbsite groups. Thus a high sillimanite content indicates thorough weathering in the source area of the original sediment. Complete weathering, including leaching of silica, gives bauxite or other laterites, and on metamorphism these would give corundum plus magnetite with or without sillimanite. Emery is the fine-grained corundum-magnetite rock derived from laterite.

Kyanite. Kyanite forms from potash-deficient sediments in lieu of sillimanite in some environments. Generally, kyanite is indicative of either higher pressure or lower temperature than sillimanite and is thus generally considered as indicative of a somewhat lower grade of metamorphism. In gneisses, kyanite forms by the breakdown of muscovite to yield potash feldspars and is therefore normally accompanied by abundant feldspar. The crystals of kyanite are typically bladed and relatively euhedral. A blue, pale-blue, or white color is characteristic, and cleavage cracks at two directions near 90° to each other are generally prominant. The variable hardness, 5 parallel to the length of the blades and 7 across the blades, is diagnostic.

Kyanite blades tend to lie with their long dimension in the plane of the foliation. They may be preferentially concentrated in certain bands, and some zones may contain 40 percent kyanite. The common associated minerals are feldspars, quartz, biotite ± garnet, and staurolite. With increasing temperature, kyanite may invert to sillimanite, yielding clumps of parallel fibers.

Staurolite. Large brown crude euhedra of staurolite may be present in some gneissose rocks but are far more common in schistose rocks. Staurolite is characteristic of iron- and alumina-rich potash-poor sediments and thus indicates a lateritic component in the source but not a laterite source. With increasing metamorphism, staurolite becomes unstable and yields garnet, a much more typical ferromagnesian mineral of gneisses. Staurolite and kyanite commonly occur together, but generally in these rocks, mica is still abundant so that they would be classified as schists.

Cordierite. Cordierite is the magnesium-aluminum-silicate typical of high-grade metamorphism. It is typical of highly chloritic or montmorillonitic clays and is often associated with sillimanite, representing excess alumina. Crystals are typically anhedral and have a blue or gray color. The crystals often attain a very large size, several centimeters, but generally remain anhedral. A crude hexagonal shape may be noticeable. Cordierite is often aasociated with anthophyllite, which occurs as fibrous light-brown prisms. Anthophyllite is a magnesium-rich orthorhombic amphibole which shows typical well-developed amphibole cleavage.

Garnet. Garnet is the most common ferromagnesian mineral, except biotite and hornblende, characteristic of these gneisses. The crystals are usually large and rounded, but well-defined crystal faces are much less common than in garnetiferous schists (Figs. 4-3 and 7-3). Megascopically recognizable inclusions of quartz, biotite, graphite, and sillimanite are common. The inclusions are often lined up as trains through the garnet, preserving planar fabrics characteristic of the lower grades of metamorphism. Curved or S-shaped trains are occasionally recognizable, indicating rotation of the crystal during growth. Garnet is usually in the almandite-pyrope series, and pink, red, and purplish red colors are common.

Quartzo-feldspathic minerals. Quartz and feldspars are normally dominant minerals of these rocks. Most commonly, quartz is more abundant than feldspar because of the paucity of alkalis in shale. However, metasomatic introduction of alkali may counteract this normal deficiency. Some crude banding of the quartzo-feldspathic material is generally noticeable as either distinct feldspar-rich and feldspar-poor bands or indistinct lenslike segregations of quartz. Grain size in these minerals is widely variable but is generally coarse-grained. Lenses of quartz usually have coarser grains than quartz of the groundmass mixture.

The feldspar of these rocks is dominantly orthoclase and sodic plagioclase in the oligoclase-andesine range. More highly calcic plagioclase may occur, depending on the soda to lime ratio of the initial rock. The feldspars are normally white, pink, or red, occurring as anhedral grains. Polysynthetic twinning is usually noticeable in the plagioclase of gneiss in contrast with its rarity in the plagioclase of schist. This probably reflects the reduction of shear due to the apparent plasticity of gneiss and the limitation of chemical reactions, resulting in recrystallization, which dominate metamorphism in the range of environments represented by schist. Porphyroblasts of feldspar are common in gneiss, particularly where metasomatism is involved. Large ovoid crystals of potash feldspar or plagioclase develop and often result in augen structures.

Quartz is normally the dominant mineral in gneiss developed from sediments because of initial high silica content. It is generally colorless and glassy and has anhedral outlines and highly sutured borders. However, in certain gneisses it is fine-grained and sugary in appearance. This is typical of the metamorphic products of impure quartz sandstone or of remnant chert beds, lamellae, or lenses. Typically, the sugary quartz derived from sandstone occurs in irregular polygranular patches, interstitial between feldspars and micas. The fine-sand texture may be well preserved, and occasionally all gradations from metaquartzite to gneiss can be found in one area. Recrystallized chert varies in appearance from megascopically granular mosaic masses to porcelaneous-appearing masses. Generally, the chert layers are broken up into lenticular boudins, but strings of boudins mark initial bedding.

Ferromagnesian minerals. Biotite and hornblende are the characteristic dark minerals, in addition to garnet, staurolite, etc., described above. Planar orientation of scattered biotite crystals (Fig. 7-5) and segregations dominated by hornblende or biotite or both are the most obvious features which give these rocks their gneissose structures. The segregations of the ferromagnesian-rich zones may be the result of initial difference in composition (Fig. 7-4) or metamorphic differentiation. Generally, uniform continuous bands of ferromagnesian-rich material can be interpreted as being the result of initial compositional differences. Thus hornblende-rich bands are indicative of a higher initial lime content and may represent calcareous zones. If they are relatively thick, they may represent original lava flows, mafic pyroclastics, or dikes and sills in the initial sediments. Metamorphic differentiation generally results in lenticular structures. The ferromagnesian-rich zones indicate those portions of the rock from which the more mobile constituents have migrated during shear and its resultant dilatency. The quartzo-feldspathic zones represent former shears into which these mobile constituents have migrated.

Hornblende of the gneisses is generally black, prismatic, and subhedral. Prism faces may be fairly well developed, but terminations are generally lacking. The prisms often taper out to a point, producing spindle-shaped crystals. Normally, the long dimension of the crystals lies in the plane of gneissosity. Parallelism within that plane produces a linear element, but this is not necessarily present. Hornblende rarely makes up the major bulk of the hornblende-rich zones; there will also be admixed biotite, quartz, and feldspar plus accessory minerals. Anhedral blocky crystals of hornblende are less common than prismatic crystals but may be present or dominant. If blocky crystals are present with prismatic crystals, the c axis of the stubby crystals generally is roughly perpendicular to the gneissosity.

Biotite occurs with or without hornblende in the ferromagnesian-rich zones, but it also occurs as scattered single crystals with the quartzo-feldspathic material. Biotite flakes are generally highly tabular and subhedral and have well-developed basal faces and ragged edges. The biotite is generally brown, appearing almost black in the mass of the rock, but thin flakes may show a green or greenish-brown color. Planar orientation parallel to the gneissosity is most common. Lenses of coarse thick biotite books may occur, with the lenses parallel to the gneissosity and the biotite books at a large angle to the gneissosity. This is less common than in schist, however.

Accessory minerals. Graphite may appear in gneiss and is almost a positive evidence of sedimentary origin. The carbon was originally organic, and it is typical of gneisses developed from black shales. Graphite occurs as crystal plates, often several millimeters in diameter and a fraction of a millimeter thick. The crystals may be perfectly alined with the gneissosity, but they often appear to be quite random. The intense darkening of rocks by graphite which is characteristic of graphitic phyllite and schist is not characteristic of the graphitic gneisses. The innumerable minute flakes recrystallize into a relatively small number of coarse flakes.

Pyrite or pyrrhotite are common associates of graphite in gneiss. The sulfur is ultimately derived from organic material and is common in organic-rich shale as disseminated pyrite or marcasite. During metamorphism this recrystallizes to coarse-grained material and is usually pyrrhotite.

Corundum is found in some alumina-rich silica-poor gneiss. It commonly occurs as euhedral hexagonal prisms in random orientation with respect to the gneissosity. The original sediment was one of a lateritic derivation. If very little clastic material was added, then corundum may become dominant along with magnetite derived from the iron oxide component of the laterite. Fine-grained corundum-magnetite mixtures are called emery.

Magnetite and other spinels are common accessory minerals. The typical habit is as small octahedrons, and since these minerals have a high crystalloblastic strength, the crystals are commonly euhedral. Magnetite can occur in almost any of these gneisses, but the other spinels are restricted to silica-deficient alumina-rich environments. Spinel proper, hercynite, and intermediate mixtures, pleonaste, are all fairly common. The common colors are red, green, and black, but purple, yellow, brown, and blue occur. Magnetite is black.

Gneisses Not Clearly of Sedimentary Origin

The most common gneisses are dominantly quartzo-feldspathic rocks with hornblende or biotite. Distinctive minerals indicative of initial sedimentary compositions are rare or minor. Mineralogically, these rocks approach normal igneous rocks, and therefore it cannot be recognized from hand specimens whether they represent metaigneous rocks or metasomatosed sediments. Field relations usually indicate which source is more probable through gradations into recognizable rocks, contact relations with adjacent rock masses, or local inclusions of diagnostic mineral assemblages.

These gneisses are named on the basis of either composition or structure, and commonly names applied combine both compositional and structural features; thus, an augen granite gneiss. Compositions are indicated by the use of a plutonic igneous-rock name, i.e., granite gneiss, diorite gneiss, syenite gneiss, etc. Such terms simply indicate a gneissose metamorphic rock with a mineralogy of a granite, diorite, or syenite. They should imply nothing as to the original nature of the rock. If, however, it can be determined that the parent rock from which a gneiss was developed was a granite, then the rock should be named a *gneissose granite*. The term *granitic gneiss* is improper, because "granitic" is a textural term and rocks having a granitic texture cannot be gneissose.

A number of structural names have been applied to gneisses, some of which have a genetic implication. From hand-specimen studies it is generally best to avoid terms having genetic implication and use those terms only when a regional study warrants their use. Simple structural terms usable in names include banded gneiss, lenticular gneiss, augen gneiss, rodded gneiss, etc. These structural terms were defined earlier in this chapter. Injection gneiss and lit-par-lit gneiss imply the forceful injection of magmatic material, usually granite, into a metamorphic rock. Injection gneisses are found at the margins of some batholiths. The term *migmatite* has been applied to many banded gneisses and implies a mixed rock composed of igneous and meta-morphic material in alternating bands. It is not often clear whether the mag-matic component has been forced into the rock or whether this is the product of partial melting and subsequent segregation of the melt from the solid remnant. Various names have been applied to migmatites based upon distribution and shape of the igneous component and the interpretation of the origin. Migmatite and its related terms are best reserved for regional field studies and probably should not be used in hand-specimen descriptions.

The dominant minerals of these rocks are quartz, feldspars, biotite, and hornblende, with minor amounts of hypersthene, garnet, and sillimanite. Muscovite occasionally occurs as a late, and probably metasomatic, mineral. The proportions of these minerals vary widely, as indicated by the applied rock terms in their names; thus, syenite gneiss versus granite gneiss. Within a hand specimen the proportions of the minerals may also vary due to com-positional banding, but this is less common in gneiss derived from igneous rocks than in gneiss derived from sedimentary sources.

Quartz. Inasmuch as the commonest gneiss is granite gneiss, quartz is an important constituent, generally ranging from 15 to 50 percent. A higher quartz content is almost always indicative of a sedimentary origin. Several habits and distributions of quartz are characteristic. Segregations into lenses, pods, rods, and the corners of augen are most noticeable; however, the great bulk of quartz probably occurs as individual grains and interstitial aggregates of grains in a feldspathic or ferromagnesian groundmass. In general, segre-gated quartz consists of coarse polygonal grains with a mosaic or sutured texture. Large grains are usually set in a groundmass of smaller grains, as though some granulation had taken place. Quartz grains in lenticular segre-gations tend to be lenticular and parallel to the gross structure. Similarly, in rodded structures the grains tend to be elongated or spindle-shaped. The quartz of such segregations is generally colorless and transparent so that its apparent color depends on any associated minerals.

Disseminated quartz generally is most common in the feldspathic zones and occurs as distinctly lenticular to polygonal grains. The lenticular shape contributes to the overall foliation of the rock. Where the feldspars are porphyroblastic, quartz is generally indistinct and interstitial. This is par-ticularly common where the feldspars have developed as porphyroblasts in sandstone and the quartz appears as irregular interstitial masses with a decidedly sugary texture.

In gneiss developed by the metasomatic alteration of mafic rocks, such as lava flows or greenstone, quartz is hidden in the fine-grained ferromag-nesian groundmass. In the earlier stages of augen gneiss development, it occurs as small dark grains embedded in chloritic or micaceous material. With more thorough metasomatic development of feldspar porphyroblasts, the quartz becomes coarser-grained and more noticeable as larger crystals, often almost as large as the feldspars. In thoroughly granitized specimens, quartz and the ferromagnesian minerals form an anastomosing network between clots of feldspar. The feldspar clots represent preexisting porphyroblasts which have recrystallized to a finer-grained aggregate.

Feldspars. Feldspars are the dominant minerals of these gneisses. Potash feldspar, usually orthoclase but also microcline, and sodic to intermediate plagioclase dominate. Calcic plagioclases are characteristic of gabbro gneiss. Feldspar is usually the major potash or soda mineral present, and usually most calcium is in the feldspar. Epidote, muscovite, and calcic pyroxene are generally absent to minor. The feldspars may occur as porphyroblasts, often forming augen structures, as distinct feldspar-rich bands or lenses, or as the bulk of an essentially homogeneous rock with indistinct foliation.

Orthoclase or microcline as white, pink, or red grains is characteristic of granite gneiss and syenite gneiss. The crystals are usually of a uniform size within one area of a specimen, but size varies from lens to lens. These lenticular or streaky zones of contrasting size of grain are often the main indication of a planar element. The individual grains range widely in shape, but most are anhedral and somewhat flattened parallel to the planar element. Equant to rectangular straight-sided grains are rare. Interpenetration with adjacent quartz or biotite is the rule, because except in the metasomatic environment, potash feldspar has a low crystalloblastic strength.

Porphyroblasts and augen of pink potash feldspar are very common in metasomatically derived gneiss. The crystals are large and generally ovoid in shape and have their maximum dimensions parallel to the planar element. Inclusions of quartz, magnetite, and ferromagnesian minerals are common and often show a zonal arrangement. Color zoning is also common. Mantling of the pink orthoclase ovoids by yellowish or greenish sodic plagioclase is common in gneiss of the rapikivi type. Augen structures developed around single crystals or clusters of large crystals are characteristic of augen gneiss. The corners of the augen are usually filled with quartz. Quartz-ferromagnesian mineral bands generally curve around feldspar porphyroblasts, indicating high crystalloblastic strength for the feldspar, but the ovoid shape indicates low form energy. Porphyroblasts may become so numerous that they make up the bulk of the rock with an interstitial finer-grained groundmass of the other minerals.

Plagioclase, as disseminated crystals, is generally very similar in habit to potash feldspar but tends to be white, bluish-white, or cream-colored. Except in gabbro gneiss, it is usually sodic, and striations are difficult to find. When present the striations are often bent, and cleavage surfaces are curved. In amphibolite gneiss, derived by middle-grade metamorphism of mafic igneous rocks, plagioclase is generally white oligoclase in highly irregular interstitial grains. Some epidote may be associated, but generally, excess calcium is taken up by hornblende.

Plagioclase in gabbro gneiss is in the labradorite range of composition. It may be white but is often very dark, almost black, due to minute inclusions. The crystals are highly anhedral with very ragged borders. Bent twinning lamellae and curved cleavages are common. The crystals are generally lens-shaped with the 010 cleavage subparallel, and as a result, the striations are subparallel to the planar element. Reaction zones of brown fine-grained granular garnet are common between plagioclase and the associated pyroxene. The garnet is seen only where plagioclase is in contact with pyroxene and not between adjacent plagioclases.

Biotite. Biotite, along with hornblende, is the common ferromagnesian mineral of the silicic gneisses. It is usually brown to almost black in masses in the rock, but thin sheets may be either brown or green and are translucent. It may occur in distinct zones within the rock or as disseminated single crystals or small crystal clusters. The flakes are usually small and ragged in outline, but where they occur as single crystals, they tend to be more euhedral, because their crystalloblastic strength is greater than the associated quartzo-feldspathic material. As individual crystals, biotite is generally subparallel throughout the rock and gives the most apparent planar element. Small clusters of biotite flakes tend to occur as pods or rod-shaped structures. Clusters give a planar or linear element, and individual flakes may have rather random orientation within the cluster. However, their maximum dimension remains parallel to the structure. Thus a flake in which the cleavage is not parallel to the maximum dimensions of the structure will tend to be thick and elongated in one direction.

Biotite in the ferromagnesian-rich bands of distinctly banded gneiss appears typically as small anhedra which are often so intergrown that the borders of individuals are difficult to pick out. Biotite-rich bands usually also contain other ferromagnesian minerals, such as hornblende or magnetite, and also contain disseminated fine-grained quartz. Chlorite is also common in these zones as either a retrograde product of the biotite or a primary material not yet reconstituted into the higher-grade minerals. This last occurrence is particularly common where a gneiss is the product of metasomatic development of feldspar porphyroblasts in greenstone.

Hornblende. Hornblende is as equally common as biotite. There are two main types of hornblende-bearing gneiss: the silicic gneisses, where hornblende is associated with abundant quartzo-feldspathic material, and the amphibolites or hornblende gneisses derived from mafic igneous rocks, where hornblende is associated with plagioclase with or without epidote and quartz. In the first occurrence hornblende appears as small stubby anhedra or as elongated subhedral crystals with well-developed prism faces and ragged terminations. The hornblende is consistently black with lustrous cleavage, but its hardness and splintery fracture under a probe distinguish it from similar-appearing biotite. Hornblende generally occurs in these rocks in distinct bands, lenses, or rods of segregated crystals. In separate bands the crystals are generally elongated, with their prism zone parallel to the planar element. They may be parallel or randomly oriented in the foliation plane so that a linear element may be superimposed on the planar element. In small lensoid or rodded structures, hornblende tends to be not so well oriented. Crystals whose c axes are parallel to the structure are generally slender and elongated, whereas crystals with other orientations tend to be stubby and irregular in outline.

Hornblende in amphibolites varies widely in size and habit. In many rocks the crystals tend to be essentially equidimensional and have random orientation, but a coarse crude banding gives a gneissose structure. The bands are generally alternately hornblende-rich and hornblende-poor. Each band has the same mineral assemblage of hornblende-plagioclase-quartz $\pm$ epidote but in differing proportions. Well-developed elongated hornblende crystals give rocks which may be called either hornblende schist or hornblende gneiss, depending on the extent of mineral segregation. A linear element may or may not be present as a result of parallelism of the elongated crystals.

Pyroxene. Except in gabbro gneiss and hornblendites verging on gabbro gneiss, the pyroxenes are minor constituents. At very high temperatures, biotite becomes unstable and breaks down to yield hypersthene and orthoclase. The resulting rock is hypersthene-bearing granite gneiss, or metacharnockite. In such rocks gneissosity is poorly developed; all minerals are about equigranular, but lenticular shapes are the rule. The hypersthene generally occurs as small elongated anhedra which have a fibrous appearance due to cleavage.

Diopside appears as small grains in amphibolite at temperatures near the upper limit of hornblende stability. They are generally more abundant in certain layers of the gneiss. Augite or hypersthene replaces hornblende as the dominant ferromagnesian mineral at still higher temperatures. Hypersthene is probably the more common product, because calcium and aluminum from the hornblende go to make the plagioclase more calcic. Both pyroxenes are generally black and occur as coarse anhedral crystals which are lenticular and parallel. A brown granular garnet commonly forms as a reaction product between hypersthene and neighboring plagioclase crystals.

Sillimanite and kyanite. Sillimanite is commonly developed at high grades in gneiss developed from silicic igneous rocks as a result of loss of alkali from the rock during the earlier breakdown of alkali feldspar to sericite. The extent of development of the aluminum silicates then indicates the extent to which alkali was lost. Kyanite appears as pale-bluish to white prismatic crystals of high crystalloblastic strength. The crystals are generally small, however, and are easily overlooked. They tend to be concentrated in certain zones of the gneiss, representing the more micaceous zones of the lower-grade rock. Sillimanite occurs as its typical minute fibers to slender prisms. They are generally white with a highly lustrous cleavage but may be so minute as to be overlooked. The fibers are commonly included in massive quartz and on a weathered surface give the quartz a bleached-bone-white appearance. Sillimanite may be concentrated along certain zones of the rock, and often the rock will part along these zones, giving a silky fibrous surface. Radial arrangement of the larger needles is common.

Garnet. Garnets in the almandite-pyrope series are common accessory minerals in granite gneiss, and garnets in the pyrope-grossularite-almandite series are common in gabbro gneiss. In each case the garnet is indicative of high pressure, in that it is a denser mineral than those it replaces. In general, crystals are rounded to irregular and anhedral. The garnets of granite gneiss are pink to purplish pink, with the purplish color generally indicative of a higher pyrope content. In general, they are well rounded but not bounded by crystal faces. The garnets have a higher crystalloblastic strength than their associated quartzo-feldspathic groundmass, but because of the high plasticity of the rock during their formation, they appear to have granulated edges. Inclusions of other minerals in the garnets are much less noticeable than in garnets of schistose rocks. The lime-bearing garnets of gabbro gneiss tend to be brown and finely granular. Inasmuch as they are a reaction product between calcic feldspar and pyroxene, they occur along the contacts between those two minerals and replace them with irregular boundaries. Augen structures are not commonly developed around garnets in either of these rock types because of the relatively small size of the garnet and the paucity of micaceous minerals left in the environment of garnet development.

Muscovite. Medium-sized to large porphyroblasts of muscovite are common in granite gneiss. They appear to develop at a very late stage in the metamorphism, because they commonly have completely random orientation which is unrelated to the gneissic banding. Muscovite flakes are commonly poikiloblastic, including all the minerals of the rock as large recognizable grains. Muscovite thus appears to be a retrograde product derived by reaction of remnant aqueous vapors with primary feldspars.

Common accessory minerals. Corundum as hexagonal prismatic or barrel-shaped crystals is common in syenite gneiss. The crystals are usually gray, but red or bluish casts may be noticeable. They generally are randomly oriented in the rock but are usually concentrated along selected zones parallel to the gneissosity. This is probably a result of the loss of alkalis along these zones, leaving an excess of alumina.

Tourmaline is often developed as black flattened prismatic crystals. Its presence generally indicates a boron metasomatism. The crystals may attain a maximum dimension of several centimeters. They generally are well alined in the plane of gneissosity but may show either random criss-cross orientation in that plane or radiating stellate groups within that plane.

Zircon and apatite are common accessories but generally do not attain sufficient size to be recognizable megascopically. Occasionally, large brown euhedral zircons will be present in the granite, granodiorite, and syenite gneisses. Large elongated hexagonal prisms of apatite may be present in gabbro gneiss inherited from the parent gabbro. Various sulfides may be megascopically recognizable. Pyrite or pyrrhotite are the commonest, but thin hexagonal plates of molybdenite are often present. Graphite may be present in irregular flakes a few millimeters in dimensions. Graphite is generally indicative of a sedimentary source rock derived from organic carbon. Sphene is a rather rare constituent of the quartz-feldspar gneisses, because at the temperature of most gneiss formations, the titanium present is taken up in ferromagnesian minerals.

QUARTZITES

Quartzites are metamorphic rocks consisting predominantly of quartz. Most commonly they are produced from quartz sandstone, subgraywacke, quartz conglomerate, chert, and related rocks, but they may also be produced by metasomatic removal of cations from other silica-rich rocks. The more impure sandstones and conglomerates rarely give rocks which can be called quartzite, because their textures are dominated by other minerals, such as micas, and are schistose or gneissose in spite of high quartz content.

Quartz is the dominant mineral of all quartzites. The accessory minerals depend on the nature of the admixed materials or cements in the initial sediment. Thus a carbonate-cemented sandstone yields quartzite with lime silicate or lime-magnesium silicate minerals, whereas an argillaceous sandstone yields quartzite with micaceous or feldspathic minerals. As a result quartzites can be divided into the following groups:

Pure quartzite
Micaceous quartzite
Garnetiferous quartzite
Feldspathic quartzite
Actinolitic quartzite

The texture of most quartzite is massive to weakly schistose. Augen structures and crude gneissic banding develop from some conglomerates. Cataclastic effects are common but usually not at a megascopically visible scale. However, coarse brecciation and veining by quartz are often present. Original bedding is often very difficult to recognize, except where lenses of contrasting composition or initial texture are preserved. Thus intercalated argillaceous beds become phyllitic or schistose, and pebbles of conglomerate lenses will be preserved to give evidence of initial bedding. Color banding is common but is completely unrelated to bedding and may be very confusing. These color bands are probably related to the oxidation state of minor amounts of iron and may be a liesegang structure.

Quartz. Quartz is usually present as equidimensional to slightly lenticular-shaped grains. Grain boundaries are commonly smooth to slightly irregular, giving a simple mosaic texture. Highly sutured and interlocking grains are not typical. Generally, in any one portion of the rock all quartz grains will be approximately the same size, regardless of initial variations in sand-grain size. Where grains are lenticular, their long dimension is parallel to the planar element of associated schistose or gneissose rocks. The equidimensional grains show no megascopically recognizable preferred crystallographic orientation. However, optical studies show that they generally have marked preferred orientations. Quartz generally appears colorless, white, pink, purplish, or black. The color depends on the minor accessory minerals or minute inclusions within the quartz grains. The oxidation state of iron is a controlling factor in the lighter-colored varieties.

The initial texture of the rock may be well preserved in quartzite; the outlines of sand grains can often be recognized by subtle color contrasts, by minute inclusions at the margins of the initial grain, or by the character of the conchoidal fracture. Size of the relict initial grains often varies noticeably in a single mass and can be used to detect bedding. Pebbles in metamorphosed quartz conglomerates are often preserved as lenses, rods, or pencils of coarse-grained material. Ill-defined augen structures are developed around pebbles. In many examples a noticeable change in shape from an initial subspherical shape is indicative of flattening and tectonic stretching of the pebble. The stretching is often not parallel to bedding so that the orientation of the conglomeratic lens or bed gives bedding, whereas the orientation of the deformed pebble gives data on the rock deformation. Cataclastic textures are often seen having deformed pebbles, in that the margins are granulated and the pebble core is ruptured and veined. Metamorphism of pure chert units results in an increase in grain size. The product is a fine-grained equigranular sugary textured rock composed of equidimensional polygonal crystals.

Chlorite and micas. Chlorite and sericite are very common minor minerals in quartzite developed at low temperatures. They are derived from interstitial argillaceous material or from the breakdown of feldspathic or ferromagnesian sand grains. Two occurrences are common for mica flakes: they may be either disseminated moderately uniformly through the rock or concentrated along selected thin zones. In the first occurrence the flakes are generally relatively large, and individuals can be readily recognized (Fig.

7-26). Thin plates with well-defined borders are typical. The flakes are often bent around quartz grains, but generally mica has a higher crystalloblastic strength than quartz so that mica flakes may penetrate into quartz grains. Micaceous segregation into thin zones generally results in a film composed of small ill-defined flakes which are intergrown on the edges. The segregation may represent primary bedding, in which case the films are continuous over distances of several feet. They may also represent a product of metamorphic differentiation. The films represent zones from which the most mobile silica has been removed. In such cases the films are generally discontinuous.

Chlorite in quartzites is most commonly disseminated uniformly through the rock, giving a dark color to the whole rock. The flakes, like micas, may be randomly oriented or parallel-oriented. Clusters of radiating patches of minute flakes in the interstices between quartz-sand grains are common. Equilibrium is often not attained between the micaceous minerals of quartzites because of their scattered disposition through the quartz. As a

Fig. 7-26. Muscovite and chlorite in quartzite. Note the poikiloblastic character of the larger garnets.

result any combination of the three minerals may be present in different parts of the same rock.

Garnet. A garnet in the almandite-rich compositions may be sparsely scattered through quartzite which at a lower grade included chlorite. Generally, the garnets are much coarser than the associated quartz and are highly poikiloblastic (Fig. 7-26). The margins of garnet crystals are extremely irregular, since the crystals grew around quartz grains.

Feldspars. Potash feldspar and albite are typical of many quartzites derived from feldspathic sandstone. They also develop from the metasomatic introduction of alkali into slightly argillaceous sandstone. Feldspar as recrystallized detrital grains forms a part of the mosaic texture of the rock, because the crystalloblastic strength of quartz and feldspar is about the same. Such grains appear as more or less rectangular shapes in which recrystallization has destroyed any effects of abrasion or alteration. Pink and white are the common colors. Striations are rarely developed on the plagioclase so that the presence of two feldspars can be recognized only when there is a color difference between them.

When feldspar is developed by the metasomatic introduction of alkali, the resulting crystals are larger than the associated quartz and are commonly poikiloblastic. Large bluish-white nonstriated albite crystals are common and include quartz and the minor minerals of the matrix. The crystals of albite may be either euhedral rectangular blocks or rounded ovoids. Potash feldspar developed metasomatically is generally pink orthoclase and, again, may be either rounded or rectangular euhedra. The extent of feldspar development generally depends on the amount of argillaceous material in the original sand. In a single outcrop, variations in feldspar content can be seen from bed to bed where the pure quartz sandstone beds contain little feldspar, whereas the argillaceous quartz sandstone beds may approach the composition of a granite. The sugary texture of what quartz remains indicates the initial clastic character of the rock.

Lime and lime-magnesia minerals. The metamorphosed equivalents of carbonate-cemented sandstones vary in character, depending on the ability of carbon dioxide to escape. Under closed systems, quartz and calcite or dolomite recrystallize simultaneously without reaction. Dolomite has a higher crystalloblastic strength than quartz and, as a result, usually recrystallizes into minute rhombs scattered through the quartz matrix. On a weathered surface the dolomite crystals are removed, leaving rhomb-shaped holes. In contrast, calcite has a low crystalloblastic strength, similar to that of quartz, and as a result, recrystallizes to irregular interstitial single crystals.

Under open systems of metamorphism, lime silicates and lime-magnesia silicates form. Inasmuch as some iron and magnesia are generally present in detrital grains, in interstitial clays, or in the carbonates, the common mineral is the tremolite-actinolite series or epidote or both. Epidote is typical of the lower grades of metamorphism and occurs as small granules embedded in the quartz mosaic. The crystals have higher crystalloblastic strength than quartz but are not generally euhedral. Elongation in one direction is common, giving a directional element to the rock. Generally, epidote, or its gray or white relative zoisite, is concentrated in specific zones which alternate with epidote-poor zones. Pale-green to dark-green amphibole develops at a higher grade at the expense of epidote. The crystals are commonly large and poikiloblastic, including granules of the groundmass quartzite. Parallelism of the elongated crystals in a plane is common, but within the plane, the crystals are most commonly randomly oriented. Stellate clusters are frequent.

Other minor minerals of these rocks are lime garnets, diopside, and calcic plagioclase. None is commonly found in megascopically recognizable grains. Diopside occurs in small pale-green rounded granules. Generally, the iron content of the rocks favors amphibole over diopside. Garnet appears as pale rounded poikiloblastic grains which may attain a size of a few millimeters. Calcic plagioclase is relatively uncommon, probably because of a deficiency of alumina in the rock. The crystals are not generally well striated and are therefore difficult to distinguish from other feldspars.

Accessory minerals. The ferruginous cement of the sandstones generally yields magnetite or thin metallic flakes of hematite. These minerals may make up an appreciable proportion of some rocks, particularly those derived from cherty-iron carbonate sediments. Magnetite is commonly coarsely granular. Euhedral octahedrons are common but may be replaced by hematite, yielding martite. Pyrite and limonite pseudomorphs after pyrite are common in some quartzite. Generally, the crystals are euhedral cubes with pyritohedral striations, but octahedrons are also common.

The detrital accessory minerals of the sand recrystallize to euhedra during metamorphism but generally retain their small size. Tourmaline and zircon are common in the purer quartzites. Secondary euhedral overgrowths are often of a contrasting color to the detrital core. Sphene may develop in calcareous quartzite from primary titanium minerals, but in the absence of lime, rutile is typical as minute needles.

CALCAREOUS SOURCE ROCKS

Marbles

Pure calcite or dolomite rocks recrystallize readily to marbles under the metamorphic environment without the appearance of new minerals. If the initial rock contained minor impurities, the minerals described in the section on calc-silicate rocks may be present in minor amounts and widely scattered grains. Commonly, the parental calcareous rocks vary significantly from bed to bed in the presence and character of impurities. As a result pure marbles may be intercalated with marbles containing a wide range of types and percentages of silicates. The distribution of the different assemblages is often the only evidence of bedding left after metamorphism.

The original clastic or chemically precipitated textures of limestones are rapidly destroyed by metamorphism. However, under special circumstances they may be preserved and be recognizable after low-grade metamorphism. Such an environment may develop in limestone lenses enclosed in shale, where the shale has recrystallized to phyllite. Apparently, here the phyllite acts to seal off the limestone from any gas escape, and most of the mechanical effects of metamorphism are taken up in the plastic behavior of the phyllite. The limestone beds are then commonly broken up into boudins and undergo slight mechanical and recrystallization effects. Cataclastic textures are developed in the limestone and are particularly noticeable if the parent limestone was a coarse bioclastic accumulation. Grains along the margin of the boudins will become lenticular due to shearing, and larger grains will be set in a granulated matrix. Fossils may be noticeably deformed but still recognizable. The more massive organic structures, such as crinoid plates and brachiopod shells, are the most commonly recognizable.

Normally, however, thorough recrystallization destroys all sedimentary textures. The rock becomes an equigranular crystalline aggregate which may show either sutured or simple grain boundaries. No preferred orientation of grains is usually noticeable megascopically, although some marbles are composed of parallel lenticular grains. These approach the calc-schists described above. A preferred orientation of crystallographic axes can usually be recognized microscopically. With increasing intensity of metamorphism, grain size generally increases, but the equigranular character is retained. This is brought about by the selective growth of favorably oriented grains. The driving force is directional pressure operating on the growth anisotropy of the minerals. As a result preferred orientation increases with increase of grain size.

Calcite and dolomite. Calcite or dolomite or both are the dominant minerals of pure marbles. Most commonly, one or the other greatly predominates, but occasionally both occur. There is no foolproof method for distinguishing them, but the simple acid test is most reliable. Both minerals are commonly polysynthetically twinned and show striations similar to those of plagioclase. In calcite the striations are parallel to the long diagonal of the cleavage rhomb and therefore bisect the acute angle between the cleavage cracks. In dolomite the striations may be parallel to either diagonal of the rhomb but are most common parallel to the short diagonal. They therefore bisect the obtuse angle between the cleavages. In marbles containing both minerals, dolomite tends to be much more euhedral than calcite (Fig. 7-20). The dolomite then appears as rounded rhombic cleavages with simple boundaries, whereas calcite consists of highly irregular grains partially enclosing the dolomite with sutured boundaries along contacts of adjacent calcite grains. Monomineralic rocks develop an equigranular simple mosaic or sutured texture. Veining of a rock dominated by one carbonate by the other carbonate is quite common. Weathered surfaces of marbles containing both carbonates usually show the dolomite crystals in relief as euhedral to rounded rhombs, whereas calcite is selectively leached out.

Quartz. Detrital quartz grains or recrystallized chert masses are common in marble, particularly those very low in dolomite. Quartz and calcite recrystallize in equilibrium without reaction up to high temperatures, where wollastonite becomes stable. Quartz also persists, in spite of the presence of dolomite, in a strictly closed system from which CO_2 cannot escape. Generally, sand grains retain their detrital character without apparent recrystallization. Etching by the carbonate matrix is common so that the surfaces of quartz grains are deeply pitted or corroded, giving a coarse frosted effect. Chert recrystallizes to a sugary or coarsely granular texture. Some recrystallized carbonates, sulfides, or graphite may be present, or new silicates may develop within a chert mass. Commonly, chert masses in dolomite are completely or peripherally replaced by silicate-reaction products such as tremolite, diopside, or forsterite.

Accessory minerals. The presence of clay in the parental limestone may result in the appearance of characteristic accessory minerals. Muscovite develops as scattered rounded books when the clay is illitic. Small nearly colorless to yellowish-green epidote granules are characteristic of low-grade metamorphism of kaolinitic clays. Brown, green, yellow, or white garnets of the grossularite-andradite series appear in the middle grades of metamorphism. The garnets commonly occur as small rounded granules, but well-developed euhedral crystals or crystal clusters occur. Generally, with the appearance of garnet, both muscovite and epidote disappear, and their places are taken by microcline and plagioclase. Both feldspars occur as rounded granules which are difficult to identify megascopically. Very commonly, any of these minerals, except muscovite, can be recognized only in an insoluble residue from the marble.

Magnesium and lime-magnesium silicates are generally indicative of a dolomitic component or the presence of chloritic or montmorillonitic clays. Tremolite as bladed crystals having a flattened lenticular cross section is common (Fig. 7-27). It is generally white to green or gray in color. The scattered blades may show random or preferred orientation, and stellate clusters are common. Diopside is indicative of a higher grade of metamorphism (Fig. 7-28). Usually, when diopside is present in minor amounts, it is difficult to identify because of its fine granular character. Larger crystals are characterized by gray to light-green color, nearly equidimensional

Fig. 7-27. Tremolite-dolomite marble. Randville dolomite, Felch, Michigan.

Fig. 7-28. Diopside-dolomite marble. Randville dolomite, Randville, Michigan.

prismatic habit, cleavage, and parting. Crystals are dominated by nearly equal development of (110), (100), and (010) and are terminated by (001) plus crude first- and fourth-order prisms. Parting parallel to either (001) or (100) is common. With increase in iron content, crystals become darker-colored in the diopside-hedenbergite series, and may be almost black. Uralitization of diopside to tremolite is moderately common, giving prismatic crystals with octagonal cross sections which show 60° amphibole cleavage. White to gray rounded granules of forsterite may appear at moderate to high grades of metamorphism. The granules are usually difficult to recognize because of fine grain size and lack of clearly defined cleavage. Forsterite granules are commonly serpentinized during retrograde metamorphism, producing the calcite-serpentine rock ophicalcite. The serpentine is generally green and fine-grained. It may show the patchy color distribution characteristic of olivine replacement.

The metasomatic introduction of volatiles results in a characteristic group of minerals. These are very common in lime silicate skarns but are less important here. Generally, the carbonates are minor when they appear. Vesuvianite (idocrase) involves introduction of water and is characteristic of rather highly argillaceous calcareous rocks. Abundant iron and magnesia are required. Vesuvianite most commonly appears as coarsely crystalline prismatic aggregates of greenish-brown to brown color. Individual crystals have octagonal cross sections composed of first- and second-order tetragonal prisms. Columnar aggregates are strongly striated parallel to the c axis but with poor terminations. Bright-green thin tetragonal dipyramids are also common. The scapolite group develops when chlorine or sulfate is introduced. Crystals are again large and are crudely formed or columnar aggregates. They are usually white to gray but may be greenish, pinkish, purplish, or brownish in cast. Introduction of fluorine results in the development of the chondrodite group in dolomite. The crystals are yellow, brown, or pink, have a resinous luster, and are complex monoclinic forms.

Pyrite or pyrrhotite is common. They may represent an introduction of sulfur, but more likely, they are inherited from authigenic minerals of the sediment. They generally occur as irregular masses, or stringers, or as clusters of euhedral crystals. Graphite is a common accessory where the initial limestone had a high organic content. The graphite appears as rounded flakes, occasionally showing hexagonal outlines. The flakes may show preferred orientation.

Calc-silicate Rocks

Calc-silicate rocks are those developed by metamorphism of limestone and dolomite under conditions such that very little carbonate remains. The rocks consist essentially of calcium, calcium-aluminum, calcium-magnesium, or magnesium silicates. Terms which are applied depend on texture and metamorphic environment. Calc-silicate gneiss and calc-silicate hornfels are products of highly impure limestone and dolomite such that most initial carbonate was exhausted in reaction with the impurities. These rocks are interbedded with purer marble, schist, and gneiss and grade along the strike into other metamorphic facies without appreciable change in bulk composition. The term *hornfels* is applicable to the fine-grained assemblages, whereas *gneiss* is applicable to the coarse-grained rocks. The terms *skarn* and *tactite* are applied to calcareous derivatives which have been subject to extensive metasomatic introduction of material. They are characteristic of intrusive contacts. Introduction of silica and iron is commonly very extensive. These rocks are intercalated with hornfels and schist of thermal origin and grade along the strike into purer limestone or dolomite or equivalent marbles. A change of bulk composition away from the contact with an adjacent intrusive is very noticeable.

Many of these rocks are extremely fine-grained, and little can be done with a hand lens to describe them. Generally, sufficient calcite remains, at least in parts, so that their calcareous nature can be recognized. The rocks are generally buff- to tan-colored and have a greenish or brownish cast. Bedding and other primary structures are generally obscured, although bedding may be reflected in alternations of mineral assemblages. This may be noticeable only in changes of color or texture in the fine-grained rocks. Occasionally, however, bedding is reflected in alternation of coarse- and fine-grained rocks. The introduction of metallic minerals into the fine-grained tactite and skarn is often notable, and these rocks may be of economic significance.

In the coarser-grained rocks a number of minerals are characteristic. These include diopside, tremolite, forsterite, lime garnets, vesuvianite, scapolite, wollastonite, epidote, calcic plagioclase, phlogopite, spinel, and many accessory minerals (Fig. 7-29). Generally, any two or three of these minerals are present along with either calcite or quartz. Individual crystals are often very large in size so that a hand specimen may be essentially monomineralic. A completely random orientation of grains is characteristic of the thermally metamorphosed or metasomatosed rocks, such as skarn. Regionally metamorphosed rocks, however, generally show a planar or linear element in parallel bands, lenses, or pods of silicate minerals dispersed through silicate-poorer marble. Even in such rocks the orientation of individual crystals may appear random.

Amphiboles. Tremolite is characteristic of the low to middle grades of metamorphism of dolomitic rocks (Fig. 7-27). Its absence and the presence of diopside or forsterite or both are indicative of high grades. Generally, tremolite is white to pale green and occurs as radiating bundles of fibers or bladed crystals. The crystals are typically rounded to give a lens-shaped cross section and are striated parallel to the c axis by alternation of growth of the (110) and (100) faces. Orientation of the blades is commonly completely random, particularly where the rock is composed essentially of silicates. At contacts between original dolomite and chert concretions, the

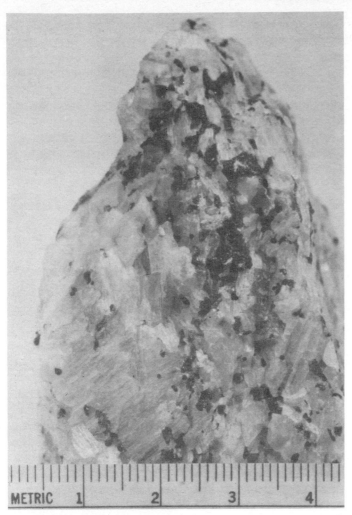

Fig. 7-29. Wollastonite-garnet skarn rock. Willsboro, New York.
(*Ward's Universal Rock Collection.*)

blades may tend to be alined perpendicular to the contact. Crystals of large size, ranging up to several inches in length, are common, and a wide range of size may be associated in a single specimen.

Darker-green actinolite may develop locally, where the original sediment contained considerable iron. Actinolite is very similar to tremolite in its habit but does not generally develop into large crystals. The darker amphiboles are generally indicative of a higher grade of metamorphism than is tremolite.

Pyroxenes. Diopside is the typical pyroxene of metamorphosed dolomitic rocks, but a pyroxene approaching hedenbergite will develop where iron has been introduced metasomatically. Diopside is typically gray to light green and occurs as nearly equidimensional prismatic crystals (Fig. 7-28). The crystals are often euhedral, particularly where there is some amount of carbonate remaining from the metamorphic reactions. In this case weathering or treatment with acid will remove the calcite, leaving diopside crystals apparently growing into cavities. The crystal faces generally have a greasy luster, are marked by pits, and are somewhat rounded. Parting parallel to the (001) pinacoid is commonly developed in diopside, probably as a result of shearing. During retrograde metamorphism, diopside may be uralitized to fibrous tremolite in parallel orientation to the original diopside. Thus light-gray to white crystals having an octagonal prismatic outline show 60° amphibole cleavage.

Dark-green to nearly black granular hedenbergite develops with the metasomatic introduction of iron. It is therefore generally associated with other iron-rich silicates, particularly andradite garnet. The hedenbergite typically consists of equidimensional rounded to polygonal granules in completely random orientation. Its very poor 90° cleavage distinguishes it from other dark minerals, such as hornblende.

Forsterite. The magnesium-rich olivine forsterite develops in dolomites which are relatively poor in silica. It is generally associated with both calcite and dolomite and may be associated with a silica-deficient mineral, commonly spinel, or be serpentinized. Forsterite is difficult to identify megascopically so that the associated minerals are indicative of its presence. The crystals are usually small rounded granules of white, yellowish, or gray color. Occasionally, the crystals are euhedral with rectangular outlines. Serpentinization is common, producing a rock of soft, green, rounded serpentine masses embedded in coarsely crystalline carbonates. Such a combination is called ophicalcite. The generally associated spinel is the magnesium-aluminum spinel and is essentially black. The crystals are rounded granules, octahedra, or massive intergrowths of granules.

Garnets. Grossularite- and andradite-rich garnets are typically developed during metamorphism of nonmagnesian calcareous rocks. Andradite-rich garnets are indicative of iron introduction, whereas grossularite develops from aluminous limestone. Grossularite is typically a light-colored yellow or green. The crystals are usually relatively coarse and are euhedral where grown against carbonates. The most common form is the dodecahedron with the trapezohedron absent or minor. Andradite is usually a cinnamon brown and most commonly is coarsely crystalline, anhedral, and granular. Well-developed crystal faces are found against carbonates and usually show subequal development of the dodecahedron and trapezohedron. Andradite masses often include other minerals developed simultaneously. Wollastonite, hedenbergite, and vesuvianite are all commonly included. Almandite and pyrope garnets are not developed from these rocks.

Wollastonite. White bladed, fibrous, or granular wollastonite is typical of high grades of metamorphism of calcite rocks (Fig. 7-29). Generally, it is indicative of an introduction of silica and is most common in contact-metasomatosed skarns. The crystals are generally anhedral, prismatic to elongated, and are strongly cracked parallel to the c axis. Three cleavages are well developed, with angles close to 45° between them. The cleavage distinguishes wollastonite from white tremolite or diopside. Coarse granular material usually includes green granular vesuvianite or brown granular andradite. Very fine-grained wollastonite may develop and is difficult to identify megascopically. Generally, it has a silky fibrous appearance due to the three well-developed cleavages, but its occurrence at the contacts of limestone with intrusive bodies is its most diagnostic feature.

Vesuvianite (Idocrase). Vesuvianite is a common product of the metamorphism of argillaceous ferruginous limestone. It occurs typically in one of two habits: thin bright-green dipyramidal crystals or greenish-brown to brown massive prismatic crystals. The bright-green variety appears to be more typical of contact skarns associated with wollastonite and andradite. The crystals are euhedral to rounded lenticular shapes with smooth greasy-appearing surfaces. The brown prismatic crystals appear to be typical of a regional-metamorphic environment. They are rarely euhedral, but prism faces are commonly developed which are markedly striated parallel to the c axis. Crystalline masses are usually coarse-grained and in a hand specimen may be dominantly parallel. In an outcrop, however, the orientation is random.

Minor or less common minerals. The epidote group is occasionally very important in aluminous limestone equivalents. Green glassy granular epidote or white to gray translucent bladed zoisite or clinozoisite are characteristic low-grade minerals. Zoisite and clinozoisite are difficult to identify megascopically. They occur as columnar prismatic masses which are generally strongly striated parallel to their length and as radiating bladed crystals.

The feldspars are normally minor in these rocks. Anorthite may appear at high grades, taking the place of epidote group or grossularite at low grades. The crystals are generally rounded granules without twinning striations. Potash feldspars may appear at medium to high grades, taking the place of muscovite at lower grade. Potash feldspar is usually microcline. Both feldspars are generally masked by other minerals in highly silicated rocks.

The typical micas are muscovite and phlogopite. Muscovite represents recrystallized clay minerals developed in a closed system, where reaction to form lime and lime-alumina minerals is inhibited. The crystals are generally hexagonal to rounded books and are nearly colorless. Phlogopite develops in argillaceous dolomite as rounded to hexagonal brown books. Thin flakes are faintly brown. When mica crystals are embedded so that only the edges of the books are visible, they appear dark-brownish and deeply grooved parallel to the cleavage.

Scapolites develop abundantly in lieu of feldspars in the presence of volatiles, particularly the halogens. Pink, purple, brown, and white to gray are typical colors. The crystals are strongly prismatic and show two distinct cleavages at 90°. The chondrodite group also develops in the presence of fluorine as complex monoclinic crystals which are yellow, brown, or red in color and have smooth resinous-lustered crystal faces.

Characteristic metallic minerals are graphite and pyrrhotite, although contact-metasomatosed rocks may have many other minerals and are important ores. Graphite appears as black flakes, often several millimeters in diameter. The flakes are thin but generally show no preferred orientation. Minute flakes of graphite may be abundant in the carbonate, giving them a gray color. Pyrite and pyrrhotite generally occur as irregular to lensoid masses, films, or crystal clusters.

GRANULITES

Granulites, as used here, are equigranular medium- to fine-grained rocks with only poorly developed planar or linear structural elements. They are dominantly quartzo-feldspathic but show their metamorphic origin through their accessory minerals and field relationships. For the most part they belong to the granulite facies, but many rocks of that facies show strongly developed gneissosity and are excluded here. Granulites may be developed in several ways. They may be the very-high-grade equivalents of argillaceous or graywacke sediments, the product of metasomatic granitization, or the product of metamorphosed igneous rocks of silicic types. The high-grade equivalents of mafic igneous rocks are also included as granulites if they show only poorly developed planar structure.

Mineralogically, granulites consist dominantly of quartz, plagioclase, and orthoclase in varying proportions (Fig. 7-30). Hornblende, hypersthene, and diopside are the dominant ferromagnesian minerals, and garnet, sillimanite, kyanite, and cordierite are the characteristic accessory minerals. Biotite may be present, but generally it is destroyed at the temperature of granulite formation.

Quartz or feldspar or both form a uniform granular groundmass of anhedral crystals. Generally, when quartz is abundant, both potash feldspar and sodic plagioclase are present, usually in subequal amounts. As quartz becomes minor, the feldspar becomes dominantly sodic to intermediate plagioclase. Quartz commonly occurs as lenticular to rodded grains, giving some

Fig. 7-30. Diorite granulite. Hornblende rims have formed around large garnet porphyroblasts. Gore Mountain, Warren County, New York. (*Ward's Universal Rock Collection.*)

directional fabric to the rock. Feldspar also tends to occur as flattened grains, but less noticeably so, and then without noticeable preferred crystallographic orientation. Plagioclase rarely shows twinning striations in these rocks, probably because of active recrystallization. Therefore the presence of two feldspars can be recognized only if they are of different color or have undergone different degrees of alteration. Grain boundaries among the three minerals are generally highly irregular.

Diopside and hypersthene are both characteristic of higher-grade granulites. Again, anhedral, irregular shaped, granular crystals are characteristic. Hypersthene, however, tends to be somewhat elongated and crudely prismatic. The presence of three directions of cleavage, one of which is quite good, distinguishes hypersthene and gives the crystals a fibrous appearance. Hypersthene also tends to be darker in color than associated diopside. Hornblende is characteristic of slightly lower grades of metamorphism than the pyroxenes, and it may have associated biotite. Parallel or subparallel orientation of elongated hornblende anhedra or biotite flakes often is the only readily recognizable directional fabric. Those hornblende prisms oriented in the preferred direction tend to be several times longer than they are in diameter, whereas those hornblende crystals in other orientations within the same rock tend to be stubby and more nearly equidimensional.

Garnet is a characteristic accessory mineral in granulites of all compositional ranges. The garnet is generally in the almandite-pyrope range of compositions and has a characteristic purplish-pink color. Crystals are, for the most part, anhedral, irregular, and often lens-shaped. Inclusions of quartz are common in the garnets, frequently becoming so abundant that the crystal is almost skeletal. The garnets may be considerably larger than the groundmass, forming porphyroblasts. Sillimanite is also a common accessory, particularly in those granulites rich in quartz and potash feldspar. Two habits are characteristic: as individual bladed crystals and as matted fibers. Single crystals are generally greatly elongated and essentially euhedral. They are colorless with a very perfect cleavage, giving a highly lustrous face. Fibrous sillimanite generally occurs included in quartz or in quartz-rich zones of the rock. The fibers are generally minute and randomly oriented. Less common are radiating clusters of sillimanite fibers. On the weathered surface the quartz-sillimanite aggregates weather to rounded knobs of a lusterless dead-white color. Kyanite may proxy for sillimanite under high pressure. The kyanite here is generally colorless to white and has only a trace of its typical blue color. Crystals are generally euhedral to subhedral and bladed. They show characteristic striations parallel to their length and irregular cross fractures. Cordierite develops in pyroxene-poor rocks as very irregular, distinctly blue masses which generally include the minerals of the groundmass. Minor recognizable accessory minerals include plates of coarsely crystalline graphite and granular magnetite.

SERPENTINE

Serpentine rocks develop at low temperatures from ultramafic igneous rocks. They are generally interpreted as being a product of autometamorphism, but many may have been intruded as essentially solid serpentine masses. The rocks are composed dominantly of various serpentine minerals plus possible remnants of olivine or pyroxene and commonly extensive carbonate veining. Extensive fracturing of the rock is common, often with the development of slickensides on the fractures. No systematic order for the fracturing or displacement as indicated by slickensides can usually be recognized.

The main mass of the rock is generally composed of extremely fine-grained material of various shades of green. Olivine textures may occasionally be recognized as granules cut by irregular cracks, with the borders and cracks somewhat darker than the remaining material. Original pyroxene textures may be recognizable as patches which show a bronzy luster due to minute plates of oxides parallel to the (100) face of the original pyroxene. Serpentine after orthopyroxene is also commonly finely platy and darker-colored due to disseminated fine granular iron oxides. In most serpentine, however, neither texture can be recognized megascopically. An appreciable proportion of fine-grained carbonates and iron oxides is usually admixed.

Fibrous serpentine, approaching asbestos, is common in most massive serpentine rocks as either cross-fiber veinlets or slickensides (Fig. 7-31). The asbestiform varieties are generally lighter in color than the main mass of the rock. Fracture fillings generally show fibers perpendicular to the walls of the fracture throughout most of the vein width. The fibers may be bent along the walls, however, giving a drag effect and indicating movement after development of the fibers. Slickensided surfaces are often coated with asbestiform serpentine, and the orientation of the fibers indicates the apparent direction of movement. This material is generally much coarser, more

Fig. 7-31. Carbonate veining cutting serpentine. Presque Isle Park, Marquette, Michigan.

Fig. 7-32. Asbestiform serpentine and talc developed in massive serpentine. Verde Antique quarry north of Ishpeming, Michigan.

compact, and cannot be readily shreaded into fine flexible hairs. It can often be broken into stiff fibrous units which cannot be further broken down.

Carbonate veins composed of calcite, dolomite, or magnesite are very common (Fig. 7-32). The veins generally divide the rock into polygonal blocks with zones several inches wide densely cut by small carbonate stringers, whereas the blocks are only sparsely cut. On a weathered surface the zones of abundant carbonates are much more readily weathered.

Magnetite and chromite are common accessory minerals of the serpentines. Often they occur in well-defined zones composed dominantly of fine granular oxides. The zones may be as regular as bedding, but a single zone may bifurcate, and stringers of oxide-rich material may extend diagonally from one oxide bed to another. Sulfides may be abundant in the carbonate veins, with pyrite, pyrrhotite, and chalcopyrite as the commonest minerals.

APPENDIX FUNDAMENTALS OF PHASE RELATIONSHIPS

Phase relationships as represented by phase diagrams are of prime importance in the understanding of igneous rocks. Experimental determination of the crystallization and composition of minerals is the basis of modern petrology. Through these studies the changing composition of the liquid as crystallization proceeds can be traced, and the relationship between various igneous bodies can be predicted.

The basic relationships involved in phase studies are summarized in *Gibbs' phase rule*, $F = C - P + 2$. F is the *number of degrees of freedom:* the number of independent variables which can be changed without causing the appearance of a new phase or the disappearance of an old phase. C is the *number of components* in the system: the smallest number of chemical entities which must be specified to describe chemically all the phases in the system. P is the *number of phases* in the system: the number of homogeneous, physically distinct, and mechanically separable units in the system. The phase rule is applicable under equilibrium conditions. Equilibrium is a state of chemical or physical balance between the phases of the system such that as long as the environmental factors — temperature, pressure, or composition of the system — remain unchanged, there will be no change in the composition or proportions of the phases present. A false equilibrium state may be attained when the rate of chemical or physical change is so slow that it is not detectable. In the study of rock-forming systems, this often requires the maintenance of a fixed set of conditions for several days. True equilibrium can be recognized, in that (1) it is sensitive to external change, that is, change of temperature or pressure; (2) the composition and proportions of phases are independent of time; and (3) the same state may be attained from several directions, that is, from either elevating temperature to the required conditions or reducing temperature to the same conditions.

The degrees of freedom are the number of independent variables which can be changed without destroying the given equilibrium state. The independent variables are temperature, pressure, and composition. Other variables exist and are important in certain relations, such as the presence of light in photosensitive reactions, the presence of electric or magnetic fields, the position in a gravity field, or the relationship in capillary or adsorbed films. These variables normally have little influence on equilibrium among rock-forming materials and can be ignored for the present purposes. The early silicate-phase studies were carried out under atmospheric pressures and in the absence of volatile phases. As a result, for these classical and fundamental studies pressure can be ignored, because the vapor pressure of silicate liquids and solids is very low. For the purposes here, then, Gibbs' phase rule can be reduced to $F = C - P + 1$, and only the variables of temperature and composition need be considered. In contemporary experimental work, techniques have been developed to treat systems under high pressure in the presence of volatiles, water, carbon dioxide, and oxygen. In such studies pressure must be considered, so what is said here is not completely applicable to such work.

Components of a system are the smallest number of chemical elements or compounds necessary to define the composition of all phases present. In a simple system involving water and ice in equilibrium at the freezing point, both phases, liquid water and solid ice, have compositions expressible by the compound H_2O, and therefore this is a one-component system. It could be expressed in terms of two components, hydrogen and oxygen; however, this is unnecessary, because the ratio of hydrogen to oxygen is always fixed at $2:1$. Inasmuch as components of a system have been defined as the smallest number of chemical units necessary to define the system, the water-ice system is a one-component system. This becomes particularly important when considering the complex rock-forming mineral systems. Thus in dealing with the plagioclase feldspars, there are several ways in which they can be considered. If a system consists of only albite ($NaAlSi_3O_8$) and anorthite ($CaAl_2Si_2O_8$), the components could be considered as Na-Ca-Al-Si-O or as Na_2O-CaO-Al_2O_3-SiO_2. However, the ratios of these components are fixed in plagioclase in the proportions $1:1:3:8$ for Na: Al: Si: O and $1:2:2:8$ for Ca: Al: Si: O or, in the oxide relations, $1:1:6$ for $Na_2O: Al_2O_3: SiO_2$ and $1:1:2$ for $CaO: Al_2O_3: SiO_2$. The smallest number of chemical units which expresses the composition of all possible plagioclase compositions is two, $NaAlSi_3O_8$-$CaAl_2Si_2O_8$, and this is therefore a two-component system. If, however, the ratios of the four oxides involved in plagioclase are not fixed in the plagioclase proportions, then plagioclase becomes a part of a larger more complex system with more components.

Phases in a system are the physically distinct, homogeneous, and mechanically separable units in the system. A phase may be solid, liquid, or gas, but in anhydrous silicate systems, only solid and liquid phases need be considered. Physically distinct phases may be of the same composition but of different structure, such as orthorhombic and monoclinic pyroxenes or the different structures of silica. Each would be considered a separate phase.

A phase, however, does not need to be continuous; it may consist of a number of distinct separate crystals or bodies of liquid, all having the same composition and structure. The homogeneous aspect of a phase implies complete uniformity of composition. Thus zoned or mantled crystals, discussed in Chap. 2, cannot be considered a single phase. Instead, each zone represents a different phase, and generally, zoning indicates a lack of attainment of equilibrium. Mantled crystals, in contrast, may represent equilibrium conditions, the core and mantle each representing a mechanically separable homogeneous unit.

Intermediate compounds remain to be defined. They consist of phases whose composition consists of a mixture of the components in fixed proportions. Thus in the system forsterite-silica (Mg_2SiO_4-SiO_2), clinoenstatite is an intermediate compound formed by the reaction

$$Mg_2SiO_4 + SiO_2 \rightleftharpoons 2MgSiO_3$$

The two components are represented in the intermediate compound in the fixed proportions of $1:1$. There is some variability in the composition of phases due to solid solution of one phase in another. In most silicate systems involving different mineral families, this is very limited. Thus in the system forsterite-silica, with the intermediate compound clinoenstatite, clinoenstatite will accommodate very little excess or deficiency of silica. In other words clinoenstatite will dissolve little forsterite or silica in solid solution. Within a single-mineral family there is commonly a wide range of solid solubility; some examples involve complete solid solution between the end members. Thus in the plagioclases and olivines a complete solid solution exists, whereas in the monoclinic pyroxenes only a limited solid solution exists.

TWO-COMPONENT SYSTEMS

Many two-component systems are important in petrology, and a discussion of each is impossible. Fortunately, in all diagrams there are a few basic principles on which a complete interpretation of the diagram can be based. The most important factor is that at any one temperature value and composition of the total sample, only one thing can be happening under equilibrium conditions at a time. This means that for any sample at one temperature, only one limited area of the diagram need be considered, and all the rest of the diagram can be ignored. Similarly, for any one bulk composition only that portion of the diagram crossed by that bulk-composition line need be considered.

GRAPHICAL REPRESENTATION

Under conditions assumed above, in which only temperature and composition are considered, the phase relationships can be represented in a two-dimensional area. Mixtures in all proportions of two components can be represented by a series of points on a straight line. If one end of the line is assigned as 100 percent of one component A, and the other end of the line is assigned as 100 percent of the other component B, then the line can be divided into units representing all mixtures of A and B. The midpoint represents 50 percent A and 50 percent B. A point one-third the distance from the A end toward the B end represents a mixture of $\frac{2}{3} A$ and $\frac{1}{3} B$. The relationships are illustrated in Fig. A-1. The composition line is plotted as the abscissa of the two-component diagram, and temperature is plotted as the ordinate. The variables in a system at a particular instant can be plotted as a point on a horizontal line at the temperature of the system; the point is vertically above the composition of the system.

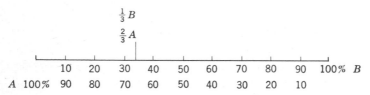

Fig. A-1. Representation of composition in a two-component system.

In a two-component system under equilibrium conditions, only three situations are allowable according to Gibbs' phase rule, the one-, two-, and three-phase situations:

$$F = C - P + 1$$

$2 = 2 - 1 + 1$	one phase, two degrees of freedom
$1 = 2 - 2 + 1$	two phases, one degree of freedom
$0 = 2 - 3 + 1$	three phases, no degrees of freedom

No fewer than one phase can exist, and there cannot be a negative number of degrees of freedom. With a basic understanding of these three relationships, any diagram may be interpreted.

ONE-PHASE SITUATION

The one-phase situation allows two degrees of freedom. This means that the composition of the phase may be changed within limits without causing the appearance of a new phase. Similarly, the temperature of the system can be changed within limits without the appearance of a new phase. The physical and chemical properties may change with changing temperature and composition, but the changes of properties will be continuous and without discontinuities. Thus in the two-component system salt-water, all degrees of salinity are possible from almost pure water to the saturation point, and the temperature of the system may vary over a wide range. Hot highly salty water is very different in its properties from cold slightly salty water, but one can be converted to the other by addition or removal of heat, addition of more salt, or dilution with more water without passing through a detectable abrupt change in the properties of the material. Thus they both belong to the same phase. In the phase diagram a one-phase situation is represented by an area bounded by two- and three-phase relations. One-phase fields are generally labeled on the diagram according to the phase which is stable through that range of composition and temperature.

If a given composition at a specified temperature falls in a one-phase field, change of temperature will move the projection point up or down in the one-phase field without change in the phase. With change of composition, the point will move to the left or right in the field. As long as the system remains in a one-phase field, no change will take place.

TWO-PHASE SITUATION

In a two-phase situation there is one degree of freedom. The variables present are the temperature and the composition of each of the two phases. Either of these variables may be changed at will, but as it is changed, the other variable must change in a fixed manner. Thus at each temperature the composition of the two phases is fixed, but if temperature is changed, the composition of the phases must adjust accordingly. Or if the composition of one phase is changed, the temperature and the composition of the other phase must adjust accordingly to maintain equilibrium.

At any given temperature, two phases of contrasting composition will be in equilibrium. One phase will of necessity be richer in one component than the bulk sample, and the other phase will be richer in the other component so that the two will average the composition of the sample. If one phase has the composition of the sample, then either only one phase is present or the two phases are two structural modifications of the same compound. In a diagram the two-phase situation can be represented by two points at the same temperature (t_1), one above the composition of each phase $(c_1$ and $c_2)$. The total system will be represented by a third point somewhere between the two and at the same temperature (see Fig. A-2).

At some other temperature with the same two phases, the situation will be represented by two more points over the compositions of the phases. When all possible temperatures are considered in the two-phase relationship, the infinite number of pairs of points generate two lines. One line represents the changing composition of one phase and the other line the changing composition of the other phase. From theoretical considerations these two lines may have any attitude except horizontal. Horizontal lines are ruled out, because a horizontal line represents an infinite number of compositions of minute differences all coexisting at the same temperature, and as such it does not represent two phases but a large number of phases.

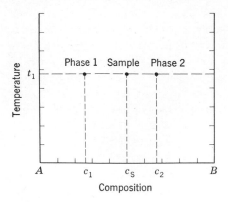

Fig. A-2. Plotting a two-phase situation at one temperature.

The two lines may be parallel, converging, diverging, vertical, or inclined in either direction (see Fig. A-3).

If the composition of both phases is changing with each increment of temperature change while the bulk composition of the sample remains constant, then of necessity the proportions of the two phases must change. In Fig. A-4 the lines representing the composition of the two phases are both sloping to the left, indicating an increase in component A in both. If in a system the bulk composition remains constant and both phases become richer in component A, then of necessity the A-poor phase must give up A and B to form more of the A-rich phase. This requires an increase in the proportion of the A-rich phase and a decrease in the proportion of the A-poor phase. In any two-phase situation the proportion of the two phases

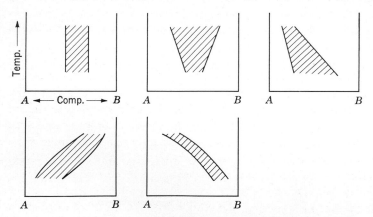

Fig. A-3. Various possible attitudes for two-phase lines. The cross-hatched areas are the two-phase fields.

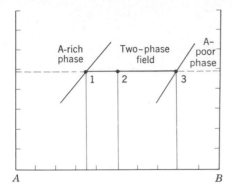

Fig. A-4. Points for the lever rule.

can be determined at any temperature by applying the *lever rule*. If the length of the line joining the composition of the two phases at any temperature is considered as 100 percent of the sample, then in Fig. A-4 the proportion of the phases may be found by measuring lines 1-2 and 2-3, where point 2 is the bulk composition of the sample; thus

$$\% \, A\text{-rich phase} = \frac{2\text{-}3}{1\text{-}3} \times 100$$

and

$$\% \, A\text{-poor phase} = \frac{1\text{-}2}{1\text{-}3} \times 100$$

Note that the amount of a given phase is proportional to the length of that segment of the line joining the composition of the two phases which terminates at the other phase. This relationship can be proved mathematically, and is known as the *lever rule*.

In Fig. A-5 the changing compositions and proportions of the two phases α and β can be followed through the temperature range t_1 to t_5 of a sample of bulk composition X. The two-phase field (the area between the two lines representing the compositions of the two phases) is bounded to the left and right by one-phase fields. At temperature t_1, bulk composition X falls in the one-phase α field. Temperature can decrease to temperature t_2 without change of the equilibrium. At temperature t_2, however, the system enters a two-phase $\alpha + \beta$ field, and from phase α (composition c_1) nuclei of β (composition c_2) begin to crystallize.

On cooling to temperature t_3, the equilibrium composition of phase α is c_3, and the equilibrium composition of β is c_4. In the process of cooling from t_2 to t_3, all preexisting α changes progressively in small increments

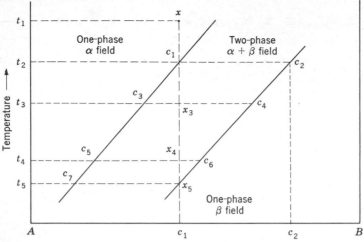

Fig. A-5. Cooling through a two-phase field.

from c_1 to c_3, and similarly, the composition of all β changes from c_2 to c_4. The first-formed β had composition c_2, and as temperature falls, even these first-formed nuclei of β must, under equilibrium conditions, completely change composition with each increment of temperature. It is the lack of adjustment of composition which produces zoning in crystals and is a non-equilibrium condition. At temperature t_3 the proportions of the two phases have changed. Application of the level rule to the three points c_3, X_3, and c_4 indicates that the proportion of phase α has decreased to $X_3\text{-}c_4/c_3\text{-}c_4$ and now composes about $\frac{2}{3}$ of the system. At the same time the proportion of β has increased to $c_3\text{-}X_3/c_3\text{-}c_4$ and now composes about $\frac{1}{3}$ of the system.

At temperature t_4 the composition of phase α has changed to c_5, with greater enrichment in component A, and phase β has changed to composition c_6. Again the composition of each phase must change in its entirety with each temperature change. The proportions of the two phases also change progressively, with a continual decrease in phase α and an increase in phase β. Finally, approaching temperature t_5, the composition of phase β approaches that of the bulk sample and increases in proportion, while phase α approaches composition c_7 and the proportion decreases. When, at temperature t_5, phase β has the composition of the bulk sample X_5, all of phase α has been used up, and the system again consists of one phase β. The last-minute quantity of phase α present had composition c_7 and was greatly enriched in component A. With further cooling, the system enters the one-phase β field, and no further changes take place.

The behavior while cooling through any two-phase field is exactly analogous. The extent to which proportions change is directly dependent on the variability of composition of the two phases. A situation in which

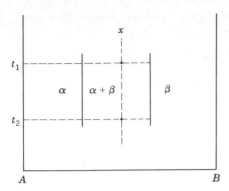

Fig. A-6. Two-phase field with no change in composition or proportions of the phases.

there is no variability is shown in Fig. A-6. Here the two phases have fixed compositions, and the two-phase field is bound by two vertical straight lines. On cooling throughout the temperature range t_1 to t_2, the composition and proportions of the two phases remain constant.

THREE-PHASE SITUATION

According to the phase rule, in a system of two components and three phases there can be no degrees of freedom. This means that the composition of the three phases and the temperature must all be fixed as long as the three phases coexist in equilibrium. This relationship appears on the phase diagram as three points, corresponding to the composition of the three phases, all on one horizontal temperature line. In the diagram the three points are joined by a horizontal line. This line becomes the limit of three associated one-phase fields and three associated two-phase fields, as developed below.

In the three-phase relationship, one of two chemical processes must be occurring: either two phases are reacting chemically to form a third, or one phase is breaking down chemically to form two others. In the first relationship of a reaction between two phases, the product must have a composition intermediate between the composition of the reactants. Thus if phase α, rich in A, and phase β, rich in B, react at a given temperature to produce phase γ, then the composition of γ must be intermediate between that of α and β. It need not be at the midpoint between them, however, because the ratio of α to β need not be $1:1$ in the reaction. At temperatures above the three-phase reaction, α and β can coexist in equilibrium. Therefore, above the three-phase line the diagram consists of a two-phase field which of necessity is limited to the right and left by one-phase fields of α and β. In the reaction $\alpha + \beta \rightarrow \gamma$, the proportions of α and β used in the reaction are fixed. However, the proportions of α and β when the three-phase temperature is reached in the system depend on the bulk composition of the sample. Therefore there may be an excess of one of the reactants over that required by the composition of γ, the reaction product. Reaction will proceed until one of the reactants is used up, and the product will be γ plus excess α or β. If α and β are present in the precise proportions required by γ, then α and β will be exhausted simultaneously, and the product will be one-phase γ. Therefore below the three-phase temperature there are three possible equilibrium assemblages: a two-phase $\alpha + \gamma$ assemblage, a one-phase γ of fixed composition, and a two-phase $\gamma + \beta$ assemblage. The two-phase fields are of necessity bounded to the right and left by one-phase fields. The diagram therefore has the relationships indicated in Fig. A-7. The proportions of the three phases during the reaction cannot be determined. However, proportions of α and β when the three-phase temperature is reached can be determined by applying the lever rule to the two-phase field, using the bulk composition. The proportions after completion of the reaction can be determined by applying the lever rule to the appropriate two-phase field below the three-phase temperature in which the bulk composition lies.

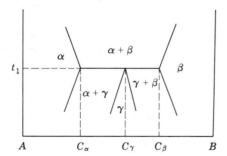

Fig. A-7. The three-phase situation at temperature t_1, $\alpha + \beta \rightarrow \gamma$. Composition of the phases are C_α, C_β, and C_γ.

The second three-phase situation involves the breakdown of one phase to yield two phases on cooling. The composition of the phase breaking down must be intermediate between the composition of the two phases produced. Phase α breaks down to yield phase β, richer in A than α, and phase γ, richer in B than α. Again, the three-phase situation has associated with it three one-phase fields, α, β, and γ, and three two-phase fields $\beta + \alpha$, $\alpha + \gamma$, and $\beta + \gamma$. The assemblage $\beta + \gamma$ is not stable above the three-phase-reaction temperature, and the phase α is not stable below that temperature. The relationship is shown in Fig. A-8. Bulk composition X at temperature t_1 is

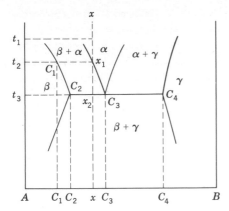

Fig. A-8. The three-phase situation at temperature t_3, $\alpha \rightarrow \beta + \gamma$. Compositions of the three phases at the three-phase reaction are $\alpha = C_3$, $\beta = C_2$, and $\gamma = C_4$.

in a one-phase α field. As temperature falls, there is no change until temperature t_2 is reached, at which point phase β begins to separate from α. The composition of the first β is c_1. With further cooling to the three-phase temperature, more β separates, and its composition shifts from c_1 to c_2. At the same time the composition of α shifts from X_1 to c_3. The proportions of α and β can be found by applying the lever rule to the line c_2-X_2-c_3. At the three-phase temperature, α of composition c_3 breaks down to form more β of composition c_2 and more γ of composition c_4. The preexisting β does not enter into this process but remains unaffected. As long as all three phases are present the temperature must remain constant, and α breaks down progressively into more β and γ. When α is exhausted, the system consists of β of composition c_2 and γ of composition c_4. The proportions can be found by applying the lever rule to the line c_2-X_2-c_4. The system now enters a two-phase $\beta + \gamma$ field, and further change of composition and proportions of the two phases follows the rules for a two-phase situation.

It should be noted that any bulk composition between the limiting compositions of c_2 and c_4 must pass through the three-phase reaction, and in each sample the composition of phase α when it reaches the three-phase temperature must be c_3. Bulk samples whose compositions lie outside the range c_2-c_4 cannot reach this three-phase reaction under equilibrium conditions. This is a general rule and holds in all equilibrium states. The only phase relationships in which a given bulk sample can be involved are those one- and two-phase fields and three-phase lines through which a vertical line drawn from the composition of the sample passes.

The three-phase reactions are named according to the nature of the reaction and the nature of the phases. Five of the ten possible situations are important in igneous petrology and are named as follows:

Liquid $\alpha \rightarrow$ solid $\beta +$ solid α Eutectic
Solid $\alpha \rightarrow$ solid $\beta +$ solid α Eutectoid
Liquid $\alpha \rightarrow$ liquid $\beta +$ solid α Monotectic
Liquid $\alpha +$ solid $\beta \rightarrow$ solid α Peritectic
Solid $\alpha +$ solid $\beta \rightarrow$ solid γ Peritectoid

The first three are of the second type discussed above. The eutectic is particularly important, in that it involves the crystallization of two phases from a molten mixture of those two phases. The reaction between olivine crystals and silica-enriched liquid to form clinoenstatite is a simple peritectic reaction.

Figures A-9 and A-10 are both two-component systems, each involving all five important three-phase reactions. They differ only in that each phase in Fig. A-9 has some variability in composition, whereas each phase in Fig. A-10 has fixed composition, except for the liquid phases. The three-phase reactions are (1) a peritectic, (2) a monotectic, (3) a eutectic, (4) a eutectoid, and (5) a peritectoid. The cooling history of liquid X involves the following sequence of phase assemblages:

X to a	A one-phase-liquid field.
a to b	A two-phase-liquid $+ \alpha$ assemblage, involving progressive crystallization of α from the liquid.
At temp. b	A three-phase reaction, liquid $+ \alpha \rightarrow \beta$; the temperature remains fixed until one phase is used up, in this case the liquid, leaving two solid phases α and β.
b to c	Cooling through a two-phase field, with the composition and proportions of the two phases changing.
At c	A three-phase situation; phase β becomes unstable and breaks down to form two solid phases α and γ; the preexisting α is not involved; at the end of the reaction there is a large proportion of α and relatively minor γ.
c to d	A two-phase $\alpha + \gamma$ field; on cooling there is minor adjustment of composition and proportions of the phases.
At d	A three-phase reaction; α and γ are no longer stable in association and react chemically to form δ; γ is rapidly used up in the reaction so that the product is a mixture of α and δ.
d to e	A two-phase field $\alpha + \delta$; the composition of α changes toward an enrichment in component B and approaches the bulk of the sample; the proportion of phase δ decreases progressively, as indicated by the diminishing gap between the α-composition curve and the bulk-sample line.

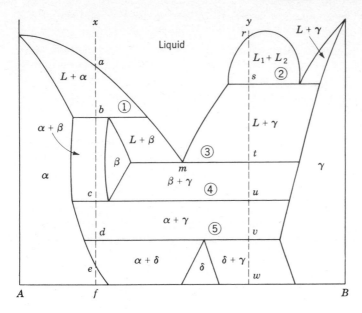

Fig. A-9. Two-component system with variable compositions of the phases. See text for discussion.

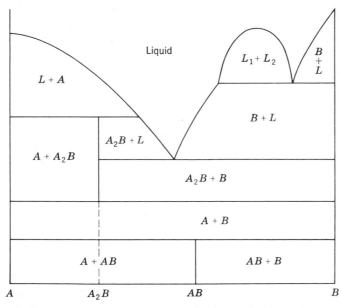

Fig. A-10. Two-component system like Fig. A-9 but with no variation allowed in the composition of any phase.

At *e*	Phase α has the composition of the sample, and phase δ is exhausted; the system enters a one-phase α field.
e to *f*	Cooling through a one-phase field of the entire sample composed of α the same composition as the original example X.

The cooling history of liquid Y involves the following relationships:

Y to *r*	A one-phase-liquid field.
At *r*	The liquid begins to break down into two immiscible liquids, one richer in A, the other richer in B.
r to *s*	As the liquids cool, each changes in composition, exsolving immiscible droplets of the other; both liquids change continuously, and the proportions of the two change accordingly.
At *s*	The three-phase monotectic is reached; the liquid richer in component B breaks down, forming crystals of γ and producing a liquid identical in composition to the A-rich liquid present when the monotectic was reached.
s to *t*	A two-phase-liquid $+ \gamma$ field involving progressive crystallization of γ; the composition of the liquid moves progressively toward composition m.
At *t*	Liquid of composition m crystallizes at a fixed temperature to yield two crystalline phases β and γ; the liquid composition m is known as the *eutectic composition*; preexisting solid γ does not enter into the eutectic crystallization.
t to *u*	A two-phase solid $\beta +$ solid γ field; the compositions and proportions of the two phases shift with each increment of temperature change.
At *u*	Phase β becomes unstable and breaks down to form two solid phases α plus more γ; preexisting γ does not enter into the breakdown.
u to *v*	A two-phase $\alpha + \gamma$ field; compositions and proportions of the phases shift slightly on cooling.
At *v*	Phases α and γ can no longer coexist stably but react to form δ; the proportions are such that α is exhausted and considerable γ is left over.
v to *w*	A two-phase $\delta + \gamma$ field; compositions and proportions of both phases shift while cooling through the field.

Two additional important terms must be defined and can be related to Fig. A-9. The *liquidus* is that assemblage of lines in the phase diagram above which all phases present are liquid. The two-phase field involving two immiscible liquids above the monotectic is the only two-phase field above the liquidus. The remainder of the liquidus is composed of the liquid composition line in the two-phase fields $L + \alpha$, $L + \beta$, $L + \gamma$, and $L + \gamma$ above the monotectic. The *solidus* is that assemblage of lines below which all phases are solid. In Fig. A-9 it is composed of the lines representing the composition of the solids in the two-phase fields $L + \alpha$, $L + \beta$, and $L + \gamma$ plus a portion of the peritectic line and the eutectic line. Below this assemblage of lines, all phases are solid.

THREE-COMPONENT SYSTEMS

The three-component systems are of great importance to igneous petrology, because with them, a reasonable first approximation of simple magma systems can be investigated. As in two-component systems, there are many different important diagrams, but in each, a simple group of basic relationships exist. Under equilibrium conditions only one thing can occur at a time so that interpretation is much simpler than the complexity of the diagrams would indicate. To reduce the complexity of the diagrams, only those relationships involving crystallization from a liquid will be considered here. This simplification is justifiable, because the main concern in igneous petrology is the processes involved in the crystallization of the magma. Many important relationships exist in the subsolidus portion of the diagrams, but these do not influence differentiation of the magma.

GRAPHICAL REPRESENTATION

The variation of the proportions of three components can be represented on an equilateral triangle. If the three apices of the triangle represent 100 percent of components A, B, and C, then any mixture of the three components which totals 100 percent will be represented by a point within the triangle. In Fig. A-11 let the top apex represent a sample composed of 100 percent component A, the lower-left apex represent a sample composed of 100 percent component B, and the lower-right apex represent a sample composed of 100 percent component C. Then the base of the triangle represents a two-component B-C system free of A. This line can be divided to represent various proportions of B and C in the same manner as in two-component systems. Similarly, the A-B side of the triangle represents the two-component A-B system, and the A-C side represents the two-component A-C system. Thus each three-component system is limited by three two-component systems.

If the three altitudes of the triangle are drawn from each apex to the midpoint of the opposite base, they intersect at a point one-third the distance from the base to the apex. The altitudes are divided into units of 100 percent at the apex to 0 percent at the base. The percentage of each component in a three-component mixture is first picked on the appropriate altitude. Three lines drawn at right angles to each altitude at the appropriate percentage for each component will intersect at a point. This point of

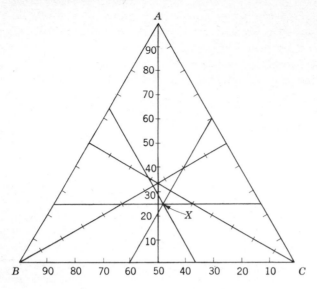

Fig. A-11. Plotting a three-component composition on an equilateral triangle. Point X represents 25 percent A, 35 percent B, and 40 percent C. See text for discussion.

intersection represents the composition of the mixture. Thus, given a sample composed of 25 percent A, 35 percent B, and 40 percent C, the plot of this mixture can be found in Fig. A-11 as follows:

1. On the altitude from apex A to side B-C, measure 25 percent of the length from B-C toward A.
2. On the altitude from apex B to side A-C, measure 35 percent of the length from side A-C toward B.
3. On the altitude from apex C to side A-B, measure 40 percent of the length from side A-B toward C.
4. From the 25-percent point on the A altitude, draw a line parallel to base B-C.
5. From the 35-percent point on the B altitude, draw a line parallel to side A-C.
6. From the 40-percent point on the C altitude, draw a line parallel to side A-B.

The three lines will intersect in a point which represents the total sample. Any other mixture of A, B, and C will plot at a different point. If $A + B + C$ is not equal to 100 percent, then there must be an additional component or several additional components. In such a case the lines perpendicular to the altitudes will not intersect in a point but will form a triangle, the size of which will depend on the proportion of the additional components.

In addition to composition, the variable temperature must be indicated in the diagram. This requires the introduction of a third dimension and requires the use of a trigonal prism having composition as the base and temperature as the altitude. Within this prism the important phase relationships are represented by inclined surfaces and lines. It is therefore possible to reduce the prism to a two-dimensional figure by contours of equal temperature — isotherms. Each horizontal plane in the prism represents the phase relationships at one temperature. The lines of intersection between a temperature plane and all the surfaces in the prism become an isotherm in the two-dimensional representation (see Figs. A-12 and A-13).

In a three-component system under equilibrium conditions, only four

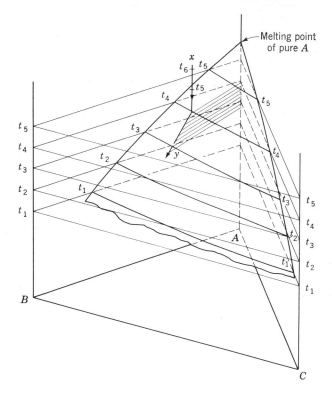

Fig. A-12. A three-component trigonal prism with an interior surface. Five temperature planes intersect the surface. Above temperature t_4, liquid X is in a one-phase volume. At t_4, A begins to crystallize out and the liquid composition moves along the surface toward Y.

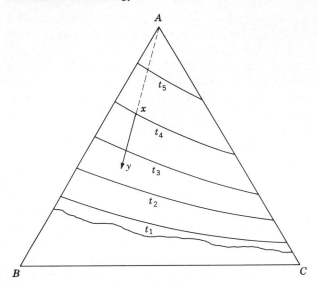

Fig. A-13. Trigonal prism of Fig. A-12 reduced to a plane with the inclined surface contoured as isotherms. Liquid X begins to crystallize out solid A at temperature t_4, and the composition of the remaining liquid moves directly away from A along line X-Y.

fundamental situations exist. According to Gibbs' phase rule, where $F = C - P + 1$:

$$3 = 3 - 1 + 1 \quad \text{A one-phase relation}$$
$$2 = 3 - 2 + 1 \quad \text{A two-phase relation}$$
$$1 = 3 - 3 + 1 \quad \text{A three-phase relation}$$
$$0 = 3 - 4 + 1 \quad \text{A four-phase relation}$$

No fewer than one phase may exist, and negative degrees of freedom cannot exist. In the following discussion, situations in which the solid phases have fixed composition and no appreciable solid solution will be discussed first. Subsequently, relationships involving solid solutions will be considered.

ONE-PHASE SITUATION

With three components and only one phase, three degrees of freedom are allowable. This means that temperature and composition of the phase are variable within limits without causing the appearance of a new phase. The third degree of freedom is meaningless here, because only two variables are being considered. In the trigonal prism representing such a system, the one-phase situation must occupy a volume within the prism, because composition can be varied, giving a horizontal area, and temperature can be varied, a vertical component, generating a volume. In the following discussions only one one-phase volume is required, and that is the one-phase liquid region. That one-phase volume occurs at all temperatures above that indicated by the isothermal line appropriate for the given sample composition. One-phase solid conditions may also exist at lower temperatures if there is mutual solubility between components in the solid state. Such a condition is not considered yet.

TWO-PHASE SITUATION

The two-phase situation is encountered where one component or intermediate compound is crystallizing from the liquid. It may also be encountered in an immiscibility gap between two liquids, but such gaps do not occur in compositions approaching igneous conditions and will not be considered. With two phases and three components, two degrees of freedom exist; this means that composition can be varied within limits without required temperature change. That is, all liquid compositions lying on an isotherm are in equilibrium with the same crystals at the same temperature. The two-phase situation is represented by a surface inclined downward into the prism from the melting temperature of the pure solid phase involved and by a vertical line through the composition of the solid. The surface represents all compositions of the mixed liquid in equilibrium with the crystal phase at the temperature of equilibrium. Thus in Fig. A-12 the inclined surface represents all compositions of a liquid composed of components A, B, and C which are in equilibrium with crystals of A only and gives the temperature of that equilibrium. Similar surfaces will be present in the prism, representing all composition of liquids in equilibrium with crystals of B and C (Fig. A-14).

In Fig. A-13, a planar reduction of the trigonal prism (Fig. A-12), when a mixture of composition X is cooled from a high temperature, a two-phase surface is reached at t_4, and crystals of component A begin to separate. The liquid, as a result, begins to be depleted in component A, and the composition of the liquid begins to move directly away from A. This illustrates a basic principle in the interpretation of three-phase diagrams, in that the composition of the liquid always moves directly away from the composition of the solid material then separating. As crystallization proceeds, composition changes in the direction of falling temperature but not necessarily in a direction at right angles to the isotherms. To repeat, the composition of the liquid shifts directly away from the composition of the phase separating. At any time in the two-phase situation the proportions of the phases can be found by applying the lever rule to the line joining the present composition of the liquid and the composition of the solid phase; the point representing the bulk composition of the sample is the intermediate point of the lever.

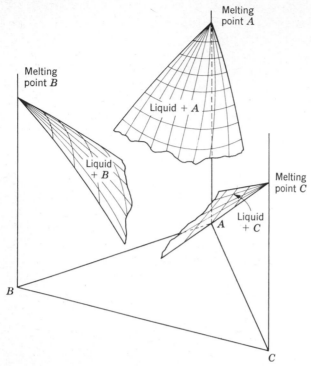

Fig. A-14. Three partial surfaces within the trigonal prism representing liquids in equilibrium with crystals of *A*, *B*, and *C*.

Thus as cooling proceeds, the proportion of solid *A* increases and the proportion of liquid decreases.

COTECTIC THREE-PHASE SITUATION

A three-phase situation allows one degree of freedom. This means that temperature can be changed within limits and still retain the same three phases; however, at each temperature the composition of each of the three phases is fixed. Similarly, the composition of the liquid may be changed within limits, but the temperature and composition of the other two phases must change a specified amount.

In the two-phase situations as *A* crystallizes from a liquid, the composition of the liquid changes along the liquidus surface, becoming progressively richer in *B* and *C*. Ultimately, this liquid must become saturated in one or the other of these components, and it must also crystallize from the liquid as temperature decreases. The composition of a liquid from which *A* has been crystallizing when it first becomes saturated with *B* is the point of intersection of the *A* liquidus surface with the *B* liquidus surface and the line of changing liquid composition. The two surfaces representing liquids saturated in A and B are inclined into the prism and must intersect along a curve. This curve represents all compositions of liquid which are saturated in both A and B so that both A and B will crystallize simultaneously from any such liquid. Such curves are called *cotectics* and exist between any pair of components or intermediate compounds which have a eutectic relationship between them in a simple two-phase diagram.

In Fig. A-15 liquid X first becomes saturated with component A. As A crystallizes, the liquid composition moves toward point e. At e the cotectic is reached, and A is joined by B. With three phases the system is fixed to one degree of freedom, and as cooling proceeds, the liquid composition must move along the cotectic in the direction of declining temperature toward point f. Again the liquid must be changing in a direction directly away from the composition of the solid being removed from it. Therefore the composition of the mixture of A and B can be found by drawing a tangent to the cotectic at the liquid composition and extending the tangent to intersect a line joining the composition of the two solid phases A and B at point g. Applying the lever rule to the line A-g-B gives the composition of the solids separating from a liquid of composition f.

At any point in the three-phase situation the proportions of the three phases present may be found by applying the lever rule. A line is drawn from the present composition of the liquid through the composition of the sample

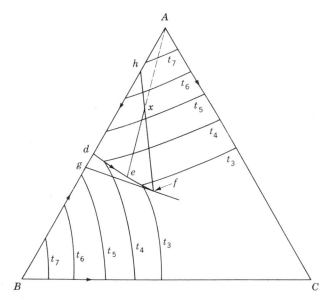

Fig. A-15. Three-phase cotectic crystallization; the cotectic boundary is curve d-e-f. See text for explanation.

and is extended to intersect the line joining the composition of the two solid phases. Thus in Fig. A-15 the line *f-X-h* is drawn, intersecting the line joining the two solid phases *A-B* at *h*. The point *h* represents the total composition of the solid at the time the liquid has the composition *f*. Applying the lever rule to the line *f-X-h*, X-h/f-h represents the percent liquid, and f-X/f-h represents the percent total solid. In the solid, A-h/A-B represents the percent *B*, and h-B/A-B represents the percent *A*. Thus in this example equilibrium is established with approximately

> 30% liquid of composition *f*
> 70% total solid of composition *h*

In the solid there is

> 83% solid *A*
> 17% solid *B*

Therefore the system consists of

> 30% liquid of composition *f*
> 58% solid *A*
> 12% solid *B*

At each temperature along the cotectic line, the proportions will be different for any bulk composition of the sample and examination will show that, on cooling, the proportion of liquid progressively decreases, and the relative amounts of solid *A* and solid *B* change.

THREE-PHASE REACTION BOUNDARY

In two-component systems the peritectic reaction involves the reaction of crystals and liquid to form a set of crystals of contrasting composition. Such a relationship is not destroyed by adding a third component to the system which does not enter into the reaction. Thus in Fig. A-16 at the three-phase reaction (1), crystals of *A* react with a liquid enriched in *B* to form the intermediate compound *AB*. Liquid, *A*, and *AB* have a reaction relationship, whereas liquid, *AB*, and *B* have a eutectic relationship. When a third component *C* is added, these relationships are not destroyed but are simply modified as to temperature of reaction and composition of the liquid. The relationship is now represented by a line *d-e* of Fig. A-17, which represents the composition of the liquid involved in the reaction of A + liquid $\rightarrow$ *AB*. The curve is the intersection of the two two-phase fields liquid + *A* and liquid + *AB*.

Two features of the three-phase diagram identify a *reaction boundary*. First, the composition point of the phase formed by the reaction *AB* does not fall in the two-phase field in which that phase is in equilibrium with the

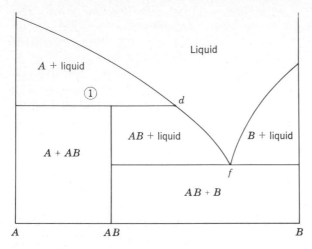

Fig. A-16. Two-component system with peritectic reaction $A + \text{liquid} \to AB$. This diagram forms the $A\text{-}B$ side of the three-component system of Fig. A-17.

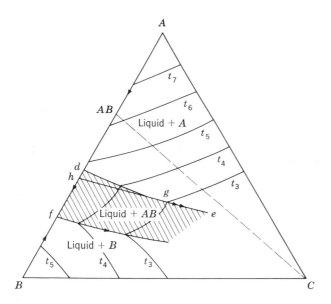

Fig. A-17. Three-component diagram with reaction boundary $d\text{-}e$. Field of $AB \to \text{liquid}$ is cross-hatched. See text for discussion.

liquid. In the case of the cotectic relationship, the composition of the phase always falls in the two-phase field in which that phase is in equilibrium with the liquid. Second, a tangent to the reaction boundary does not intersect the line joining the composition of the two solid phases involved in the reaction at a point intermediate between the two phases. Thus in Fig. A-17 the tangent g-h intersects the line A-AB at point h, which does not fall between A and AB. Point h does represent the composition of the solid separating from liquid g, but this solid (40 percent A + 60 percent B) includes B being removed from the liquid for the reaction A + liquid $\rightarrow$ AB and simultaneous crystallization of some AB as a direct crystallization product and not the product of the reaction.

The changing proportions of phases can be followed in two situations in Fig. A-18a and b, both of which are enlarged portions of the A corner of the triangle. In both diagrams the dashed line is a straight line joining the intermediate compound AB and the third component C. In each, a liquid of composition X begins crystallization with the separation of A, and the liquid moves directly away from A on the line X-d. At point d the reaction boundary is reached, and reaction commences between liquid of composition d and crystals of A to form AB. The composition of the liquid now begins to move along the reaction boundary from d toward e. At e the proportions of

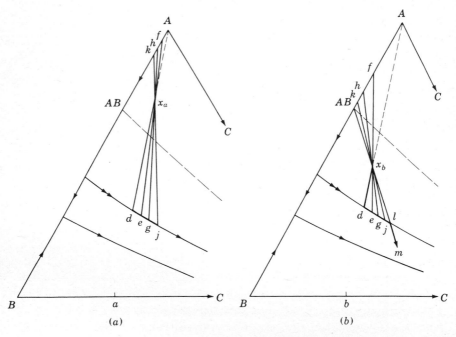

Fig. A-18. Crystallization of two liquids X_a and X_b on a reaction boundary. See text for discussion.

the phases present can be found by drawing a line from the present composition of the liquid (e) through the composition of the sample (X) and extending it to intersect the line joining the composition of the two solid phases A-AB at f. Then f-X is the proportion of liquid, and X-e is the proportion of total solid. In the total solid, A-f is the proportion of AB and f-AB the proportion of solid A.

As reaction continues, the liquid moves from e to g. Again proportions can be found by drawing line g-X-h. By comparing the proportions of solid A with solid AB on the line A-h-AB, it can be seen that in Fig. A-18b the proportion of solid A in the total solid is decreasing much more rapidly than in Fig. A-18a. When the liquid has reached composition j, the solid has proportions indicated by k. In Fig. A-18b very little solid A is left, whereas in Fig. A-18a a major proportion of solid A is left. In Fig. A-18b all A will be used up when the liquid reaches the composition indicated by the line from AB to X extended to reach the reaction boundary. At this point the reaction is complete, and only two phases exist, liquid and solid AB. With further cooling, AB crystallizes in a two-phase relationship, and the composition of the liquid moves directly away from AB along the line l-m.

FOUR-PHASE TERNARY EUTECTIC

During crystallization along a cotectic boundary (Fig. A-15), crystals of A and B are separating simultaneously, and the liquid is continuously enriched in C. Ultimately, the liquid will become saturated with C, and it must join A and B in crystallizing. At this point four phases are present: solid A, solid B, solid C, and liquid. According to the phase rule $F = C - P + 1 = 3 - 4 + 1$, there can be no degrees of freedom, and therefore the composition of all four phases and the temperature are fixed until one phase disappears. In a cooling situation, the phase to disappear would be the liquid.

The composition of the liquid while three phases are crystallizing from it is fixed at a single point in the diagram. This point is known as a *ternary eutectic*. At this point the three solids crystallize out in the proportions indicated by the composition of the point, and temperature remains constant until the liquid is exhausted. In Fig. A-19 A and B have been crystallizing out prior to reaching the ternary eutectic M, and the total solid has the composition indicated by point d. At the eutectic, A, B, and C crystallize in the proportion of point M, and the composition of the total solid moves toward M on the line d-X. When the total solid has the composition X, the composition of the sample, the liquid is exhausted and crystallization is complete.

The ternary-eutectic points occur at the intersection of three cotectic boundaries and represent the minimum temperature in the diagram, or in a portion of a diagram, at which liquid can exist. In Fig. A-20 any liquid in the triangle changes composition toward the ternary eutectic M. Thus liquid U begins with crystallization of A, is joined by C at the cotectic, and moves

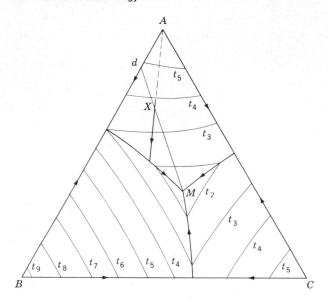

Fig. A-19. The ternary eutectic point M, the minimum melting point of a system. See text for discussion.

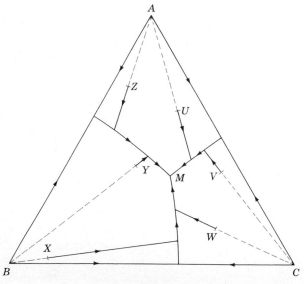

Fig. A-20. Six possible different crystallization histories, each completing crystallization at the ternary eutectic point M:

Liquid $U \rightarrow A, A+C, A+C+B$ Liquid $X \rightarrow B, B+C, B+C+A$
Liquid $V \rightarrow C, C+A, C+A+B$ Liquid $Y \rightarrow B, B+A, B+A+C$
Liquid $W \rightarrow C, C+B, C+B+A$ Liquid $Z \rightarrow A, A+B, A+B+C$

along the cotectic to M, where A and C are joined by B. Similarly, liquid W begins with crystallization of C, which is joined by B at the cotectic. The liquid composition moves along the cotectic to M, where C and B are joined by A. The six possible types of paths are illustrated by liquids $U, V, W, X, Y,$ and Z.

Figure A-21 illustrates more complicated systems in which intermediate compounds form between the components. In each case a simple eutectic relationship exists among all compounds. Thus in Fig. A-21a the intermediate compound AB forms, which has eutectic relationships with A, B, and C. In effect the system consists of two simple three-component systems: A-AB-C and AB-B-C. The line AB-C divides the system into these two subsystems, and a ternary eutectic exists in each subsystem. Figure A-21b shows the relationship when intermediate compounds form between each pair of components. This system in effect consists of four simple three-component subsystems: A-AB-AC, AB-B-BC, C-AC-BC, and AB-BC-AC. Each can be treated as a simple system, and in each a ternary eutectic exists. In each of these systems, temperature along a cotectic which connects two eutectic points falls toward both eutectics from a high-temperature point. The high-temperature points are the eutectics in the two-component systems AB-AC, BC-AB, and AC-BC. Such points are known as *equilibrium thermal divides*.

FOUR-PHASE REACTION POINT

In Fig. A-18a as reaction proceeds with the composition of the liquid moving along the reaction boundary, the liquid becomes progressively richer in component C. Ultimately, the liquid will become saturated with C and must

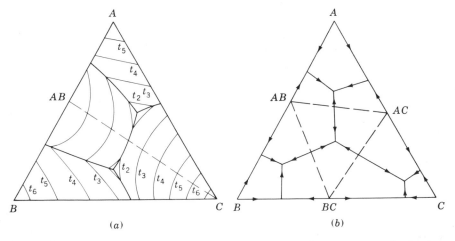

Fig. A-21. Three-component systems with intermediate compounds having eutectic relationships. Temperature shown as isotherms in (a); direction of falling temperature shown by arrows in (b).

crystallize. This means that there will now be four phases present: liquid, A (which is reacting with liquid), AB, and C. Therefore there will be no degrees of freedom; the composition of all four phases and the temperature are fixed. The composition of the liquid is therefore invariable and is represented by a point in the diagram, the *reaction point*. Temperature remains constant as crystallization and reaction proceed simultaneously until the liquid is used up.

In Fig. A-22 the reaction point is at the intersection of the reaction boundary between A and AB and the cotectic between A and C. A second cotectic between AB and C begins at this point, and temperature declines along it to the ternary eutectic between AB, B, and C. Thus these three-phase boundaries intersect at the reaction point, but temperature along one of them declines toward a lower-temperature four-phase point. Any sample whose composition falls in the triangle A-AB-C will complete crystallization at the reaction point, whereas any sample whose composition falls in the triangle AB-B-C will complete crystallization at the ternary eutectic.

In Fig. A-22 liquid of composition X begins crystallization with the separation of crystals of A. The reaction $A + \text{liquid} \rightarrow AB$ begins when the composition of the liquid reaches the reaction boundary at point d. Reaction proceeds with the liquid composition moving along the reaction boundary from d to e. At e, a composition slightly off from the reaction point, the proportions of the phases are indicated by the application of the lever rule to lines e-X-f and A-f-AB, and it is apparent that A and liquid are still both

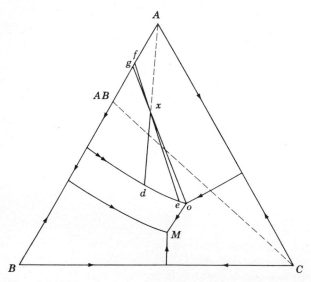

Fig. A-22. The four-phase reaction point o and related phase assemblages. See text for discussion.

present. When the liquid composition reaches the reaction point, C appears as a solid phase and the reaction forming AB from A continues. The composition of the total solid now moves directly toward the composition of the reaction-point liquid from g to X. When the total solid has composition X, the liquid is exhausted and crystallization is completed. In contrast, liquids falling between the C-AB join and the reaction boundary will behave as in Fig. A-18b and will never reach the reaction point. After A is used up, AB crystallizes and the liquid moves directly away from AB until the AB-B cotectic is reached and crystallization is complete at the ternary eutectic.

In some systems a reaction boundary may become a cotectic boundary, which in turn moves to a reaction point. Figure A-23 illustrates such a system. The three-phase curve between B and BC is a reaction curve from d to e and a cotectic from e to the reaction point o. Point e is the point of tangency of a line drawn from BC tangent to the curve. Tangents to points on the curve between d and e do not intersect a line joining the composition of the two solid phases B and BC at a point intermediate between them. This is a criterion for recognizing reaction boundaries. However, tangents to points on the curve between e and o do intersect the line B-BC at points intermediate between B and BC. This is a criterion for recognizing cotectic boundaries.

THREE COMPONENTS WITH SOLID SOLUTION BETWEEN TWO COMPONENTS

Many important rock-forming minerals form solid-solution series in which the end members and all intermediate compositions have eutectic relations

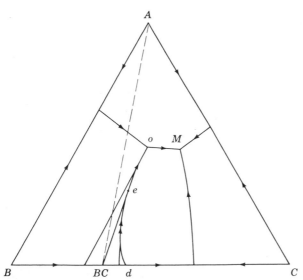

Fig. A-23. A reaction boundary (d-e) that becomes a cotectic (e-o). Point o is a four-phase reaction point.

with a third important component. A three-component system is thus set up as illustrated in Fig. A-24 in which A and B form a solid-solution series, whereas A and C and B and C have simple eutectic relationships. The three-component diagram then consists of two two-phase fields, A-B solid solution + liquid and C + liquid, and a three-phase cotectic. No four-phase point is possible. The major problem in the system is that the composition of the liquid and of one of the solid phases is changing continuously with changing temperature.

The compositions of the two variable phases can be determined graphically in this system. Given a liquid of composition X which lies on the cotectic (Fig. A-25), the composition of the solid solution AB in equilibrium with it can be found as follows: First, graphically remove component C from the system by drawing line C-X-d, a straight line. Point d now represents the proportions of A and B in the liquid. Second, draw a vertical line down into the A-B two-component diagram to the liquidus line at point e. Liquid composition e is in equilibrium with crystals of composition f, as determined by drawing a horizontal temperature line. Third, point f is projected vertically to the base of the three-component diagram and represents the composition of the solid solution in equilibrium with liquid X. Line X-f is drawn and is termed a *three-phase join*, in that it joins the compositions of the two phases of variable composition when three phases exist in equilibrium. A separate three-phase join can be constructed for each point on the cotectic curve. Thus g-h, i-j, and k-l are all constructed three-phase joins. They are not necessarily parallel, but they will not cross. Thus with a few constructed joins, intermediate joins can be interpolated with reasonable accuracy.

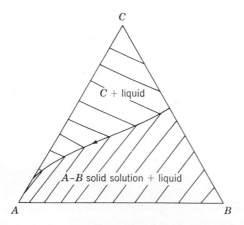

Fig. A-24. Three-component system with a complete solid-solution series between A and B and a cotectic relationship with C.

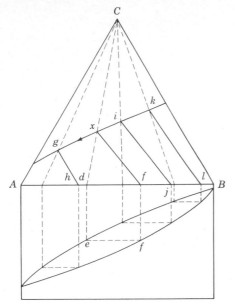

Fig. A-25. Construction of three-phase joins. See text for discussion.

The crystallization history of a liquid which falls on the solid-solution side of the cotectic may be followed by construction of a series of three-phase joins. The first crystals to form will be rich in component B and can be found graphically by drawing line c-x-e (Fig. A-26) and then dropping vertically into the two-component diagram along line e-f. The composition of the first solid-solution crystals is then at point g in the two-component diagram, or its equivalent point g' in the three-component diagram. The composition of the liquid now moves directly away from g', but as this occurs, the composition of the solid-solution crystals in equilibrium with the liquid changes. Therefore the composition of the liquid moves away from a moving point and moves along a curved line x-r-s. The liquid reaches the cotectic at the end of the three-phase join which passes through the composition of the sample, the join s-h. The liquid now has composition s and the solid-solution crystals have composition h. Component C now joins the solid solution in crystallizing, and the composition of the liquid moves along the cotectic from s to t. When the liquid reaches the t, the solid-solution crystals have composition i at the other end of the three-phase join t-i. The proportions of the phases can be found in the usual way by drawing the line t-x and extending it to intersect the line C-i at o; o-X is proportional to the remaining liquid, and x-t is proportional to the total solid. In the solid, C-o is proportional to the amount of solid solution and o-i is proportional to the crystals of C. The final composition of the solid solution is represented by point e and the last

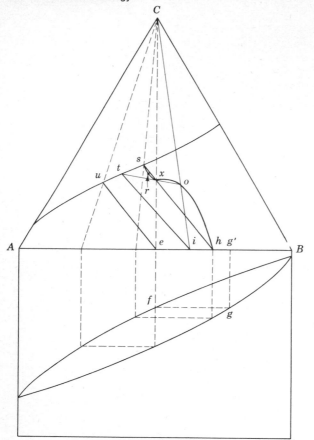

Fig. A-26. Crystallization of a liquid in the solid-solution field. See text for discussion.

liquid by the other end of the three-phase join *e-u*. During crystallization the composition of the total solid has followed the line *g'-h-o-X*.

It is important to note that no two liquids falling in the solid-solution two-component field will follow the same exact lines of composition change. This is illustrated in Fig. A-27 where liquids *e*, *f*, and *g* all fall on the same three-phase join and therefore reach the cotectic with the same composition. However, the composition of the first solid-solution crystals to form will be different for each, and the composition of the final phases will be different, as indicated by the lines *C-e-h*, *C-f-i*, and *C-g-j*. Similarly, liquids *m*, *n*, and *o* lie on the same line radiating from *C* and will therefore begin crystallization with the same composition of solid-solution crystals *r'* and complete crystallization with the same composition of the phases (last liquid at *s'* and final solid solution at *t'*). However, each falls on a different three-phase join and will therefore reach the cotectic with different liquid and solid-solution crystal compositions.

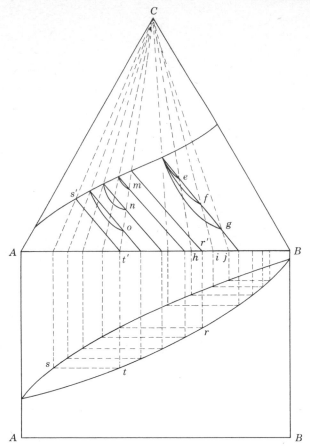

Fig. A-27. Figure demonstrating that no two liquids in the solid-solution field follow the same path during crystallization.

Systems occur in which an intermediate compound AB forms a solid solution with component C but has eutectic relationships with components A and B. This is illustrated in Fig. A-28. Two cotectics then appear, one separating a solid-solution two-phase field from an A two-phase field and the other separating the solid-solution two-phase field from the B two-phase field. Such diagrams can be treated as two three-phase diagrams A-AB-C and B-AB-C joined along a two-component diagram AB-C.

THREE COMPONENTS WITH SOLID SOLUTION
BETWEEN A REACTION PRODUCT AND THE THIRD COMPONENT

In the two-component system involving a peritectic reaction $A + $ liquid $\rightarrow$ AB, a solid solution may be developed between AB and a third component C. Eutectic relationships exist between A and C and between B and C. Within the three-component system (Fig. A-29) there is therefore a reaction

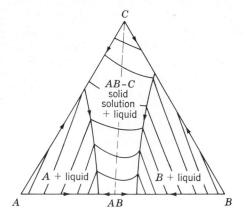

Fig. A-28. Solid solution between inter-mediate compound *AB* and component *C*.

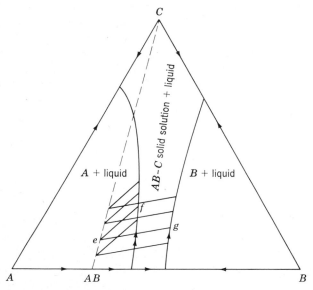

Fig. A-29. Solid solution between an intermediate com-pound (*AB*) formed by reaction (*A* + liquid → *AB*) and the third component (*C*). Empirically determined three-phase joins given as *ef* and *eg*.

boundary along which *A* and liquid react to form a solid-solution series *AB-C*, and a cotectic along with the *AB-C* solid solution and *B* crystallize simul-taneously. As the reaction boundary approaches the *A-C* eutectic, it converts into a cotectic boundary. No four-phase points occur in the diagram. The reaction boundary joins the composition of the liquid at the peritectic to the composition of the liquid at the *A-C* eutectic. The cotectic joins the *AB-B* eutectic point with the *B-C* eutectic point.

In this system two sets of three-phase joins are necessary inasmuch as there are two situations, each involving two phases of variable composition. One set of three-phase joins connects the composition of the solid solution in equilibrium with liquids on the reaction boundary, and the other set connects the composition of solid-solution crystals with liquids on the co-tectic. Neither set can be determined graphically from a two-component diagram, because the *AB-C* join is not a simple two-component system. Certain mixtures of *AB* and *C* begin with the crystallization of *A*. Thus inas-much as the composition of all phases which appear cannot be expressed in terms of the two components, the system cannot be a simple two-component system. Three-phase joins in this situation must be determined empirically and given in the diagram. In Fig. A-29 the two sets are shown; line *e-f* and similar lines are the three-phase joins between solid-solution crystals and liquids on the reaction boundary, and line *e-g* and similar lines are two-phase joins between solid-solution crystals and liquids on the cotectic boundary.

The crystallization histories of three liquids are of particular interest. These are liquids which begin crystallization with crystals of *A* and are in-volved in the reaction. The first is a liquid such as composition X (Fig. A-30) in which the final solid phases are *A* and *AB-C* solid solution. The crystalli-zation in this instance is completed while the liquid is on the reaction bound-ary. The second is a liquid such as *Y* which falls in the *AB-B-C* triangle, and complete crystallization yields solid solution plus *B*. In this instance crystalli-zation is complete when the liquid is on the cotectic. The third condition is met when the bulk-sample composition falls on the *AB-C* join, and the final

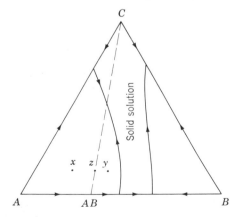

Fig. A-30. Positions of liquids considered in Figs. A-31, A-32, and A-33. Arrows indicate direction of declining tempera-ture.

solid consists of only solid-solution crystals. Liquid here is exhausted while on the reaction boundary. To follow each of these a careful construction of the lines required to determine proportion by the lever rule is necessary, and therefore only an enlarged portion of the total diagram will be used in the following discussion.

The crystallization of liquid X can be traced in Fig. A-31. The first crystals to form are A, and the composition of the liquid moves directly away from A along line X-d. At point d the liquid reaches the reaction boundary, and A reacts with liquid to form AB which takes up C in solid solution. The first solid-solution crystals will have composition e; the line d-e is a three-phase join. The composition of the liquid now moves down temperature along the reaction boundary from d to f to h. With each temperature change and liquid-composition change, the solid-solution crystals must change in composition. Thus when the liquid is of composition f, the solid-solution crystals have composition g, and when the liquid has composition h, the solid-solution crystals have composition i. At this point the proportions of liquid and solid must be considered. Proportions are found by drawing lines joining the composition of the two solid phases A and AB-C solution. These lines are represented by A-g and A-i. The lines from the successive compositions of the liquid through X (f-X-o and h-X-p) intersect A-g and A-i at o and p, respectively, and these points represent successive compositions of the total solid. Applying the lever rule to lines f-X-o and h-X-p, the proportions of the remaining liquids are represented by the lines X-o and X-p. It can be seen that the proportion of liquid is decreasing rapidly. Construction of further three-phase joins and lever-rule lines shows that the liquid will be exhausted when it reaches composition j, the solid-solution crystals having

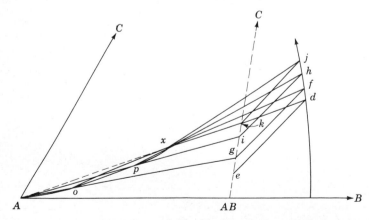

Fig. A-31. Crystallization of liquid X of Fig. A-30. Liquid composition follows line X-d-f-h-j, whereas solid composition follows line A-o-p-X.

composition k. The lines j-X and k-X-A intersect at X, and all the solid has the composition X. The point k represents the composition of the solid solution in the initial sample X found by removing excess A graphically with the line A-X-k. During crystallization the composition of the total solid remained at A while the liquid moved from X to d. It then moved along the curve A-o-p to X as the liquid moved along the reaction boundary.

Liquid Y lies in the AB-B-C triangle and will complete crystallization with the solid phases B and solid solution. The composition of the final solid-solution crystals can be found by graphically removing excess B (Fig. A-32) with the line B-Y-o, o being the composition of the final mixed crystals. The first crystals to form, however, are A, and the liquid composition moves from Y to d, a point on the reaction boundary. When the liquid reaches point d, reaction begins producing solid-solution crystals of composition e and the liquid moves along the reaction boundary to points f and h. Simultaneously, the composition of the mixed crystals moves from e to g to i as required by the three-phase joins d-e, f-g, and h-i. Construction of the lever-rule lines A-g and f-y-r shows that the proportion of A to mixed crystals is changing rapidly, and the proportion of A is decreasing as indicated by r-g. All A will be used up when the liquid reaches point h, and the mixed crystals have composition i. The line h-i is the three-phase join passing through the bulk composition Y. Application of the lever rule shows that line h-Y extended to intersect the line joining the composition of the two solid phases A-i intersects it at i, and A-i is proportional to the percent of solid solution in the total solid, that is, 100 percent solid solution. Liquid still remains, however, as indicated by the lever i-Y.

At this point only two phases are present, liquid and solid-solution crystals. The composition of the liquid now moves across the two-phase

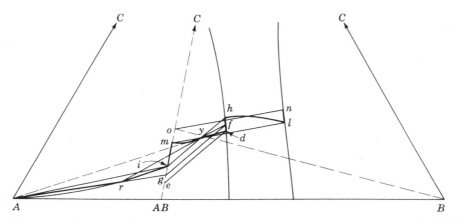

Fig. A-32. Crystallization of liquid Y of Fig. A-30. Liquid composition follows lines Y-d-h-l-n, whereas solid composition follows lines A-r-i-m-Y.

solid-solution field along the curved line h-l and reaches the cotectic boundary at the end of the three-phase join to the cotectic which passes through the bulk composition of the sample Y. During this two-phase crystallization, the composition of the solid-solution crystals changes from i to m. When the liquid reaches the cotectic, crystals of B begin to separate also, and the liquid moves along the cotectic from l to n, whereas the composition of the solid-solution crystals moves from m to o. The line n-o is the three-phase join to the cotectic through the composition of solid solution in the initial sample. To recapitulate, during crystallization the composition of the liquid moves from Y to d, then from d to h along the reaction boundary, then along the curved line from h to l, and finally, along the cotectic from l to n, at which point all liquid is used up. Simultaneously, the solid begins with crystals of A; at the reaction boundary AB-C solid solution appears; and the total solid moves along the curved line A-r to i, where all A is used up. The total solid then moves from i to m in a two-phase crystallization. While the liquid is moving along the cotectic, the total solid moves along a curved line from m to Y, where crystallization is complete.

Liquid Z falls on the AB-C solid-solution join and therefore winds up with only one solid phase, the solid solution of composition Z. The liquid and crystals of A are used up simultaneously during reaction on the boundary. Again (Fig. A-33), crystals of A separate first, and the liquid moves directly away from A to the reaction boundary at d. There A plus liquid react to form AB-C solid solution of composition e. The liquid moves along the reaction boundary, and the composition of the solid-solution crystals adjusts according to the three-phase joins d-e, f-g, h-i, and j-k. Construction of the lever-rule lines allows interpretation of proportions. Lines A-g, A-i, and A-k join the compositions of the successive solid phases, and f-Z-o, h-Z-p, and j-Z-q join the present composition of the liquid and present composition of

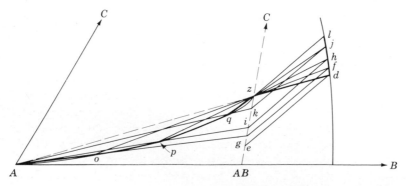

Fig. A-33. Crystallization of liquid Z of Fig. A-30. Liquid composition follows line Z-d-h-l, whereas solid composition follows line A-o-p-q-Z.

total solid through the bulk composition. The successive lengths of levers Z-o, Z-p, and Z-q show that the liquid is being rapidly used up, whereas levers g-o, i-p, and k-q show that A is being rapidly used up. When the liquid reaches the end of the three-phase join through Z, both A and liquid are exhausted, because lines joining A and the composition of the solid solution (A-Z) and the line joining the present composition of the liquid and bulk composition (l-Z) intersect at Z. The liquid-solid lever is all solid, and the A to AB-C solution lever is all AB-C.

FOUR-COMPONENT SYSTEMS

The basic principles of four-component systems can be outlined, but the detailed crystallization histories cannot be followed through at this time. Very few such systems of petrologic significance have been worked out in sufficient detail to warrant a detailed discussion. However, the fundamentals of such systems are important in approaching the complex systems of magmas. To represent four components in all proportions, a three-dimensional figure is required to indicate composition alone. Therefore the second variable, temperature, can only be considered as specific temperatures of fixed points. Accurate temperature for all phase situations cannot be shown graphically.

The three-dimensional figure used to represent composition is the tetrahedron, where each apex represents 100 percent of one component. The tetrahedron is then bounded by four three-component systems representing the faces and six two-component systems representing the edges. The composition of any four-component mixture is represented by a point within the tetrahedron. Given a mixture of A, B, C, and D, a point representing this mixture can be determined by recalculating A, B, and C to 100 percent and plotting this point on the A-B-C face of the tetrahedron. A line is constructed from this point to the D corner of the tetrahedron, and a point on this line proportional to the proportion of D is the plot of the total sample. Thus in Fig. A-34 samples with composition 20 percent A, 40 percent B, 20 percent C, 20 percent D, and 20 percent A, 10 percent B, 25 percent C, 45 percent D have been plotted as points X and Y, respectively.

Each face of the tetrahedron consists of a three-component system, and therefore, under the simplest conditions each face will contain three cotectics and a ternary eutectic. With the addition of the fourth component a surface will rise into the tetrahedron from each cotectic along which the composition of the liquid will move while two phases are crystallizing simultaneously. These surfaces will divide the tetrahedron into four two-phase volumes, one at each corner. Liquids falling in these volumes will crystallize out one phase at appropriate temperature, and the composition of the liquid will move directly away from that phase until it reaches a three-phase surface.

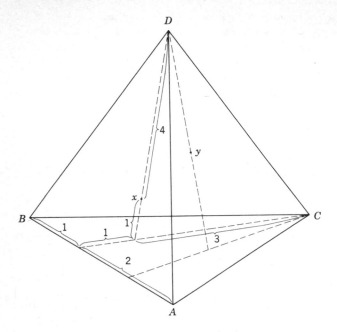

Fig. A-34. Plotting composition in a tetrahedron. Point x represents 20 percent A, 40 percent B, 20 percent C, and 20 percent D. Ratios are $A:B = 1:2$; $C:A+B = 1:3$; $D:A+B+C = 1:4$. Point Y represents 20 percent A, 10 percent B, 25 percent C, and 45 percent D, with ratios $A:B = 2:1$; $C:A+B = 45:55$; and $D:A+B+C = 45:55$.

The three-phase surfaces will intersect along lines which move into the tetrahedron from the ternary eutectics. These lines will represent four-phase situations of a cotectic type. In the four-component system there will be six three-phase planes intersecting along four four-phase lines. The four-phase lines will intersect at a five-phase eutectic point which will occur at the lowest possible temperature for the presence of a liquid phase.

Figure A-35 illustrates this system and the crystallization of a melt. Points E_1, E_2, E_3, and E_4 are the ternary eutectic in the three-component systems, and point o is the five-phase eutectic point for the entire system. A liquid of composition X falls in the two-phase A plus liquid volume. Crystallization begins with the separation of A, and the liquid composition moves directly away from A, reaching the three-phase surface at point e. At this point crystals of B join A, and the composition of the liquid moves along the three-phase surface. A line tangent to the surface at the liquid composition intersects the A-B edge at the composition of the crystals separating from that liquid. At point f the liquid reaches the four-phase line, and C joins A

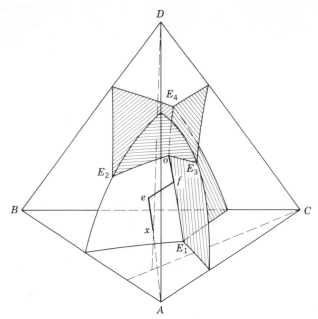

Fig. A-35. Crystallization of a liquid in a four-component eutectic situation. Liquid X has composition 65 percent A, 7 percent B, 13 percent C, and 15 percent D. The liquid composition follows the line X-e-f-o.

and B in crystallizing. The composition of the liquid moves along the four-phase line toward the five-phase point. A line tangent to the curve through the liquid composition intersects the A-B-C face at a point representing the composition of the crystals separating from that liquid. The liquid ultimately reaches the five-phase point o, at which point D joins A, B, and C in crystallizing. No degrees of freedom exist; therefore the composition of the liquid and the temperature remain constant until the liquid is exhausted. The proportion of all phases at any point can be found by applying the lever rule to appropriately drawn lines. During three-phase crystallization a line from the liquid composition through X will intersect the A-B edge at a point proportional to the ratio of solid A to solid B. During four-phase crystallization a line from the liquid composition through X will intersect the A-B-C face at a point proportional to the amounts of solid A, B, and C. In both situations application of the lever rule to the line from liquid through X to solid gives the proportion of liquid to total solid.

Solid solution between two components is important. Such diagrams involve only three two-phase volumes and one four-phase line. There is no five-phase point. Figure A-36 illustrates such a system. A and B form a continuous solid-solution series. For liquids lying in the solid-solution two-

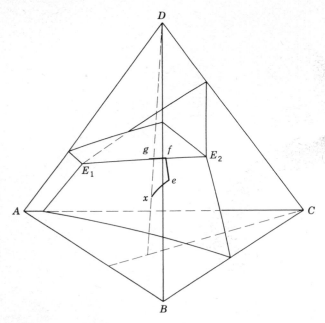

Fig. A-36. Four-component system with solid solution between A and B. E_1 and E_2 are ternary eutectics; the line E_1-E_2 is a four-phase line. Liquid X contains 24 percent A, 36 percent B, 15 percent C, and 25 percent D. During crystallization, liquid X follows the line X-e-f-g, with final liquid composition near g.

phase field, the first crystals formed have a composition lying on the A-B edge. The proportions can be determined from the composition of the liquid with respect to A and B and from a given two-component diagram for A and B. The liquid follows a curve such as X-e-f-g during the complete crystallization. At e a three-phase surface is reached and A-B solid solution is joined by c. The liquid moves along this surface from e to f. At f a four-phase line is reached and D joins in the crystallization. The liquid then moves along the four-phase line until crystallization is complete.

INDEX

INDEX